I0888538

AI and Intelligent Systems for Sustainable Industrial and Environmental Applications

Kamalakanta Muduli
https://orcid.org/0000-0002-4245-9149
Papua New Guinea University of Technology, Papua New Guinea

Sujoy Kumar Jana
https://orcid.org/0000-0003-0901-5316
Papua New Guinea University of Technology, Papua New Guinea

Rashmi Prava Das
CV Raman Global University, Odisha, India

Indrajit Pal
Asian Institute of Technology, Thailand

IGI Global
Scientific Publishing
Publishing Tomorrow's Research Today

Vice President of Editorial	Melissa Wagner
Director of Acquisitions	Mikaela Felty
Director of Book Development	Jocelynn Hessler
Production Manager	Mike Brehm
Cover Design	Jose Rosado

Published in the United States of America by
IGI Global Scientific Publishing
701 East Chocolate Avenue
Hershey, PA, 17033, USA
Tel: 717-533-8845 | Fax: 717-533-7115
Website: https://www.igi-global.com E-mail: cust@igi-global.com

Copyright © 2027 by IGI Global Scientific Publishing. All rights reserved. No part of this publication may be reproduced, stored or distributed in any form or by any means, electronic or mechanical, including photocopying, without written permission from the publisher. Use of this publication to train generative artificial intelligence (AI) technologies is expressly prohibited. The publisher reserves all rights to license its use for generative AI training and machine learning model development.

Product or company names used in this set are for identification purposes only. Inclusion of the names of the products or companies does not indicate a claim of ownership by IGI Global Scientific Publishing of the trademark or registered trademark.

Library of Congress Cataloging-in-Publication Data

LCCN: (CIP Data Pending)
ISBN13: 9798260023839
Isbn13Softcover: 9798260023846
EISBN13: 9798260023853

British Cataloguing in Publication Data
A Cataloguing in Publication record for this book is available from the British Library.

All work contributed to this book is new, previously-unpublished material.
The views expressed in this book are those of the authors, but not necessarily of the publisher.
This book contains information sourced from authentic and highly regarded references, with reasonable efforts made to ensure the reliability of the data and information presented. The authors, editors, and publisher believe the information in this book to be accurate and true as of the date of publication. Every effort has been made to trace and credit the copyright holders of all materials included. However, the authors, editors, and publisher cannot assume responsibility for the validity of all materials or the consequences of their use. Should any copyright material be found unacknowledged, please inform the publisher so that corrections may be made in future reprints.

Table of Contents

Detailed Table of Contents

the causal interlinkages of DI with MS. FDI, trade openness, population density, population growth and gender inequality are included as control variables (CVs). Log-linear functional forms of models are formulated to examine the coefficients of DI and CVs with MS, and MS, and CVs with DI using a panel data during 2006 – 2022. DI has a positive impact on MS, and MS is found suitable to increase DI. This study makes a well-intentioned contribution in theoretical and empirical literature, and provides effective policy suggestions to increase the sustainability of MS and DI. It suggests research direction to undertake further research in DI and MS.

Increasingly, traditional waste management systems are struggling to keep up due to rising global waste generation and an associated demand for intelligent automation solutions, which are quickly rising to meet that need. The current chapter details the development and operation of an autonomous smart waste sorting system. This autonomous system uses artificial intelligence along with multisensor technologies to improve the sorting performance of the various materials that it sorts. The machine learning model used in the system supports visual identification of the materials being sorted as well as separate detection of all metals, glass and organic waste by means of unique physical and chemical properties. The operational data for this system is communicated to a centralized digital platform, which provides real-time visibility and management of the system through a user interface. The experimental results of the prototype presents an encouraging avenue for growth of scalable smart waste management solutions.

Integrating wind energy into the power grid offers significant benefits in reducing reliance on fossil fuels, but challenges such as intermittent generation impact system stability and power quality. This chapter explores advanced control techniques for optimizing grid-connected wind energy systems, focusing on Maximum Power Point

Floods in India cause significant socio-economic, infrastructure, and environmental damage. Traditional monitoring systems are inadequate for real-time flood prediction and risk mitigation. This chapter introduces an AIoT-based flood warning system that utilizes IoT sensors for continuous data acquisition on rainfall and river stream levels, combined with machine learning models for accurate flood prediction. The system uses edge and cloud computing for low-latency processing, ensuring timely alerts. Custom regional models cater to Assam's riverine floods and Kerala's flash floods, improving lead times and emergency responses. Integrated with disaster management platforms, the system enhances flood resilience and offers a scalable, intelligent solution for climate-sensitive areas.

This chapter presents a multi-hazard prediction model that addresses the complexity of contemporary disaster scenarios, where multiple hazards interact and cascade in real time. Traditional disaster risk assessments, focused on single hazards and non-dynamic modeling, fail to capture the dynamic interdependencies in such situations. Leveraging advancements in data availability and computational intelligence, this model integrates machine learning, deep learning, and ensemble-based predictions to forecast cascading disaster mechanisms. By combining past data, real-time sensor measurements, satellite images, and socio-economic data into an adaptive, scalable pipeline, the model enhances forecasting accuracy, reduces latency, and increases sensitivity to risk trends. The approach enables proactive disaster risk mitigation and supports emergency management with actionable insights, advancing both theoretical and practical disaster management frameworks.

In many regions of Türkiye, land use development is chosen and managed without consideration to land capability, which results in disinvestment and lowers ecological capacity. In order to support the local authority's decision-making process, this article presents an analytical method that uses individual suitability maps to determine if a region is suitable for developing various economic sectors. Using a Multi-Criteria Decision Analysis (MCDA) integrated with Geographic Information Systems (GIS), the method is made up of the Analytic Hierarchy Process (AHP), a deterministic and fuzzy logic approach, the latter two were used for the standardisation of the criteria maps. The land-use suitability study took into account a number of criteria, such as geophysical characteristics, accessibility, built-up area and infrastructure, vegetation, and green and blue infrastructure. When compared to alternatives based on conventional approaches, the integration of fuzzy logic, AHP, and GIS is an advanced methodology that can enhance suitability evaluations.

Preface

The rapid advancement of Artificial Intelligence (AI) and intelligent systems is transforming industries, environmental management practices, and decision-making processes across the globe. As societies confront pressing challenges related to sustainability, climate resilience, industrial efficiency, resource optimization, and disaster preparedness, AI-driven technologies are emerging as powerful enablers of innovative and sustainable solutions. The book *AI and Intelligent Systems for Sustainable Industrial and Environmental Applications* brings together diverse scholarly contributions that explore the intersection of intelligent technologies, industrial sustainability, renewable energy systems, environmental monitoring, and smart decision-support frameworks.

This volume presents a multidisciplinary perspective on how AI, machine learning, the Internet of Things (IoT), edge computing, predictive analytics, and geospatial intelligence can collectively address contemporary industrial and environmental challenges. The chapters illustrate the transformative role of intelligent systems in areas such as predictive maintenance, renewable energy optimization, waste management, flood forecasting, disaster prediction, food distribution, smart agriculture, and industrial land-use planning.

A central theme throughout the book is sustainability. The contributing studies demonstrate how AI-enabled technologies can improve operational efficiency while reducing environmental impact and resource waste. Several chapters focus on intelligent industrial systems that support predictive maintenance, digital inclusion, and energy optimization within the framework of Industry 4.0. Others examine AI applications in environmental sustainability, including smart waste sorting systems, renewable energy integration, flood alert mechanisms, fire activity recognition, and multi-hazard forecasting models designed to enhance resilience against climate-related disasters.

The volume also emphasizes the importance of data-driven decision-making and interdisciplinary innovation. Advanced computational approaches such as deep learning, Long Short-Term Memory (LSTM) forecasting, computer vision, fuzzy logic, scientometric analysis, and hybrid control systems are explored in practical contexts that highlight their real-world relevance and scalability. By integrating intelligent technologies with sustainable development goals, the chapters collectively contribute to ongoing efforts toward building resilient industrial systems and environmentally conscious societies.

Another distinguishing feature of this book is its strong practical orientation and global relevance. The contributions reflect experiences and case studies from a variety of regional contexts, particularly from developing and emerging economies where technological innovation can significantly accelerate sustainable development. The studies provide valuable insights for researchers, practitioners, policymakers, engineers, environmental planners, and students seeking to understand the evolving role of AI and intelligent systems in sustainable transformation.

This book aims to serve as both a scholarly reference and a catalyst for future research. It encourages collaboration across disciplines and highlights the importance of integrating technological innovation with ethical, environmental, and societal considerations. The editors hope that the ideas presented in this volume will inspire further advancements in intelligent systems and foster sustainable solutions for industrial and environmental applications worldwide.

CHAPTER OVERVIEW

1. The first chapter examines the evolution of predictive maintenance within Industry 4.0 environments, focusing on the integration of IoT sensors, artificial intelligence, edge computing, and machine learning for intelligent asset monitoring, fault diagnosis, and maintenance optimization.
2. The second chapter investigates the relationship between digital inclusion and the sustainability of the manufacturing sector across multiple countries. It presents empirical findings on how digitalization contributes to industrial sustainability and economic development.
3. The third chapter presents an intelligent waste-sorting system that combines artificial intelligence and multisensor technologies to improve automated waste classification, operational efficiency, and scalable smart waste management practices.
4. The fourth chapter explores advanced control and optimization strategies for grid-connected wind energy systems using Maximum Power Point Tracking (MPPT) and hybrid controllers to enhance renewable energy efficiency, reliability, and grid stability.
5. The fifth chapter provides a comprehensive scientometric evaluation of AI-driven innovations in Indian agriculture and allied sectors. It analyzes research

trends, collaboration networks, thematic developments, and future opportunities in sustainable agricultural intelligence.

6. The sixth chapter introduces an AIoT-based flood alert system designed for flood-prone regions. By integrating rainfall and river sensors with machine learning models, the system supports real-time flood prediction, early warning, and disaster resilience.
7. The seventh chapter discusses dynamic multi-hazard forecasting approaches that use machine learning and ensemble modeling to predict cascading disasters. The framework integrates sensor data, satellite imagery, and socio-economic indicators to support proactive disaster management.
8. The final chapter presents a GIS- and Multi-Criteria Decision Analysis-based framework for industrial land suitability assessment. The study demonstrates how fuzzy logic, AHP, and spatial analysis techniques can support sustainable industrial land-use planning and environmental decision-making.

The editors sincerely hope that this volume will contribute meaningfully to the growing body of knowledge on AI and intelligent systems for sustainable development. By bridging industrial innovation and environmental responsibility, this book aspires to support the advancement of resilient, efficient, and sustainable technological ecosystems for the future.

Chapter 1
Advancing Industrial Maintenance Using Predictive Maintenance in the Era of Industry 4.0

K. Angayarkkani
Dwaraka Doss Goverdhan Doss Vaishnav College, India

H. Vignesh Ramamoorthy
Amity Institute of Information Technology, Amity University, India

N. Priyadharshini
Shri Krishna Arts and Science College, India

T. Gopalakrishnan
http://orcid.org/0000-0003-0923-4076
Vels Institute of Science, Technology, and Advanced Studies (VISTAS), India

S. Karunya
Dwaraka Doss Goverdhan Doss Vaishnav College, India

A. V. Senthil Kumar
http://orcid.org/0000-0002-8587-7017
Nehru Institute of Information Technology and Management, India

Mini Srivastava
http://orcid.org/0000-0002-9002-5126
Shri Guru Ram Rai University, India

ABSTRACT

Predictive Maintenance (PdM) has emerged as a critical enabler of Industry 4.0, overcoming the limitations of traditional corrective and time-based preventive maintenance strategies. Through the integration of IoT-enabled sensors, advanced condition monitoring, edge computing, and artificial intelligence, PdM supports real-time health assessment, early fault detection, and data-driven maintenance decisions.

DOI: 10.4018/979-8-2600-2383-9.ch001

2027, IGI Global Scientific Publishing. Copying or distributing in print or electronic forms without written permission of IGI Global Scientific Publishing is prohibited. Use of this chapter to
ative artificial intelligence (AI) technologies is expressly prohibited. The publisher reserves all rights to license its use for generative AI training and machine learning model development.

This chapter outlines core PdM architectures, the role of machine learning in fault diagnosis and remaining useful life estimation, and key system integration challenges influencing intelligent asset management in modern industrial environments.

BACKGROUND

Industrial maintenance practices have evolved continuously in response to rising system complexity, tighter production interdependencies, and growing economic exposure to unplanned downtime. Early industrial facilities relied almost exclusively on corrective maintenance—waiting for equipment to fail before intervening. While this reactive posture minimised short-term expenditure, it introduced prolonged stoppages, secondary damage to adjacent components, safety hazards, and significant production losses. As facilities expanded and machines became more tightly coupled, these consequences became operationally and financially untenable (Jardine et al., 2006).

The response was preventive maintenance: scheduled inspections and component replacements based on manufacturer recommendations or historical failure statistics. Reliability improved, yet the strategy carried a structural flaw. Maintenance actions were executed according to a calendar rather than the actual condition of equipment, which led to premature replacement of healthy components in some cases and missed degradation in others. The cost of this mismatch was particularly visible in high-throughput environments, where both over-maintenance and under-maintenance eroded lifecycle efficiency (Zonta et al., 2020).

The arrival of Industry 4.0—characterised by cyber-physical systems, pervasive connectivity, and data-driven decision-making—altered this picture fundamentally. Industrial assets that had once been opaque mechanical entities became continuously observable data sources. This shift created the conditions for condition-based and predictive maintenance: strategies that evaluate actual equipment health rather than elapsed time. Predictive Maintenance (PdM) continuously monitors operating data to detect degradation early and anticipate failure before it occurs, enabling maintenance actions that are proactive, targeted, and economically justified (Lee et al., 2015).

The rapid growth of the Industrial Internet of Things has been central to this transformation. Dense networks of embedded sensors now capture vibration, temperature, pressure, acoustic emissions, and electrical signals at high resolution, revealing subtle deterioration that periodic inspections cannot detect (Pech et al., 2021). At the same time, advances in machine learning have made it possible to extract actionable insight from these data streams, while edge computing has enabled low-latency diagnostics to be performed close to the asset—reducing bandwidth demands, improving responsiveness, and limiting unnecessary data exposure (Lu et al., 2023).

Despite its promise, PdM deployment continues to face real challenges. Data heterogeneity, scarce failure labels, legacy system interoperability, cybersecurity risks, and organisational resistance each constrain adoption. More critically, even accurate predictions fail to deliver operational value if they are not communicated effectively to maintenance personnel and embedded within actionable workflows. Predictive maintenance must therefore be understood as a systemic transformation of maintenance philosophy—not merely an analytical upgrade—if it is to fulfil its potential as a strategic enabler within the Industry 4.0 ecosystem (Zonta et al., 2020).

Table 1. Comparison of Traditional Maintenance and Predictive Maintenance Strategies

Aspect	Traditional Maintenance	Predictive Maintenance
Maintenance Trigger	Periodic schedules or post-failure response	Condition-based, using real-time operational data
Downtime Pattern	Often unplanned and disruptive	Planned and minimised through early intervention
Cost Impact	High long-term cost from reactive repairs and wasted resources	Reduced lifecycle cost through optimised scheduling
Data Utilisation	Limited use of historical records	Continuous use of real-time IoT-driven analytics
Decision Support	Manual, experience-based judgement	AI-assisted, data-driven decision support
Asset Life	Shortened by late detection of degradation	Extended through early fault detection and timely action

Source: Adapted and synthesised from Mobley (2002), Jardine et al. (2006), and Zonta et al. (2020).

A comparative summary of traditional approaches and predictive maintenance is provided in Table 1. The closed-loop interaction between industrial assets, IoT-based sensing, data engineering, decision support, maintenance execution, and continuous feedback that defines the PdM framework is illustrated conceptually in Figure 1.

Figure 1. Conceptual framework of predictive maintenance architecture in Industry 4.0

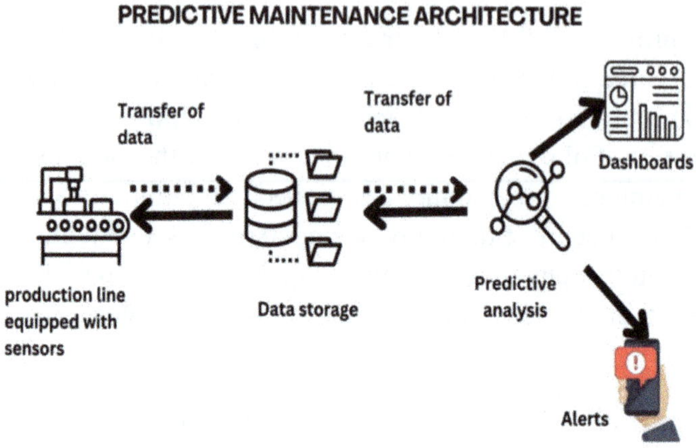

Source: Adapted from Jardine et al. (2006), Lee et al. (2015), and Zonta et al. (2020).

The following sections contextualise this evolution within the broader Industry 4.0 transformation before addressing lifecycle implementation, sensor ecosystems, data engineering, analytics, integration, and human-centred execution.

SCOPE AND CONTRIBUTION OF THIS CHAPTER

Published work on predictive maintenance remains fragmented across sensing, analytics, and proof-of-concept deployments. Many contributions focus narrowly on a particular sensor modality, a specific algorithm, or a single industrial case without explaining how these components must be combined, governed, and sustained in real environments. Organisations attempting deployment frequently discover that technical accuracy alone does not translate into operational value unless PdM is embedded within maintenance workflows, decision authority, and enterprise systems (Jardine et al., 2006; Zonta et al., 2020).

This chapter addresses that gap by adopting a system-level, operationally grounded perspective. Rather than cataloguing algorithms, it treats PdM as a closed-loop socio-technical system spanning asset selection, instrumentation, data engineering, analytics, decision support, work execution, and continuous improvement. Predictive signals create value only when they trigger the right action at the right time—and when outcomes feed back into the system to sharpen future predictions.

The chapter makes four targeted contributions. First, it situates PdM within the Industry 4.0 ecosystem, explaining how IoT networks, edge–cloud architectures, and AI-based diagnostics and prognostics function as interdependent rather than standalone elements. Second, it foregrounds maintenance data engineering as a decisive success factor, covering sampling strategy, windowing, feature extraction, noise filtering, order tracking, and feature store governance—elements routinely underrepresented in surveys but strongly determinative of deployment reliability (Pech et al., 2021). Third, it organises analytical methods within a clear diagnostic–prognostic–prescriptive taxonomy, reducing conceptual ambiguity and supporting method selection aligned with operational objectives (Serradilla et al., 2022). Fourth, it connects PdM with classical reliability engineering—FMEA, RCM, Weibull analysis, and fault tree reasoning—to strengthen interpretability and industrial legitimacy in safety-critical environments (Jardine et al., 2006).

Beyond prediction, the chapter covers system integration through APIs, MES, CMMS, and EAM platforms, and examines Augmented Reality as a human-centred execution interface. Five structured case studies illustrate performance across manufacturing, process, energy, rail, and maritime sectors. The chapter closes with a critical assessment of deployment challenges and future research directions. It is intended to serve two roles simultaneously: a structured reference for graduate students and researchers seeking conceptual clarity, and a practical roadmap for engineers and maintenance managers designing PdM systems aligned with Industry 4.0 objectives.

PdM SYSTEM LIFECYCLE AND IMPLEMENTATION ROADMAP

Effective PdM deployment requires a structured lifecycle perspective rather than isolated deployment of sensors or analytics tools. Empirical evidence consistently shows that many initiatives stall not because of weak algorithms, but because of poor alignment with asset criticality, operational workflows, data governance, and decision authority (Jardine et al., 2006; Zonta et al., 2020). A lifecycle framework addresses this by ensuring that predictive insights are actionable, economically justified, and sustainable. The framework is iterative rather than linear, enabling continuous refinement as assets age, conditions change, and organisational maturity grows.

Asset Selection and Maintenance Criticality Assessment

The lifecycle begins with systematic asset selection rather than indiscriminate sensor deployment. Not all equipment justifies predictive monitoring, and poor prioritisation is a primary cause of low return on investment (Mobley, 2002). Criticality assessment evaluates failure impact on production continuity, safety and environmental consequences, repair cost, failure frequency, and process coupling. Reliability-Centred Maintenance (RCM) and Failure Modes and Effects Analysis

(FMEA) are applied at this stage to identify dominant failure modes and rank assets by operational risk, ensuring that investment is driven by value rather than technological enthusiasm.

Baseline Study and Operational Context Definition

Following asset selection, a baseline study characterises normal operating behaviour and historical performance trends. This phase is frequently underemphasised in PdM literature, yet it is essential for distinguishing genuine degradation from routine operational variability (Zonta et al., 2020). Analysis of maintenance records, failure logs, operating regimes, load profiles, and environmental conditions establishes reference states against which future deviations can be interpreted with confidence. Without this grounding, predictive models generate excessive false alarms that erode user trust.

Sensor Planning and Instrumentation Strategy

Sensor selection and placement should be failure-mode driven rather than technology-driven. FMEA outcomes guide the choice of sensing modalities to capture early indicators of specific degradation mechanisms (Selcuk, 2017). Bearing fatigue, for example, is best monitored through high-frequency vibration and acoustic emission sensors; electrical insulation degradation calls for temperature monitoring and current signature analysis. Practical constraints—installation access, environmental exposure, calibration stability, and lifecycle cost—are factored into every instrumentation decision, with scalability and interoperability prioritised to ensure long-term compatibility across heterogeneous equipment fleets (Lee et al., 2015).

Maintenance Data Engineering and Data Pipeline Design

Raw industrial sensor data are rarely suitable for direct analytical use. Noise, incompleteness, and non-stationarity make data engineering a decisive determinant of PdM success (Tsui et al., 2015). This stage addresses signal synchronisation, timestamp alignment, noise filtering, normalisation, missing-data handling, and feature extraction across time, frequency, and time–frequency domains. Architectural decisions on edge versus cloud processing are also finalised here. Latency-sensitive diagnostics run at the edge; long-term trend analysis, model training, and fleet-level analytics are centralised in cloud environments.

Predictive Model Development and Validation

Model development is guided by data characteristics, failure behaviour, and operational objectives rather than algorithmic novelty. Classical statistical methods remain effective for well-understood degradation processes, while deep learning is reserved for complex, non-linear patterns where sufficient data exist (Si et al., 2011). Time-series-aware validation—rolling or blocked cross-validation—prevents information leakage. Cost-sensitive evaluation distinguishes false alarms from missed failures, reflecting their asymmetric operational impact. Performance is assessed at both asset and fleet level to confirm robustness and generalisability.

Deployment, Integration, and Closed-Loop Operationalisation

Deployment marks the transition from analytical experimentation to operational impact. Predictive outputs must integrate into Manufacturing Execution Systems (MES), Computerised Maintenance Management Systems (CMMS), or Enterprise Asset Management (EAM) platforms to enable closed-loop execution (Mobley, 2002). Through standardised Application Programming Interfaces, predictive insights can automatically trigger work orders, adjust preventive schedules, support spare-parts planning, and inform workforce allocation. Without this integration, PdM risks becoming an isolated dashboard with limited organisational influence.

Monitoring, Model Drift, and Lifecycle Governance

Predictive models degrade over time as equipment ages, processes change, and sensors are replaced. Continuous performance monitoring is therefore essential to maintain reliability and credibility (Zonta et al., 2020). Lifecycle governance encompasses drift detection, retraining policies, periodic audits, version control, and documentation of model changes—practices that align PdM with MLOps principles and support accountability in regulated industries.

Continuous Improvement and Enterprise-Scale Expansion

The final lifecycle stage focuses on scaling successful pilots across similar assets, production lines, or facilities. Lessons from initial deployments inform standardised architectures, reusable analytical components, and best-practice workflows. Clear ownership of data, models, and decision authority prevents fragmentation and ensures consistency across sites. Through iterative refinement, PdM evolves from a localised technical initiative into a strategic capability embedded within the broader Industry 4.0 transformation (Lee et al., 2015).

EVOLUTION OF PREDICTIVE MAINTENANCE IN THE INDUSTRY 4.0 ERA

The progression of maintenance strategies from reactive correction to predictive intelligence reflects a deeper industrial transformation rather than a sequence of incremental improvements. Each shift was driven by new production realities that exposed the limitations of existing approaches.

Early industrial settings were dominated by corrective maintenance: interventions performed only after failure. This was sustainable when machinery was simple, production lines were loosely coupled, and downtime costs were modest. As automation expanded, unplanned stoppages cascaded across interconnected systems, and the unpredictability of reactive maintenance became economically and operationally untenable (Lei et al., 2018).

Preventive maintenance introduced time- and usage-based inspection schedules, reducing unpredictability but introducing a different inefficiency: maintenance actions were performed irrespective of equipment condition. Healthy components were replaced prematurely; incipient faults sometimes went undetected between scheduled visits. In high-automation environments, these mismatches accumulated into significant hidden operational costs.

Condition-based maintenance (CBM) marked a conceptual step forward by incorporating direct measurements of equipment health. Monitoring vibration, temperature, and pressure against predefined thresholds improved responsiveness, but CBM remained largely diagnostic. Static thresholds lacked sensitivity to context and could not forecast future degradation, limiting the ability to plan proactively.

Industry 4.0—combining cyber-physical systems, ubiquitous sensing, high-speed communications, and scalable computing—created the conditions for a genuinely predictive approach. Continuous data streams from embedded sensors enabled the analysis of subtle degradation patterns invisible to periodic inspection. Machine learning introduced the capacity to model non-linear relationships between operating conditions and failure mechanisms and to estimate Remaining Useful Life under variable conditions (Serradilla et al., 2022). Edge computing resolved the latency problem that had constrained cloud-only analytics, enabling near-real-time diagnostics at the asset level (Lu et al., 2023).

Contemporary PdM extends beyond failure prediction to encompass maintenance scheduling optimisation, spare parts planning, energy efficiency improvement, and risk-informed decision-making. Maintenance is no longer treated as a reactive cost centre but as a strategic function contributing to operational resilience, sustainability, and competitiveness. Predictive insights are increasingly embedded within MES, EAM, and AR-based execution interfaces, reinforcing the view that PdM represents not the latest stage in maintenance evolution, but a structural redefinition of how industrial reliability is achieved in connected, data-intensive environments.

FAILURE MODES, DEGRADATION MECHANISMS, AND RELIABILITY-CENTRED LINKAGE

The industrial credibility of predictive maintenance depends on its alignment with established reliability engineering principles. Advanced analytics and IoT connectivity are valuable only when grounded in the physical realities of how machines degrade and fail. Without that grounding, data-driven predictions may be statistically impressive yet operationally untrusted. Integrating PdM with FMEA, RCM, Weibull reliability modelling, and fault tree analysis ensures that predictive insights are interpretable, actionable, and accepted by maintenance practitioners.

Common Failure Modes in Industrial Assets

Rotating equipment—motors, gearboxes, pumps, compressors—accounts for the majority of industrial maintenance events. Bearings, among the most failure-prone components, typically fail through surface fatigue, spalling, lubrication breakdown, or contamination-induced wear. These mechanisms originate at the microstructural level and progress gradually, generating characteristic vibration and acoustic signatures well in advance of catastrophic failure.

Gear failures include pitting, scuffing, tooth cracking, and misalignment-induced wear. Early-stage defects manifest as subtle harmonic distortions in vibration spectra; advanced damage produces impulsive responses and elevated noise. Electric motors present failure modes spanning multiple physical domains—stator winding insulation degradation, rotor bar cracking, eccentricity, and thermal overload—each detectable through combinations of current signatures, temperature gradients, and vibration patterns. Pumps and compressors add process-coupled modes: cavitation, seal wear, impeller erosion, and flow-induced vibration require correlation of mechanical indicators with pressure, flow, and fluid property data.

Degradation Trajectories and Progressive Failure Behaviour

Most industrial failures are progressive rather than instantaneous. Incipient defects generate weak, intermittent signals masked by operational noise; as damage accumulates, amplitude, frequency content, and statistical properties change in systematic ways. Recognising these patterns allows PdM systems to distinguish genuine deterioration from normal variability and to extend the lead time available for planned intervention.

Degradation paths are seldom linear. Load variations, environmental changes, and partial maintenance actions can accelerate, decelerate, or temporarily suppress observable indicators. Predictive models must therefore accommodate stochastic, non-monotonic behaviour rather than assuming smooth trajectories—a requirement that underscores the value of physics-informed constraints alongside data-driven learning.

Weibull Reliability Modelling and Failure Statistics

Weibull analysis characterises time-to-failure behaviour and failure rate evolution through a flexible shape parameter that accommodates early-life failures, random failures, and wear-out phenomena. Integrating Weibull-based population statistics with individual asset condition data allows maintenance teams to contextualise observed behaviour within broader failure distributions, supporting risk-informed prioritisation. Hybrid frameworks combining reliability statistics with machine learning are increasingly adopted in industrial PdM programmes, balancing predictive accuracy with physical interpretability (Jardine et al., 2006).

FMEA as a PdM Design Instrument

Failure Modes and Effects Analysis serves as a critical bridge between traditional maintenance engineering and predictive analytics. In PdM system design, FMEA is not a documentation exercise but a structured method for aligning sensors, features, and models with known failure mechanisms. Each failure mode is assessed for severity, occurrence probability, and detectability, guiding sensor selection toward the physical parameters most sensitive to early-stage degradation. Mapping FMEA outputs to predictive models reduces false alarms, improves diagnostic relevance, and builds trust among maintenance personnel. Systems designed without FMEA grounding frequently suffer from poor interpretability and limited operational acceptance.

Reliability-Centred Maintenance and PdM Integration

Reliability-Centred Maintenance provides a decision framework that assigns maintenance strategy—corrective, preventive, or condition-based—according to functional importance, failure consequences, and risk tolerance. PdM fits naturally within RCM as the preferred strategy for assets where failure consequences justify proactive intervention and monitoring is technically feasible. Integrating PdM outputs into RCM workflows ensures that predictive alerts translate into maintenance policies rather than isolated warnings, reinforcing PdM as a strategic component of enterprise asset management rather than a standalone analytical tool.

Fault Tree Analysis and Causal Interpretation

Fault tree analysis provides a causal representation of how component-level failures propagate to system-level outcomes. Associating sensor-derived indicators with fault tree nodes gives predictive systems explanatory capability: maintenance teams can distinguish symptoms from root causes, reducing unnecessary interventions and improving diagnostic precision. This causal structure directly addresses explainability—one of the most frequently cited barriers to PdM adoption in safety-critical environments.

NEXT-GENERATION IoT SENSOR ECOSYSTEMS FOR PREDICTIVE MAINTENANCE

The effectiveness of any PdM system is bounded by the quality and interpretability of the data it acquires. Sensing has evolved from a peripheral monitoring activity into a core engineering function that determines diagnostic accuracy, prognostic reliability, and decision confidence. Modern PdM implementations depend on sensor ecosystems that combine physical robustness, high-resolution measurement, embedded intelligence, and interoperable communication.

Role of Intelligent Sensing in PdM Architectures

Traditional industrial sensors were designed for threshold-based alarm generation. Adequate for detecting gross abnormalities, they are ill-suited for early-stage degradation detection, where fault signatures are weak and easily masked by operational noise. Predicting failure requires sensors that capture subtle variations in physical behaviour over extended periods, often under harsh and variable conditions. IoT-enabled sensors address this by providing continuous, high-frequency measurements that reveal incipient fault development. Inadequate sensing at this stage cannot be compensated by sophisticated analytics downstream.

MEMS-Based Sensors and Distributed Monitoring

Micro-Electro-Mechanical Systems (MEMS) sensors have become a cornerstone of contemporary PdM implementations. Compact size, low power consumption, and cost-effectiveness make MEMS accelerometers and gyroscopes practical for vibration and motion monitoring across motors, gearboxes, and pumps. Their scalability enables dense, spatially distributed deployment rather than reliance on a small number of measurement points (Sabato et al., 2017). MEMS sensors offer sufficient sensitivity to capture frequency components associated with bearing defects, imbalance, and misalignment while maintaining mechanical robustness under industrial operating conditions.

Complementary Sensor Modalities for Multi-Physics Monitoring

PdM systems increasingly employ multi-modal sensing strategies to capture degradation mechanisms that vibration data alone cannot reveal. Thermal sensors provide non-contact measurement of surface temperature distributions, enabling early detection of overheating from friction, electrical faults, or lubrication failure. Acoustic and ultrasonic sensors detect high-frequency emissions from micro-cracks, cavitation, or early bearing damage—often before vibration changes become detectable.

Current and voltage transducers play a critical role in monitoring electric motors and drives. Current signature analysis enables detection of rotor bar defects, stator insulation degradation, and load anomalies without direct mechanical access. Combined with mechanical sensors, electrical measurements support cross-domain fault validation, reducing false alarms and improving diagnostic confidence. The roles of commonly adopted sensor technologies are summarised in Table 2.

Table 2. Advanced IoT Sensor Technologies and Their Roles in Predictive Maintenance

Sensor Type	Measured Parameter	Typical PdM Application	Engineering Advantage
MEMS Accelerometers	Vibration, acceleration	Rotating machinery diagnostics	Compact, low power, scalable for fleet deployment
Piezoelectric Sensors	Dynamic strain, vibration	Bearing and gearbox fault detection	High sensitivity to impulse events
Thermal / Infrared Sensors	Surface and component temperature	Overheating and insulation fault detection	Non-contact; no process interruption
Acoustic Emission Sensors	Sound and acoustic emission	Early-stage fault diagnosis and crack detection	Detects micro-level defects before vibration changes
Ultrasonic Sensors	Thickness, leakage, flow	Corrosion monitoring and leakage detection	Subsurface fault identification
Current / Voltage Transducers	Electrical signatures	Motor rotor and stator fault detection	No mechanical access required

Source: Adapted from Jardine et al. (2006), Sabato et al. (2017), and Pech et al. (2021).

Embedded Intelligence and Edge-Aware Sensing

A defining characteristic of next-generation IoT sensors is integrated processing capability. Rather than functioning as passive data collectors, modern sensors increasingly perform preliminary signal conditioning, feature extraction, and anomaly screening at the device level. This design reduces transmission volumes, filters irrelevant data before it reaches analytics platforms, and enables near-real-time response in latency-sensitive applications such as high-speed production lines or safety-critical machinery.

Interoperability and Industrial Integration Considerations

Sensor ecosystems operate within heterogeneous environments that span legacy equipment, modern control systems, and enterprise software. Interoperability is therefore a prerequisite rather than a secondary feature. Industrial IoT sensors increasingly support standardised communication protocols and data formats that facilitate seamless integration with analytics engines, MES, and maintenance management platforms. Interoperable architectures reduce vendor lock-in, simplify expansion, and support long-term sustainability of PdM programmes.

EDGE COMPUTING AND INTELLIGENT DIAGNOSTICS IN PREDICTIVE MAINTENANCE

High-frequency sensor streams generated by vibration, acoustic, thermal, and electrical monitoring quickly overwhelm cloud-centric analytics architectures. Transmitting all raw data to centralised servers introduces latency, bandwidth congestion, and cybersecurity exposure that are incompatible with real-time maintenance decision-making. Edge computing has emerged as the architectural response to these constraints.

Rationale for Edge Computing in PdM Architectures

Early fault indicators frequently manifest as transient signal anomalies requiring immediate interpretation to prevent cascading failures. Cloud-only architectures cannot reliably meet these responsiveness requirements, particularly in geographically distributed plants or bandwidth-constrained environments. Relocating computation to the machine, controller, or gateway level reduces end-to-end latency and supports near-real-time anomaly detection. This capability is especially valuable in tightly coupled production systems where a delayed response can propagate a fault across multiple assets.

Edge–Cloud Partitioning and Functional Allocation

Effective PdM systems distribute computation according to latency sensitivity, computational complexity, and data aggregation requirements rather than routing everything to one tier. Edge nodes handle signal preprocessing, feature extraction, anomaly detection, and threshold-based diagnostics—tasks that benefit from asset proximity and require modest resources. Cloud platforms carry computationally intensive workloads: deep model training, fleet-level trend analysis, historical data mining, and long-term optimisation. This functional partitioning keeps systems both fast and analytically robust. The architectural interaction between edge diagnostics and cloud analytics is illustrated conceptually in Figure 2 (Lee et al., 2015; Lu et al., 2023).

Figure 2. Edge–cloud processing framework for predictive maintenance.

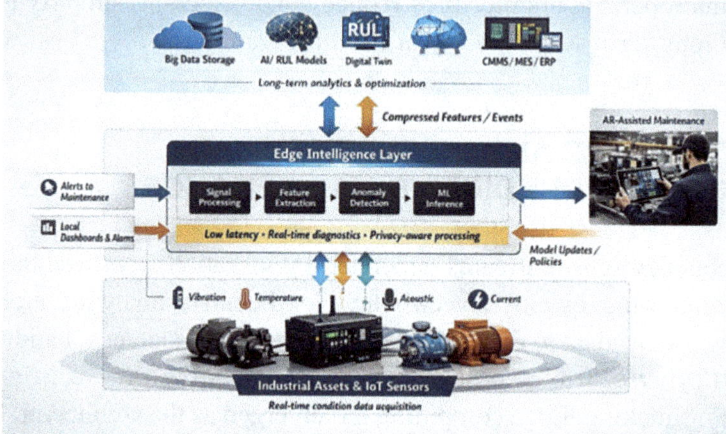

The figure illustrates sensor-level data acquisition, edge-based preprocessing and anomaly detection, and cloud-level model training, prognostics, and enterprise integration. Source: Adapted conceptually from Lee et al. (2015) and Lu et al. (2023).

Intelligent Diagnostics at the Edge

Edge platforms increasingly incorporate lightweight machine learning models optimised for resource-constrained environments—limited memory, processing power, and energy budgets. Typical edge-level analytics include statistical anomaly detection, compact neural networks, and rule-augmented classifiers. Intelligent diagnostics at the edge filter noise, prioritise critical events, and suppress irrelevant operational variability before data reaches higher-level platforms. This selective handling reduces false alarms and improves the interpretability of predictive outputs, addressing one of the most persistent barriers to PdM adoption: technician distrust of automated alerts. Edge diagnostics also ensure operational continuity during network disruptions, preserving maintenance functionality and safety assurance when cloud access is unavailable.

Data Reduction, Security, and Privacy Advantages

Processing raw sensor data locally minimises the volume of sensitive operational information transmitted across networks. Only summarised features, anomaly flags, or encrypted insights are forwarded to centralised systems, reducing exposure to cyber threats. This model aligns with regulatory and proprietary constraints in energy, aerospace, and defence manufacturing, and supports defence-in-depth cybersecurity strategies by limiting attack surfaces and isolating critical diagnostics from external networks.

Engineering Trade-Offs and Design Considerations

Edge computing introduces trade-offs that must be managed explicitly. Limited computational resources constrain model complexity, requiring careful algorithm and feature selection. Updating edge-deployed models across distributed nodes demands robust lifecycle management mechanisms for version control and performance monitoring. Modern PdM architectures address these challenges through modular edge software stacks that support remote updates and continuous drift monitoring, ensuring that edge intelligence evolves alongside equipment behaviour and process changes.

MAINTENANCE DATA ENGINEERING AND SIGNAL PROCESSING

Practical PdM effectiveness depends on the quality, structure, and engineering integrity of the data supplied to analytical models. Raw industrial sensor data are inherently noisy, non-stationary, and influenced by variable operating regimes. Without disciplined data engineering and signal processing, sophisticated AI models produce unstable predictions, excessive false alarms, and limited operational trust

(Tsui et al., 2015). Data engineering must therefore be regarded as a foundational engineering layer within the PdM lifecycle rather than a preprocessing afterthought.

Sampling Strategy and Data Acquisition Design

Sampling rates must be selected based on the characteristic frequencies of expected degradation mechanisms, not on arbitrary data availability. For rotating machinery, vibration signals often require sampling frequencies several times higher than the highest fault-related frequency to capture transient impulsive events accurately. Undersampling causes aliasing, where high-frequency fault signatures appear as lower-frequency components, producing misleading diagnostic indicators. Anti-aliasing filters are applied prior to digitisation, and oversampling is commonly adopted to preserve signal fidelity under variable operating conditions. These choices embody a fundamental trade-off between data resolution, storage requirements, and computational burden that must be managed deliberately when scaling across large fleets.

Windowing, Segmentation, and Temporal Context

Industrial sensor signals are segmented into temporal windows to enable localised analysis of system behaviour. Window length and overlap selection directly affect diagnostic sensitivity and response time. Short windows improve responsiveness to transient anomalies but increase variance; longer windows improve statistical stability at the cost of detection delay. Windowing functions such as Hamming or Hann windows reduce spectral leakage during frequency analysis. Overlapping windows ensure continuity in monitoring and prevent blind spots in degradation tracking.

Frequency-Domain Analysis and Spectral Feature Extraction

Frequency-domain analysis remains a cornerstone of PdM for rotating equipment, where mechanical faults manifest as characteristic spectral signatures. The Fast Fourier Transform converts time-domain vibration signals into frequency spectra revealing harmonics associated with imbalance, misalignment, bearing defects, and gear damage. Spectral features including dominant frequency amplitudes, band-limited energy, harmonic ratios, and sideband structures isolate fault-related components from background noise. However, FFT assumes signal stationarity within the analysis window—an assumption frequently violated in real environments with variable load and speed.

Time–Frequency Analysis for Non-Stationary Behaviour

Short-Time Fourier Transform and wavelet transforms handle non-stationary signals by examining frequency content over time. STFT is effective for tracking how fault signatures evolve during gradual degradation. Wavelet transforms adapt time and frequency resolution to signal content, making them particularly sensitive to impulsive events from early-stage bearing and gear faults. Wavelet-based energy coefficients and entropy measures are now routinely integrated into PdM pipelines for their ability to capture localised degradation phenomena invisible in conventional spectral representations.

Order Tracking for Variable-Speed Machinery

Many industrial assets operate under variable-speed conditions, rendering fixed-frequency analysis unreliable. Order tracking analyses vibration signals relative to rotational speed rather than absolute frequency, synchronising sensor data with rotational reference signals to enable consistent fault diagnosis across changing operating regimes. This technique is essential in automotive manufacturing, aerospace propulsion, and process industries where speed variation is intrinsic. Incorporating order tracking into PdM pipelines reduces false alarms caused by benign speed fluctuations and significantly improves robustness.

Feature Families, Health Indicator Construction, and Feature Governance

Following preprocessing, quantitative features representing equipment condition are extracted from three families: time-domain statistics (capturing amplitude and impulsiveness), frequency-domain features (characterising spectral distribution), and time–frequency features (describing evolving degradation). Combining features across domains produces richer health representations. Feature selection is guided by failure-mode relevance rather than algorithmic convenience, reinforcing the engineering foundation of the PdM pipeline.

As PdM systems scale, managing extracted features becomes a non-trivial challenge. Feature stores provide centralised repositories that standardise feature definitions, ensure consistency across models, and support version control. In safety-critical industries, feature traceability supports regulatory auditing and accountability, and aligns PdM implementations with emerging MLOps principles.

ARTIFICIAL INTELLIGENCE ALGORITHMS FOR PREDICTIVE MAINTENANCE

Artificial intelligence has become a central analytical enabler of PdM, yet its role is frequently misunderstood. Algorithms are often presented as isolated technical solutions rather than components within a broader maintenance decision framework. The value of AI in PdM lies not in algorithmic sophistication alone, but in its capacity to support diagnostic understanding, prognostic forecasting, and prescriptive decision-making under real operational constraints (Carvalho et al., 2019). PdM analytics are best organised into three functional layers: diagnostic, prognostic, and prescriptive intelligence, each serving a distinct purpose and imposing different requirements on data, models, and validation. Table 3 summarises widely adopted AI techniques and their functional roles.

Table 3. Artificial Intelligence Techniques Applied in Predictive Maintenance

AI Technique	Primary Function	PdM Application	Key Consideration
Convolutional Neural Networks (CNN)	Automated feature extraction	Visual inspection; vibration-based fault detection	Requires sufficient labelled data; high compute at edge
Long Short-Term Memory (LSTM)	Sequential time-series modelling	Remaining Useful Life (RUL) estimation	Sensitive to noise; needs time-aware validation
Reinforcement Learning (RL)	Policy optimisation under uncertainty	Maintenance scheduling and decision optimisation	Limited industrial adoption; explainability challenges
Generative Adversarial Networks (GAN)	Synthetic data generation	Rare fault modelling and data augmentation	Training instability; quality validation required
Random Forest (RF)	Ensemble-based classification	Fault severity assessment; condition classification	Robust with limited data; interpretable feature importance
Transformer Architectures	Long-range dependency modelling	Multi-sensor RUL estimation; complex degradation patterns	High computational cost; best suited for cloud inference
Graph Neural Networks (GNN)	Relational and graph-structured learning	Multi-component fault propagation modelling	Requires graph-structured asset topology data

Source: Compiled and adapted from *Zonta et al. (2020)*, *Zheng et al. (2017)*, *Sánchez et al. (2018)*, *Carvalho et al. (2019)*, and *Serradilla et al. (2022)*.

Diagnostic Intelligence: Fault Detection and Classification

Diagnostic analytics determine whether an asset is operating abnormally and identify the most likely fault category. Accuracy and timeliness of fault recognition, rather than long-horizon forecasting, are the primary objectives at this layer. Classical methods such as Random Forests and Support Vector Machines remain highly effective when failure modes are well characterised and feature sets are physically meaningful. Their robustness, interpretability, and modest data requirements frequently make them preferable to complex architectures when labelled failure data are limited.

Convolutional Neural Networks have demonstrated superior performance when diagnostic inputs involve high-dimensional representations such as time–frequency maps or thermal images. By automatically extracting hierarchical features from minimally processed signals, CNNs reduce dependence on handcrafted feature engineering. However, their deployment requires sufficient data and computational capacity, particularly when models are executed at the edge where resources are constrained.

Prognostic Intelligence: Remaining Useful Life Estimation

Prognostic analytics extend beyond fault identification to estimate how long an asset can continue operating before failure—a substantially harder problem because it requires modelling degradation dynamics rather than detecting static abnormalities. LSTM networks are widely adopted for this purpose because of their capacity to capture long-range temporal dependencies in sequential sensor data. These models learn degradation trajectories from historical time-series and update predictions dynamically as new observations arrive.

Purely data-driven prognostic models must be treated with caution. RUL predictions are sensitive to noise, operating regime changes, and incomplete failure histories. Uncertainty-aware prognostics—using ensembles, Bayesian inference, or probabilistic output layers—are increasingly preferred over deterministic point estimates, as they align predictive outputs with risk-based maintenance decision-making (Si et al., 2011).

Prescriptive Intelligence: Maintenance Decision Optimisation

Prescriptive analytics address what action should be taken and when, given competing objectives such as cost, risk, and production continuity. Reinforcement Learning evaluates maintenance actions—inspection, repair, replacement—based on cumulative reward functions encoding operational objectives, learning adaptive policies that balance early intervention against unnecessary maintenance. Prescriptive models are particularly valuable in multi-asset scenarios where decisions must account for resource constraints, spare-part availability, and workforce scheduling. Industrial adoption remains limited by explainability and policy validation challenges;

RL-based approaches are therefore most effective as decision-support tools rather than autonomous controllers.

Addressing Data Scarcity through Generative and Transfer Learning

Industrial assets are designed to fail infrequently, which inherently limits labelled training data for supervised learning. Generative Adversarial Networks partially address this by synthesising realistic fault scenarios to augment training datasets. When used judiciously, synthetic data improve model robustness without requiring costly or hazardous real-world failure experiments. Transfer learning mitigates data limitations further by adapting knowledge from related assets, operating conditions, or simulated environments, reducing both deployment time and data requirements—particularly relevant for heterogeneous fleets sharing common failure characteristics.

Hybrid and Interpretable AI for Industrial Acceptance

Purely data-driven models encounter resistance in industrial environments because of their limited transparency. Hybrid approaches that combine physics-based degradation understanding with machine learning constrain predictions within physically plausible bounds and improve generalisation under changing conditions. Explainable AI techniques—SHAP values, attention maps, saliency methods—provide insight into feature relevance and decision pathways, enabling maintenance engineers to justify actions based on predictive outputs. Such interpretability is essential in safety-critical applications where traceability matters as much as prediction accuracy.

Emerging Architectures: Graph Neural Networks and Self-Supervised Learning

Graph Neural Networks (GNNs) represent a particularly promising frontier for industrial PdM. Industrial assets rarely operate in isolation: pumps feed reactors, motors drive conveyors, and compressors supply entire process lines. Failures propagate through these topological relationships in ways that independent per-asset models cannot capture. GNNs model assets and their physical or functional connections as graph structures, learning how degradation in one node affects the health of neighbouring components. This relational learning capability makes GNNs especially well-suited for predictive maintenance in interconnected process systems, energy grids, and transportation fleets where fault propagation pathways are as important as the faults themselves (Serradilla et al., 2022).

Self-supervised learning addresses the persistent challenge of labelled data scarcity by training models on the structure of unlabelled sensor data itself—learning robust representations without requiring annotated failure events. Pre-trained self-supervised models can then be fine-tuned on small labelled datasets, dramatically reducing the annotation burden while preserving diagnostic accuracy. Combined with domain adaptation techniques that transfer learned representations across asset types or operating environments, these approaches open a path toward scalable, fleet-level predictive maintenance without dependence on large, site-specific failure archives.

REMAINING USEFUL LIFE MODELLING AND PROGNOSTICS

Estimating Remaining Useful Life (RUL) represents one of the most valuable yet technically demanding objectives of predictive maintenance. Fault diagnosis answers whether degradation has occurred; RUL estimation addresses the forward-looking question of how long an asset can continue operating within acceptable performance and safety limits. This capability transforms maintenance from reactive response into proactive lifecycle management, enabling optimised scheduling, inventory planning, and risk mitigation.

Conceptual Foundations of RUL Estimation

RUL is defined as the time interval between the current observed asset state and the point at which the asset can no longer fulfil its intended function. Unlike deterministic lifetime estimates derived from design assumptions, RUL is inherently uncertain and dynamic, evolving continuously as operating conditions, loads, maintenance actions, and environmental factors change. Effective RUL estimation requires models representing degradation as a stochastic process rather than a fixed trajectory, capable of accommodating cross-asset variability, non-linear behaviour, and incomplete observation of underlying failure mechanisms. The practical objective is not absolute prediction accuracy, but reliable lead time sufficient for informed maintenance decisions (Si et al., 2011).

Degradation-Based Modelling and Health Indicators

Degradation-based models track the evolution of health indicators over time, treating failure as a continuous process rather than a binary event. RUL is estimated by extrapolating degradation trajectories until predefined failure thresholds are reached.

Common indicators include increasing vibration amplitude, rising temperature trends, declining efficiency, and changes in energy consumption. These are modelled using regression techniques, stochastic processes, or state-space formulations. Such models are most effective when failure mechanisms are well understood and degradation is quasi-monotonic. Industrial degradation is rarely smooth, however: load changes, maintenance interventions, and environmental variability can temporarily mask or reverse observable trends, requiring adaptive models that incorporate uncertainty.

Data-Driven RUL Estimation Using Sequential Learning

LSTM networks are widely used for RUL estimation because of their capacity to capture long-range temporal dependencies in multivariate time-series data. These models learn mappings between sequences of condition-monitoring features and corresponding RUL targets, updating predictions dynamically as new data arrive. Strong performance has been demonstrated across rotating machinery, aerospace components, and energy systems. Transformer architectures, relying on attention mechanisms that selectively weight the most informative portions of input sequences, have emerged as a powerful alternative for scenarios involving multi-component systems or irregular operating patterns. Their computational complexity makes them best suited for cloud-based prognostics rather than real-time edge inference (Serradilla et al., 2022).

Uncertainty Quantification and Risk-Aware Decision Thresholds

A critical limitation of many RUL approaches is their reliance on point predictions that convey no uncertainty. Decision-makers need to understand not only when failure is likely, but how confident the prediction is. Overoptimistic estimates risk catastrophic failure; overly conservative predictions drive unnecessary maintenance and cost. Uncertainty-aware RUL modelling addresses this through prediction intervals generated by ensemble learning, Bayesian neural networks, Monte Carlo dropout, or probabilistic state-space models. These methods distinguish epistemic uncertainty (model uncertainty) from aleatory uncertainty (inherent variability), enabling risk-informed maintenance planning aligned with asset criticality and organisational risk tolerance. Engineers are also more likely to act on predictions that explicitly acknowledge uncertainty than on outputs that appear unjustifiably precise.

Calibration, Validation, and Model Drift

RUL models must be continuously validated against observed outcomes. Calibration techniques compare predicted RUL distributions with actual failure times and adjust parameters accordingly. Poorly calibrated models may appear accurate on average while systematically over- or under-estimating remaining life. Evolving operating conditions, equipment upgrades, and process changes introduce model drift, which must be addressed through continuous performance monitoring, retraining policies, and periodic performance audits—key elements of lifecycle governance aligned with MLOps principles.

SYSTEM INTEGRATION: APIs, MES, CMMS, AND CLOSED-LOOP EXECUTION

Predictive maintenance generates value only when analytical insights are translated into timely, coordinated maintenance actions. Many PdM initiatives fail not because of weak prediction accuracy but because predictive outputs remain isolated from operational decision-making systems. System integration is therefore the decisive factor separating a functioning strategic capability from a disconnected analytical experiment.

Need for Closed-Loop Integration

Industrial environments are heterogeneous, comprising legacy equipment, modern IoT-enabled machines, enterprise software platforms, and human operators. Without integration into execution systems, predictive insights typically manifest as standalone dashboards or alerts that lack authority, traceability, and operational follow-through. Closed-loop integration ensures that predictive intelligence flows bidirectionally: from sensors to analytics, and from analytics back to maintenance execution and performance feedback. This loop enables continuous model improvement and transforms PdM into a living operational system.

Role of Application Programming Interfaces (APIs)

APIs serve as the foundational mechanism for interoperability in PdM architectures. They enable standardised data exchange between sensor platforms, analytics engines, MES, CMMS, and EAM platforms. Through API-based integration, predictive insights such as anomaly flags, RUL estimates, and risk scores are transmitted automatically to downstream systems, reducing manual intervention, minimising latency, and ensuring consistent, traceable maintenance triggering. API-based architectures also offer scalability, modularity, and vendor independence—critical attributes in environments characterised by rapid technological evolution.

Integration with Manufacturing Execution Systems (MES)

MES coordinates shop-floor activities, monitors production performance, and manages resource allocation. Connecting PdM with MES aligns maintenance actions with production schedules, preventing unplanned disruptions and optimising throughput. Predictive alerts evaluated in the context of production priorities and batch schedules allow maintenance to be scheduled during planned windows rather than triggered by failures. This alignment directly enhances Overall Equipment Effectiveness and supports coordinated decision-making between maintenance and production teams.

CMMS and EAM Integration for Maintenance Execution

While MES integration aligns maintenance with production, CMMS and EAM platforms operationalise execution by managing work orders, spare parts inventory, labour allocation, and maintenance history. Predictive outputs integrated into CMMS/EAM automatically generate work orders, recommend inspection or replacement actions, and update schedules dynamically. RUL predictions can trigger conditional work orders that escalate in priority as failure probability rises, reducing emergency repairs, improving spare-parts planning, and supporting evidence-based maintenance prioritisation across the asset portfolio.

Closed-Loop Feedback and Continuous Learning

Mature PdM systems capture maintenance outcomes and feed them back into analytical models. Work-order completion data, inspection findings, and failure confirmations provide ground truth that improves model calibration and reduces future prediction error. This feedback mechanism supports accountability: maintenance actions are traceable to predictive triggers, enabling performance auditing and continuous improvement. Over time, closed-loop learning transforms PdM from a static predictor into an adaptive decision-support platform that evolves alongside equipment behaviour.

Engineering and Governance Considerations

Integration introduces challenges related to data consistency, cybersecurity, and system governance. APIs must be secured through authentication, encryption, and access control to prevent unauthorised intervention. Version control and documentation are essential for traceability of predictive decisions, particularly in regulated or safety-critical industries. Successful deployments adopt standardised data models, clearly defined system interfaces, and governance frameworks aligned with MLOps principles, ensuring that PdM systems remain reliable, auditable, and sustainable.

ROI/KPI Framework for Predictive Maintenance Programmes

Demonstrating business value is as important as technical performance. Predictive maintenance programmes should be measured against a structured set of Key Performance Indicators that span operational, financial, and reliability dimensions. Core KPIs include: Overall Equipment Effectiveness (OEE), which captures the combined impact of availability, performance, and quality; Mean Time Between Failures (MTBF), reflecting improvement in asset reliability; Mean Time to Repair (MTTR), measuring the efficiency of fault detection and intervention; unplanned downtime as a percentage of scheduled production time; maintenance cost ratio (total maintenance cost as a percentage of asset replacement value); and spare-parts inventory turns, which indicate the effectiveness of condition-based stocking. Return on investment should be calculated over a three-to-five-year horizon that accounts for implementation cost, ongoing operational cost, and the quantified value of avoided failures—including production losses, emergency labour, and secondary damage. Industry evidence cited in the case studies of this chapter indicates that well-implemented PdM programmes achieve ROI of 8–12 times the programme cost over a five-year period, driven primarily by reductions in unplanned downtime and emergency repair expenditure (Zonta et al., 2020; Rakholia et al., 2025).

GIS Integration for Spatially Distributed Asset Management and Sustainable Maintenance

Geographic Information Systems (GIS) provide a spatial dimension that is particularly valuable for predictive maintenance programmes spanning distributed infrastructure. In sectors such as energy transmission, water utilities, rail networks, pipeline systems, and urban facilities management, assets are physically dispersed across large geographic areas. GIS enables the spatial overlay of predictive health indicators—RUL estimates, anomaly flags, and criticality scores—onto geographic maps, allowing maintenance planners to optimise inspection routing, technician dispatch, and parts logistics based on both asset condition and physical location. A pump approaching its failure threshold in a remote area may warrant different intervention timing than an identical pump accessible from a central depot; GIS-based PdM integration captures this context explicitly.

GIS also contributes directly to sustainable manufacturing and maintenance practices. By rationalising travel routes for maintenance personnel and vehicles, GIS-integrated PdM reduces fuel consumption, fleet emissions, and logistics cost. Condition-based maintenance planning guided by spatial analytics minimises unnecessary site visits—a significant sustainability gain in geographically extensive operations such as offshore wind farms, railway infrastructure, or national pipeline networks. Furthermore, GIS supports environmental compliance by enabling spatial correlation between asset health data and environmental sensitivity maps, allowing maintenance decisions to account for proximity to protected areas, floodplains, or populated zones when scheduling interventions with potential environmental impact. Integrating GIS within the PdM data architecture therefore strengthens both operational efficiency and the environmental stewardship objectives of Industry 4.0-aligned manufacturing and infrastructure organisations (Lee et al., 2015; Tao et al., 2019).

AUGMENTED REALITY-ASSISTED MAINTENANCE AND HUMAN-CENTRED EXECUTION

Predictive maintenance derives value from the quality and timeliness of human action, not from predictions alone. Maintenance execution remains a human-intensive activity involving diagnosis, intervention, verification, and documentation. Augmented Reality (AR) has emerged as a critical interface layer bridging predictive analytics and physical maintenance work, ensuring that insights are translated into correct, timely on-site actions.

Role of AR in the Predictive Maintenance Workflow

AR overlays digital information onto the physical workspace, allowing maintenance personnel to access contextual guidance while directly interacting with equipment. Within PdM systems, AR delivers predictive outputs—fault location, severity, and recommended actions—at the point of intervention. Rather than replacing human expertise, AR reduces cognitive load, minimises interpretation errors, and improves procedural consistency. This is especially valuable in environments characterised by complex machinery, limited documentation, or variable workforce skill.

Integration of Predictive Insights with AR Interfaces

When connected to PdM platforms, AR applications dynamically display asset health indicators, historical trends, and RUL-based urgency cues. Components approaching failure thresholds can be visually highlighted, guiding technicians to critical intervention points. Step-by-step instructions, exploded component views, and safety warnings are synchronised with predictive alerts to ensure that maintenance actions align with analytical recommendations. This tight coupling addresses a common failure mode of PdM systems, where technically accurate predictions are operationally ignored because they lack clarity or contextual grounding (Tao et al., 2019). The AR-assisted PdM workflow is illustrated in Figure 3.

Figure 3. Workflow of an AR-assisted predictive maintenance process.

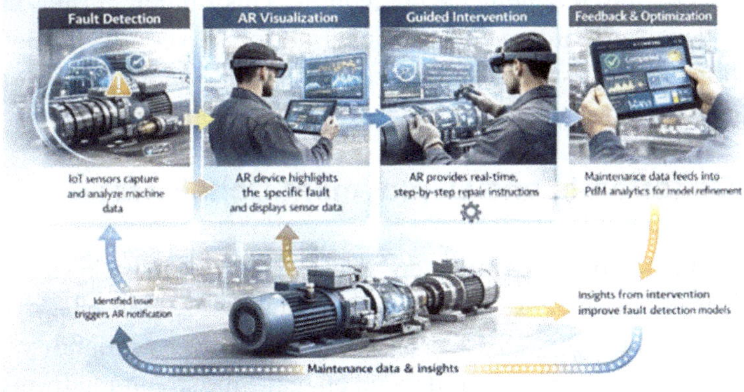

The figure integrates predictive analytics, contextual AR visualisation, guided maintenance execution, and feedback into the maintenance management system. Source: Adapted from Lee et al. (2015), Tao et al. (2019), and Zonta et al. (2020).

Human Factors, Trust, and Alarm Fatigue

A significant barrier to PdM adoption is technician scepticism of algorithmic recommendations. Maintenance personnel rely on experiential judgement developed over years of practice, and opaque AI outputs can be perceived as intrusive or unreliable. AR-assisted maintenance mitigates this by making predictive insights interpretable and contextually grounded. Visual explanations, confidence indicators, and historical comparisons allow technicians to understand why an intervention is recommended rather than merely being instructed what to do. This transparency fosters trust and supports gradual cultural adoption.

Alarm fatigue presents a related challenge: excessive or poorly prioritised alerts overwhelm technicians and reduce compliance. AR interfaces designed with criticality-based alert prioritisation—displaying only relevant, actionable information calibrated to asset risk level—prevent cognitive overload and improve decision quality under time pressure (Zonta et al., 2020).

Training, Knowledge Retention, and Workforce Enablement

Beyond real-time execution, AR plays a strategic role in workforce development. Advanced diagnostics may exceed the experience level of junior technicians; AR-guided workflows enable consistent execution across skill levels by embedding expert knowledge directly into maintenance procedures. This supports faster onboarding, reduces dependency on scarce expert personnel, and preserves organisational knowledge. Over time, AR-assisted execution contributes to a more resilient workforce capable of handling complex predictive insights with confidence.

Practical Constraints and Deployment Considerations

AR deployment introduces practical constraints that must be acknowledged honestly. Hardware ergonomics, field-of-view limitations, industrial lighting conditions, and device durability all affect usability. Connectivity reliability is a further critical factor when AR applications depend on real-time streaming from predictive platforms. Successful implementations address this through hybrid designs where core guidance is available locally and advanced analytics are synchronised when connectivity permits, ensuring continuity without over-reliance on ideal infrastructure.

STRUCTURED INDUSTRIAL CASE STUDIES IN PREDICTIVE MAINTENANCE

Industrial case studies validate predictive maintenance concepts by demonstrating how analytical methods translate into measurable operational outcomes. The five cases below use a consistent analytical framework—asset context, sensing strategy, analytical approach, deployment architecture, integration, and performance outcomes—to enable cross-domain comparison. Where possible, performance figures are grounded in published benchmark data.

Case Study 1: Rotating Machinery in Discrete Manufacturing

Asset Context: High-speed induction motors and gearboxes in an automotive assembly line.
Failure Modes: Bearing fatigue, shaft misalignment, lubrication degradation.
Sensing Strategy: Triaxial MEMS vibration sensors, surface temperature probes, motor current signature analysis.
Analytical Approach: Time–frequency feature extraction with LSTM-based anomaly detection and RUL estimation.
Deployment Architecture: Edge-based diagnostics with cloud-based trend analysis and periodic model retraining.
Integration: Predictive alerts linked to CMMS-generated work orders with automatic priority escalation.
Outcomes: Early fault detection enabled maintenance during scheduled production pauses, reducing emergency repairs. Literature benchmarks for comparable automotive deployments report unplanned downtime reductions of 15–25%, maintenance cost savings of 10–20%, and MTTR improvements of 20–30% relative to preventive baselines (Zonta et al., 2020; Rakholia et al., 2025). Bearing replacement intervals were optimised by condition rather than fixed schedules, reducing premature part replacement and improving Overall Equipment Effectiveness.

Case Study 2: Process Industry Equipment in Chemical Manufacturing

Asset Context: Centrifugal pumps and compressors in a continuous chemical processing plant.
Failure Modes: Seal leakage, cavitation, impeller erosion.
Sensing Strategy: Pressure fluctuation monitoring, acoustic emission sensors, flow rate sensors.
Analytical Approach: Statistical degradation modelling combined with survival analysis for failure probability estimation.
Deployment Architecture: Centralised cloud analytics integrated with plant historian systems.
Integration: Predictive insights embedded into EAM for maintenance prioritisation and spare-parts planning.
Outcomes: Early identification of cavitation-related degradation prevented catastrophic pump failures that had previously caused multi-day production losses. Maintenance actions were aligned with process constraints, reducing safety incidents and environmental risk. Industry studies on chemical and

process plant PdM report pump failure rate reductions of 20–40% and spare-part inventory reductions of up to 25% through improved failure anticipation (Carvalho et al., 2019).

Case Study 3: Wind Turbine Gearboxes in the Energy Sector

Asset Context: Wind turbine gearboxes and generators in an offshore wind farm.
Failure Modes: Gear tooth wear, insulation degradation, thermal stress.
Sensing Strategy: Vibration monitoring, thermal imaging, electrical signal analysis.
Analytical Approach: Hybrid physics-informed models combined with deep learning-based RUL estimation with uncertainty quantification.
Deployment Architecture: Edge computing for real-time fault detection; cloud-based prognostics with confidence interval outputs.
Integration: Predictive outputs integrated with centralised asset management platforms for maintenance window scheduling.
Outcomes: Catastrophic gearbox failures—among the costliest events in wind energy operations, with individual replacement costs exceeding £200,000—were substantially reduced. RUL-based maintenance planning enabled coordination with favourable weather windows, minimising revenue loss from turbine unavailability. Published reviews report that PdM in wind energy achieves availability improvements of 1–3 percentage points and reduces operations and maintenance costs by 10–15% compared with time-based strategies (Lei et al., 2018).

Case Study 4: Traction Systems in Rail Operations

Asset Context: Traction motors and braking systems in electric passenger trains.
Failure Modes: Thermal overload, brake pad wear, electrical faults.
Sensing Strategy: Temperature sensors, vibration sensors, onboard diagnostic data recorders.
Analytical Approach: Time-series classification models combined with rule-based safety thresholds.
Deployment Architecture: Onboard edge analytics with periodic synchronisation to central maintenance servers.
Integration: Predictive alerts interfaced with depot scheduling systems and regulatory compliance records.
Outcomes: In-service failures and unscheduled stoppages decreased, with maintenance interventions planned proactively during depot visits. Fleet availability improved. The system also supported regulatory compliance through

traceable maintenance records linked to predictive triggers. Railway PdM deployments in published case literature report in-service failure reductions of 30–50% and improved fleet availability of 1–2% for high-utilisation rolling stock (Serradilla et al., 2022).

Case Study 5: Diesel Engines in Maritime and Offshore Operations

Asset Context: Diesel main engines and auxiliary systems on offshore support vessels.
Failure Modes: Fuel system degradation, cooling system fouling, bearing wear.
Sensing Strategy: Engine performance monitoring, oil particle analysis, vibration sensors.
Analytical Approach: Ensemble learning combining statistical and machine learning models for anomaly detection and degradation forecasting.
Deployment Architecture: Hybrid edge–cloud system designed for intermittent satellite connectivity.
Integration: Predictive insights used for voyage-based maintenance planning and port call scheduling.
Outcomes: Unplanned failures during offshore operations—where access for emergency repair is limited and standby vessel day rates exceed £50,000—were reduced through earlier fault detection and improved voyage maintenance planning. The hybrid architecture ensured diagnostic continuity under connectivity constraints. Maritime industry benchmarks indicate that engine PdM programmes reduce unplanned off-hire events by 20–35% and lower annual maintenance expenditure by 8–15% (Rakholia et al., 2025).

Cross-Case Synthesis and Lessons Learned

Table 4 presents a structured cross-case comparison of the five deployments, enabling direct comparison of sensing strategies, analytical methods, deployment architectures, and quantified outcomes across sectors.

Table 4. Cross-Case Comparison of Predictive Maintenance Deployments

Sector	Asset	Key Sensors	AI Method	Architecture	Key Outcome	Citation
Discrete Manufacturing	Motors, gearboxes	MEMS vibration, thermal, current	LSTM + time-frequency	Edge + cloud	15–25% downtime reduction; 20–30% MTTR improvement	Zonta et al. (2020); Rakholia et al. (2025)
Chemical Processing	Pumps, compressors	Pressure, acoustic emission, flow	Statistical + survival analysis	Cloud + plant historian	20–40% pump failure reduction; 25% spare-part inventory saving	Carvalho et al. (2019)
Wind Energy	Turbine gearboxes, generators	Vibration, thermal imaging, electrical	Physics-informed hybrid + deep learning RUL	Edge + cloud (uncertainty-aware)	1–3 pp availability gain; 10–15% O&M cost reduction	Lei et al. (2018)
Rail Operations	Traction motors, brakes	Temperature, vibration, onboard diagnostics	Time-series classification + rule-based safety	Onboard edge + central server	30–50% in-service failure reduction; 1–2% fleet availability gain	Serradilla et al. (2022)
Maritime/ Offshore	Diesel engines, auxiliaries	Performance monitoring, oil particle, vibration	Ensemble learning (statistical + ML)	Hybrid edge–cloud (intermittent connectivity)	20–35% off-hire reduction; 8–15% maintenance spend reduction	Rakholia et al. (2025)

Source: Synthesised from case evidence and published industry benchmarks cited within each case.

Several consistent patterns emerge across the five cases. PdM delivers the greatest value when asset selection is guided by criticality, sensing is failure-mode driven, and analytical outputs are integrated directly into execution systems. Data engineering quality and human-centred execution consistently influence outcomes more strongly than algorithmic sophistication alone. Uncertainty-aware decision-making and closed-loop governance are further distinguishing features of deployments that deliver sustained rather than pilot-only performance. These findings reinforce the core argument of this chapter: PdM is best understood as an operational system, not a standalone analytical tool.

CHALLENGES, RESEARCH GAPS, AND FUTURE DIRECTIONS

Despite substantial advances in sensing, analytics, and computing infrastructure, PdM deployment continues to face significant technical, organisational, and methodological constraints. A structured examination of these issues is necessary to position predictive maintenance as a mature and reliably deployable industrial capability.

Data Quality, Ground Truth, and Label Scarcity

One of the most persistent constraints is the quality and completeness of maintenance data. Industrial sensor streams are noisy and non-stationary; reliable ground-truth failure labels are scarce because maintenance logs are frequently inconsistent, incomplete, or recorded for administrative rather than analytical purposes. This scarcity limits supervised learning approaches and increases the risk of overfitting. Future work must therefore prioritise robust data engineering practices, self-supervised and weakly supervised learning strategies, and standardised maintenance data schemas that improve data reliability across heterogeneous assets and sites.

Evaluation Metrics and Cost-Sensitive Validation

Many published PdM studies report classification accuracy or prediction error while neglecting operational consequences. The cost of a false alarm differs substantially from the cost of a missed failure, and these trade-offs vary across asset types and contexts. Future research should prioritise cost-sensitive evaluation frameworks that explicitly incorporate downtime costs, safety implications, and resource constraints. Time-aware validation strategies—rolling-window and event-based evaluation—are necessary to confirm that models perform reliably under real operating conditions rather than idealised laboratory settings.

Model Drift, Lifecycle Management, and MLOps Governance

Predictive models degrade as equipment condition, operating regimes, maintenance practices, and process configurations change. This drift is a significant threat to long-term reliability if left unmanaged. Addressing it requires lifecycle governance practices aligned with MLOps principles: continuous performance monitoring, drift detection mechanisms, retraining policies, version control, and traceable decision documentation. Research is needed to develop lightweight, interpretable drift detection techniques suitable for industrial environments with limited computational resources and labelled data.

Scalability, Federated Learning, and Fleet-Level Generalisation

Many PdM solutions perform well at pilot or single-asset level but struggle to scale across large fleets or multiple facilities. Cross-asset variability in design, operating conditions, and maintenance practices reduces model generalisability. Federated learning offers a promising path forward: models can be trained collaboratively across distributed assets without centralising sensitive data, respecting data sovereignty and cybersecurity constraints. Transfer learning and domain adaptation also show promise for enabling knowledge sharing across assets with shared failure characteristics (Serradilla et al., 2022).

Cybersecurity and System Resilience

As PdM systems become more interconnected through IoT networks and cloud platforms, the cybersecurity attack surface grows correspondingly. Unauthorised access, data manipulation, or model tampering can compromise maintenance decisions and create safety hazards. Ensuring resilience requires secure communication protocols, robust access control, and continuous monitoring for anomalous system behaviour. The intersection of predictive maintenance and industrial cybersecurity remains underexplored and represents a critical area for future investigation.

Human Factors and Organisational Readiness

Technical sophistication alone does not guarantee PdM success. Human trust, organisational culture, and decision-making authority play decisive roles in adoption and sustained use. Resistance commonly arises when predictive outputs are perceived as opaque, unreliable, or in conflict with practical experience. Future work must emphasise explainability, human-centred design, and training strategies that align PdM with existing maintenance practices. Research on alarm fatigue, usability of decision-support interfaces, and human–machine collaboration is essential to ensure that predictive systems enhance rather than hinder maintenance effectiveness.

Future Research Outlook

Looking forward, PdM research must shift from demonstrating technical feasibility to addressing deployment sustainability. Priority areas include: (i) uncertainty-aware decision-making frameworks that align predictions with operational risk tolerance; (ii) standardised industrial benchmarks beyond NASA CMAPSS and PHM08 datasets that reflect real-world heterogeneity; (iii) hybrid physics-informed machine learning that constrains predictions within physically plausible bounds; (iv) digital twin integration for closed-loop simulation and model validation; and (v) cross-disciplinary collaboration across reliability engineering, data science, cybersecurity, and human factors research. Predictive maintenance is becoming a core component of autonomous and semi-autonomous industrial systems; its future depends not on analytical innovation alone, but on thoughtful integration with organisational processes, governance structures, and human expertise.

CONCLUSION

This chapter has presented predictive maintenance as a closed-loop, socio-technical system embedded within the Industry 4.0 ecosystem. Reliability engineering foundations, advanced sensing, disciplined data engineering, AI-based analytics, enterprise system integration, and human-centred execution have been treated as interdependent rather than separate concerns. The central argument is that PdM functions as a strategic capability—rather than a standalone technological solution—only when these elements are coherently combined, governed, and aligned with operational reality.

Structured lifecycle analysis has shown that asset criticality assessment, baseline definition, failure-mode-driven sensing, and closed-loop CMMS/EAM integration are as consequential as algorithm selection. Five cross-sector case studies have illustrated how this systemic perspective translates into measurable outcomes: reduced unplanned downtime, lower maintenance costs, improved MTTR, and extended asset life. A critical examination of deployment challenges—data scarcity, model drift, cybersecurity, human trust, and scalability—has identified the research directions most likely to advance PdM from pilot success to enterprise-scale sustainability.

The chapter is intended to serve practitioners and researchers alike. For engineers and maintenance managers, it provides a deployment roadmap grounded in operational evidence. For graduate students and researchers, it offers conceptual clarity, a structured analytical taxonomy, and an honest assessment of where significant technical and organisational challenges remain. The overarching message is that predictive maintenance is not a product to be installed but a capability to be built, sustained, and continuously improved alongside the industrial systems it serves.

REFERENCES

Carvalho, T. P., Soares, F. A. A. M. N., Vita, R., Francisco, R. P., Basto, J. P., & Alcalá, S. G. (2019). A systematic literature review of machine learning methods applied to predictive maintenance. *Computers & Industrial Engineering*, *137*, 106024. Advance online publication. DOI: 10.1016/j.cie.2019.106024

Jardine, A. K. S., Lin, D., & Banjevic, D. (2006). A review on machinery diagnostics and prognostics implementing condition-based maintenance. *Mechanical Systems and Signal Processing*, *20*(7), 1483–1510. DOI: 10.1016/j.ymssp.2005.09.012

Lee, J., Bagheri, B., & Kao, H. A. (2015). A cyber-physical systems architecture for Industry 4.0-based manufacturing systems. *Manufacturing Letters*, *3*, 18–23. DOI: 10.1016/j.mfglet.2014.12.001

Lei, Y., Li, N., Guo, L., Li, N., Yan, T., & Lin, J. (2018). Machinery health prognostics: A systematic review from data acquisition to RUL prediction. *Mechanical Systems and Signal Processing*, *104*, 799–834. DOI: 10.1016/j.ymssp.2017.11.016

Lu, S., Lu, J., An, K., Wang, X., & He, Q. (2023). Edge computing on IoT for machine signal processing and fault diagnosis: A review. *IEEE Internet of Things Journal*, *10*(13), 11093–11116. DOI: 10.1109/JIOT.2023.3239944

Mobley, R. K. (2002). *An introduction to predictive maintenance* (2nd ed.). Butterworth-Heinemann.

Pech, M., Vrchota, J., & Bednář, J. (2021). Predictive maintenance and intelligent sensors in smart factory. *Sensors (Basel)*, *21*(4), 1470. Advance online publication. DOI: 10.3390/s21041470 PMID: 33672479

Rakholia, R., Suárez-Cetrulo, A. L., Singh, M., & Carbajo, R. S. (2025). Integrating AI and IoT for predictive maintenance in Industry 4.0 manufacturing environments: A practical approach. *Information (Basel)*, *16*(9), 737. Advance online publication. DOI: 10.3390/info16090737

Sabato, A., Niezrecki, C., & Fortino, G. (2017). Wireless MEMS-based accelerometer sensor boards for structural vibration monitoring: A review. *IEEE Sensors Journal*, *17*(2), 226–235. DOI: 10.1109/JSEN.2016.2630008

Sánchez, R.-V., Lucero, P., Vásquez, R.-E., Cerrada, M., Macancela, J.-C., & Cabrera, D. (2018). Feature ranking for multi-fault diagnosis of rotating machinery using random forest and k-nearest neighbors. *Journal of Intelligent & Fuzzy Systems*, *34*(6), 3463–3473. DOI: 10.3233/JIFS-169526

Selcuk, S. (2017). Predictive maintenance, its implementation and latest trends. *Proceedings of the Institution of Mechanical Engineers. Part B, Journal of Engineering Manufacture, 231*(9), 1670–1679. DOI: 10.1177/0954405415601640

Serradilla, O., Zugasti, E., Rodriguez, J., & Zurutuza, U. (2022). Deep learning models for predictive maintenance: A survey, comparison, challenges, and prospects. *Applied Intelligence, 52*(10), 10934–10964. DOI: 10.1007/s10489-021-03004-y

Si, X.-S., Wang, W., Hu, C.-H., & Zhou, D.-H. (2011). Remaining useful life estimation: A review on the statistical data driven approaches. *European Journal of Operational Research, 213*(1), 1–14. DOI: 10.1016/j.ejor.2010.11.018

Tao, F., Qi, Q., Wang, L., & Nee, A. Y. C. (2019). Digital twins and cyber-physical systems toward smart manufacturing and Industry 4.0. *Engineering (Beijing), 5*(4), 653–661. DOI: 10.1016/j.eng.2019.01.014

Tsui, K. L., Chen, N., Zhou, Q., Hai, Y., & Wang, W. (2015). Prognostics and health management: A review on data driven approaches. *Mathematical Problems in Engineering, 2015*, 793161. Advance online publication. DOI: 10.1155/2015/793161

Zheng, S., Ristovski, K., Farahat, A., & Gupta, C. (2017). Long short-term memory network for remaining useful life estimation. In *Proceedings of the IEEE International Conference on Prognostics and Health Management (ICPHM)* (pp. 88–95). IEEE. DOI: 10.1109/ICPHM.2017.7998311

Zonta, T., da Costa, C. A., da Rosa Righi, R., de Lima, M. J., da Trindade, E. S., & Li, G.-P. (2020). Predictive maintenance in Industry 4.0: A systematic literature review. *Computers & Industrial Engineering, 150*, 106889. Advance online publication. DOI: 10.1016/j.cie.2020.106889

Chapter 2

Implications of Digital Inclusion on the Sustainability of the Manufacturing Sector

Ajay Kumar Singh
http://orcid.org/0000-0003-0429-0925
Department of Humanities and Social Sciences, Graphic Era University, Dehradun, India

S. Srinivasan
http://orcid.org/0009-0002-0179-9849
Department of Humanities and Social Sciences, Graphic Era University, Dehradun, India

Arakhita Behera
http://orcid.org/0000-0002-6761-9313
Regional Institute of Education (NCERT), Ajmer, India

ABSTRACT

This study creates a digital inclusion index (DII) and it is applied as a representative variable for digital inclusion (DI) for 45 countries. It highlights the contribution of digitalization for sustainability of the manufacturing sector (MS). It examines the causal interlinkages of DI with MS. FDI, trade openness, population density, population growth and gender inequality are included as control variables (CVs). Log-linear functional forms of models are formulated to examine the coefficients of DI and CVs with MS, and MS, and CVs with DI using a panel data during 2006 – 2022. DI has a positive impact on MS, and MS is found suitable to increase DI.

DOI: 10.4018/979-8-2600-2383-9.ch002

© 2027, IGI Global Scientific Publishing. Copying or distributing in print or electronic forms without written permission of IGI Global Scientific Publishing is prohibited. Use of this chapter to erative artificial intelligence (AI) technologies is expressly prohibited. The publisher reserves all rights to license its use for generative AI training and machine learning model development.

This study makes a well-intentioned contribution in theoretical and empirical literature, and provides effective policy suggestions to increase the sustainability of MS and DI. It suggests research direction to undertake further research in DI and MS.

1. INTRODUCTION

Manufacturing sector (MS) has a vital share of GDP in many countries (Singh et al., 20220; Halkos et al., 2021; Wan et al., 2022; Qiao & Niu, 2024; Mangla et al., 2024). The performance of this sector is useful to absorb the disguised unemployment, and enhance industrial and business development. Moreover, MS has the potential to reduce poverty and income inequalities through increasing employment, physical infrastructure and money flow (Singh et al., 20220; Erumban & de Vriers, 2024). MS also enhances the contribution of a nation in exports of goods, and improves its global value chain. The growth of this sector has greater reliance on many variables in social, economic, innovation, technological and environmental factors (Singh et al., 20220; Cunningham et al., 2023). MSMEs are working as a backbone to increase social life and welfare of people in many nations of the world (Zen & Faizin, 2021; Rosyidiana & Narsa, 2024).

Performance of MSMEs and SMEs are supportive to increase foreign investment and create capital formation. Further, economic growth is supportive to increase production and growth of MSMEs. In South Asian countries, Chowdhury (2025) suggested that a growing economy helps to create ample opportunities for entrepreneurship. MSMEs absorb disguised employment from the agricultural and other sectors (Mukherjee, 2018). MSMEs and SMEs are the significant components of MS that have a significant potential to reduce unemployment, poverty and income inequalities. MS also creates physical infrastructure and it increases gross capital formation. The growth and performance of MSMEs depends on multiple factors like technology and innovation, human skills, capital formation, foreign trade, infrastructural development and government policies (Mukherjee, 2018; Surya et al., 2021; Singh & Kumar, 2021; Larios-Francia & Ferasso, 2023; Qiao & Niu, 2024). MS is a significant component of MSMEs and SMEs. Thus, development of MSMEs and SMEs help to increase the growth of MS. It is uniformly accepted that MS help to enhance economic growth and it works as an engine of growth (Li et al., 2018; Qiao & Niu, 2024).

MS is important to produce innovative products for the people. Further, MS enhances the digital inclusion (DI) as it develops and discovers digital tools and DTs. For example, ICT industries are developing new technologies like multimedia mobile, GIS technology, AI, blockchain, apps, etc. which are helping to increase favourable digital infrastructure. Oppositely, MS also accounts around 12 – 25% of greenhouse gas emission globally. Moreover, MS promotes industrial development which is responsible for environmental degradation and environmental issues (Ahmed et al., 2022). High growth of MS is also accountable to increase wastage in the environment. While environmental degradation is caused by increased climate change which is highly harmful for humanity and their dependence sectors (Ahmed et al., 2022). Hence, MS requires sustainable practices to increase the sustainable development. Sustainability of MS can be achieved through applying three bottom line approaches of sustainability (Mangla et al., 2024). Three bottom line approaches include all drivers like social, economic, environmental and technologies to make a sustainable future of society, and enhance sustainable development.

It is stated earlier that growth and development of the MS depend on technological advancement, innovation, technology transfer, financial development, capital formation, trade openness, infrastructural development and government policies. Further, digitalization is bringing innovation in the MS for improving its extensive growth and development (Qiao & Niu, 2024). Therefore, manufacturing industries are moving towards digital transformation and sustainability globally (Wen et al., 2024). Developed countries have achieved a leading position in MS due to application of DTs and digitalization, while developing countries should also promote transformation of MS through digitalization (Wen et al., 2024). Many developed countries have a greater share of MS in their GDP due to extensive technological advancement and use of DTs in this sector. Thus, extensive investment in digitalization would be a vital approach to boost robustness and competitive edges of MS (Galvani et al., 2025; Zhang et al., 2025). Qiao and Niu (2024) claimed that the industrial internet is helping to create digital transformation, and improving efficiency and reducing operating cost in MS. Kupriyanova et al. (2023) reported that digitalization showed a positive impact on productivity and efficiency of MS in Russia.

Digital transformation is making the fundamental technological shift in industries (Boffa & Maffei, 2024). Digital transformation is helping to increase competitiveness of MS (Yang & Jin, 2024). Digitalization is found positive to increase high-quality development of MS (Chang et al., 2025). Digital transformation promotes green innovation in MS (Mu et al., 2024). More specifically, 3D printing is crucial to increase the sustainability of MS (Embia et al., 2023). Use of IoT is supportive for improving the performance of MS (Galgal et al., 2023; Chhotaray et al., 2023). GIS is a digital tool which can be applied to increase the efficiency of MS. Use of GIS is crucial to improve the optimization process, resource efficiency and decision-making capacity of MS (Qing et al. 2019). MS can increase the operational efficiency through GIS technologies (Karyakarte, 2024). GIS can be used as advanced tools to analyse the data in MS (Prah & Gajšek, 2024). GIS is working as a mediator to increase the efficiency in transportation of products in MS. Monitoring of environmental issues can be identified through GIS technologies. Accordingly, GIS is supportive of increasing the efficiency of resources and suggesting alternatives to increase environmental efficiency. Hence, GIS maintains a conducive ecosystem to increase the SDGs through contributing to good environmental management and environmental regulation.

Existing researchers argued that digitalization and DTs, and GIS technologies are increasing the sustainable practices in MS (Shah et al., 2020; Shang & Zhang, 2022; Ahmadi-Gh & Bello-Pintado, 2024). Accordingly, DTs and digitalization are favourable to increase the sustainability of MS in the highly industrialized nations (Ahmadi-Gh & Bello-Pintado, 2024). DTs are increasing the movement of MS towards smart manufacturing. Smart manufacturing is likely to produce environmentally friendly products and increase environmental sustainability. Sustainability of MS includes multiple activities like creation of employment, operating the production process and minimizing its negative impact on the environment. DTs are developing resource-efficient industries (Shang & Zhang, 2022). Consequently, sustainable practices in MS are valuable to increase sustainable development. Maffei et al. (2019); Shang and Zhang (2022) opinionated that digitalization and DTs can be an effective and powerful instrument in MS to increase their sustainability. Although, few studies noted the negative implications of extensive use of digitalization in MS (Chen et al., 2020). For instance, Chen et al. (2020) noted that extensive digitalization of MS is caused to increase greenhouse gas emission (GHGE) and wastage in the environment and increasing more demand for energy and resources. Thereupon, the answer of following research questions must to be addressed by the existing researchers to understand the causal implications of DI with MS:

- How can I be supportive to increase the sustainability of MS?
- How sustainability of MS would be positive to increase sustainable development?
- Why does MS require sustainability?
- How do I enhance the growth of MS?
- Does MS is positive to increase DI?
- Does there exist a causal relationship of DI with MS?
- How can global countries increase the growth of MS through DI?

2. THEORETICAL AND EMPIRICAL REVIEW

2.1. MSMEs and Manufacturing Sector (MS)

MSMEs and SMEs are facing multiple problems in mostly developing nations. For instance, lack of managerial skills, low financial accessibility, and low investment in technologies are creating barriers for further development of MSMEs in many countries (Zen & Faizin, 2021). Progress of MSMEs could not be improved due to low infrastructure development and capital formation in the developing countries. Technological imitation rate is high due to flexibility in the IPR regime in developing nations. MSMEs are unable to increase technological upgradation due to lack of advanced technologies in many developing countries. MSMEs of the developing nations could not improve product quality as per the international requirements. R&D ecosystem is unable to provide the latest technologies to MSMEs. Research organizations and universities are not working with MSMEs to develop advanced technologies and innovation. Also, these are not doing discovery of new technologies and innovation as per the current need of MSMEs. Extensive resource use, poor financial conditions and extensive wastage are important challenges for MS (Sankar et al., 2024). MS is facing the problem due to rising population, scarcity of resources, and pollution and environmental issues (Mangla et al., 2024). Moreover, extensive use of DTs is increasing diversity in production, productivity, efficiency and sustainability of MS across nations (Wu et al., 2024). Developing countries are unable to create digital infrastructure due to lack of finance and digital skills of people. Hence, MS in these countries is not getting ample prospects to increase its performance and sustainability.

2.2. Digitalization and MSMEs

DTs and digitalization are creative ways to increase business models, entrepreneurship innovation and new opportunities for entrepreneurs (Zhang et al., 2023; Tan et al., 2025; Autio et al., 2025; Arif & Herawan, 2025). Digitalization is creative for MS to adopt a sustainable business model (Maffei et al., 2019). Entrepreneurial innovation cannot be developed without technological and innovative advancement. Use of DTs is highly valuable for creating entrepreneurial innovation. Moreover, DTs help to increase high quality institutions, and resource accessibility and allocation (Zhang et al., 2023). Application of DTs reduce all barriers which create crises in business, and help to increase better business models (Vrontis et al., 2022). DTs and digital platforms are helping MSMEs to increase their performance effectively in terms of profits, financial output, sales pattern, consumer's satisfaction and revenue generation, (Mukherjee, 2018; Rosyidiana & Narsa, 2024; Faiz et al., 2024). DTs has a significant reflection on economic sustainability and social value of enterprises.

Advanced DTs like IoT. AI and Blockchain, and GIS can be used for data-driven decision making in SMEs (Muduli et al., 2023; Karyakarte, 2024; Almeida & Okon, 2025). Blockchain technologies are crucial to improve supply chain management (Muduli et al., 2023). Big-data analytical technologies can be applied to predict the demand of new products, sale patterns, profits and resource requirements in

MSMEs. Automation DTs are reducing the human efforts in the SMEs. Previous studies applied diverse variables like internet users, social media, digital marketing, digital strategic orientation and digital transformation to examine their implications on performance, production and size of MSMEs (Zhang et al., 2023; Díaz-Pelaez et al., 2024; Almeida & Okon, 2025). Manufacturing firms can use DTs for different purposes like marketing, advertisement of new products (Zhang et al., 2023). DTs is helping to increase well-developed service industries worldwide (Zhang et al., 2023). Ochoa and Nonnecke (2019) found that DI strategies are helping to increase state-level competitiveness in Mexico. Vrontis et al. (2022) investigated the implications of modern DTs to create values in SMEs in India.

Entrepreneurial orientation maintains a link between social and economic value creation with the performance of SMEs. Sharabati et al. (2024) reported that digital marketing is found positive to increase the effectiveness of SMEs in Jordan. This study proposed that digital transformation develops digital marketing strategies in SMEs. Rosyidiana and Narsa (2024) observed the implication of digitalization on financial performance of MSMEs in Indonesia using primary data of 35 business owners. Díaz-Pelaez et al. (2024) noticed that small businesses are struggling to receive benefits from digitalization. MSMEs and SMEs use different practices of DTs, thus, the impact of digitalization of firm's performance is found significant. MSMEs are not fully ready to adopt digital infrastructure due to inadequate internet infrastructure, lack of skilled talent and low technological awareness among the workers in India (Mittal et al., 2025). Almeida and Okon (2025) examined the significance of digitalization for promoting the performance of SMEs in Portugal. Digital skills and transformation strategies help to increase the performance of firms. SMEs should integrate digital infrastructure and digital value chain to increase their performance.

2.3. Interrelatedness between Digitalization and MS, and Challenges of MS

Digitalization increases the multiple challenges for MSMEs and SMEs especially in developing countries like India (Buteau, 2021; Autio et al., 2025; Mittal et al., 2025). MSMEs do not have proper internet connectivity, skilled workers and digital tools in many developing countries. Extensive investment is required to digital infrastructure in MS. Digital illiteracy of workers is creating a barrier to increase the digital landscape in MSMEs of the developing countries. Low digital security is a challenge for MSMEs in developing countries. Thus, digitalization is increasing the skills gap among the workers in MS. Sharabati et al. (2024) found that lack of money, and digital and technological illiteracy of workers are creating berries for small businesses to adopt digital infrastructure in Jordan. Faiz et al. (2024) noted that digital adaptation capacity of small firms is insignificant due to numerous reasons. Chowdhury (2025) noted that limited financial accessibility, digital divide and poor institutional frameworks are creating a barrier for entrepreneurship development in South Asia. Moreover, low cyber security is creating a barrier to adopt digital infrastructure in MS.

Digitalization creates ample opportunities to increase the growth of MS through improving their efficiency and productivity. Digitalization is helping to increase female's participation rate in MS and enhancing wage equality, and productivity in MS (Wu et al., 2024). Digitalization is working as a crucial driver to increase the transformation of manufacturing industry through technological advancement and upgradation (Boffa & Maffei, 2024). Digital transformation promotes green innovation which help to increase sustainability of MS. DTs are promoting smart and green manufacturing which are helping to increase sustainability of MS (Chang et al., 2025). DTs are improving the operation process of manufacturing firms globally (Sankar et al., 2024). For instance, DTs are helping to increase marketing, advertisement, and consumer satisfaction in the manufacturing firms. DTs help to increase innovation and develop new business models which promote growth of MS.

Dissemination of scientific output and knowledge are easily improving due to use of DTs. Consequently, DTs is nurturing an appropriate R&D ecosystem to discover new innovation for MS. Past studies examined the impact of various drivers of digitalization (i.e., DTs, digital economy, digital inclusion, digital infrastructure, digital devices, etc.) on different aspects (e.g., production efficiency, green innovation, internationalization of firms, global value chain, business development, etc.) of MS across countries (Mu et al., 2024). Boffa and Maffei (2024) investigated the implications of digital transformation for manufacturing business models in Swedish industry using the primary information of MS. Wang and Shao (2024) examined the contribution of digital transformation on production efficiency of manufacturing firms in China. Production efficiency of MS is significantly improved due to implementation of digital transformation. Wen et al. (2024) explored the implications of digitization in upgrading the manufacturing firms in China.

Moreover, digitalization could be helpful to increase total factor productivity (TFP) and investment in fixed assets. Organizational resilience could be significantly improved due to extensive investment in DTs in China (Zhang et al., 2025). However, digitalization increases the technological investment challenge for MS (Shang & Zhang, 2022). MS is significant to increase the digital infrastructure. Many manufacturing industries like ICT firms create new DTs which help to increase appropriate digital infrastructure. MS are encouraged to increase the extensive investment in R&D sector to develop new DTs which can be favourable for improving their efficiency. Singh and Jyoti (2023); Singh (2024) investigated the impact of MS on digitalization across countries.

2.4. Digitalization and Sustainability of MS

In the era of technological development, digitalization and DTs are positive to increase the sustainability of MS (Shang & Zhang, 2022; Embia et al., 2023). Application of technologies has seemed positive to increase sustainability in MS (Mangla et al., 2024). Sustainability of MS includes all drivers which are essential to increase sustainable development (Galvani et al., 2025). Digitalization helps to increase sustainable practices in MS (Shah et al., 2020). Advancement of DTs could be beneficial for developing sustainable MS in China (Chang et al., 2025). Sustainable manufacturing includes extensive utilization of DTs to increase the efficiency and productivity of MS. Shang and Zhang (2022) highlighted that DTs are nurturing an ecosystem for MS to reduce costs and improve sustainability. 3D printing is a digital tool which helps to increase additive manufacturing products and enhance sustainability of MS (Embia et al., 2023). IoT and automation are creative to increase performance of MS (Galgal et al., 2023). Further, additive manufacturing is crucial in automative, food manufacturing and medical health care sector (Embia et al., 2023). Additive manufacturing led to reduced waste, and increased energy saving (Embia et al., 2023). Maffei et al. (2019) point out that digitalization is moving the transformation of MS to adopt a sustainable business model (SMB). Singh and Jyoti (2023) examined the impact of digitalization on the entrepreneurship ecosystem and noted a significant association between the mentioned variables. Yang and Jin (2024) suggested that digital transformation appeared positive to increase ESG performance of MS in China.

Few researchers noted that digitalization of MS is creating new challenges in the environmental sector. Chen et al. (2020) claimed that digitalization is improving environmental sustainability through increasing efficiency of resources and information in MS. Contradictory, digitalization is caused to increase GHGE and wastage from MS due to additional use of energy and resources. Hence, MS should understand the positive and negative impact of digitalization on the environmental sector in MS. Ahmadi-Gh and Bello-Pintado (2024) reported that DTs are helping to increase environmental sustainability in manufacturing companies. Mu et al. (2024) highlighted that digital transformation is providing significant benefits to increase green innovation and resource efficiency, and reduce waste materials, operation costs and use of energy in Chinese MS. Further, green innovation is assisting to adopt a sustainable business model and appropriate ecological environment (Mu et al., 2024). Chang et al. (2025) explored the implication of digitalization on MS in China and it is noted that digitalization could be helpful to increase high-quality development of MS. Moreover, digitalization is promoting innovation diffusion which is helping to increase sustainability of MS. Galvani et al. (2025) noted that digitalization enhances sustainable development.

2.5. Contribution of GIS towards Sustainable Manufacturing

Use of GIS would be useful to integrate all dimensions of sustainability in MS. GIS can be applied to examine the monitoring of environmental pollution in MS. It suggests an appropriate and scientific way for MS to comply with environmental regulation (Karyakarte, 2024). Accordingly, MS can make the plan to abate GHGE from the production activities and promote green initiatives in MS. GIS is an essential driver to increase the tracking location for improving the supply chains of industrial products (Qing et al. 2019; Karyakarte, 2024). GIS improves the transportation and labour requirements of MS (Prah & Gajšek, 2024; Karyakarte, 2024). GIS also improves the delivery of products for consumers on time (Karyakarte, 2024). Adoption of GIS nurtures a favourable ecosystem in MS to increase their efficiency and profitability. Thereupon, GIS provides a platform to reduce contamination in water and other inputs used by MS. Resource efficiency and management can be improved using GIS in the MS. For example, use of energy and its multiple practices can be increased through GIS technologies. Moreover, GIS can be applied to make the future predictions of various components of the environment which help to make plans for MS. GIS can be applied to increase as an industrial revolution tool in MS (Qing et al. 2019). Hence, GIS technologies can be effective to increase sustainability of MS.

3. METHOD AND MATERIALS

3.1. Study Area and Source of Data

This research is primarily based on secondary data which is taken from the online database of WDIs, WB (2026) and UNDP, HDR (2026). Accordingly, 45 countries are selected as per the availability of statistical data of most relevant variables. For instance, statistical data of all indicators associated with DI are available during 2006 – 2022 (Figure 1). Thus, the authors considered the mentioned time period for statistical and empirical investigations. Following countries are used in this chapter: Ireland, China, Thailand, Korea (Republic of), Malaysia, Mexico, Slovenia, Japan, Türkiye, Singapore, Indonesia, Germany, Switzerland, Hungary, Jordan, Poland, Austria, Italy, Finland, Argentina, India, Sweden, Ecuador, Denmark, Russian Federation, Peru, Portugal, South Africa, Croatia, Belgium, Latvia, Israel, Brazil, Colombia, Spain, Netherlands, Uruguay, France, Canada, Chile, United Kingdom, Greece, Jamaica, Norway, and Australia.

3.2. Estimation of Digital Inclusion Index

This chapter develops a digital inclusion index (DII) that is used as a core variable to examine the impact of DI on the MS. DII is the combination of 12 crucial variables which help to increase digitalization of global countries. The brief summary of these indicators is given in Figure 1. These indicators are divided in following categories i.e., digital skills, digital stability and digital usability (Jyoti & Singh, 2023). Simple descriptive method is applied to develop DII for nations undertaken during 2006 – 2022. This method is adopted from the studies of Jyoti and Singh (2023); Singh and Jyoti (2023).

Figure 1. List of indicators which are included for estimating DII

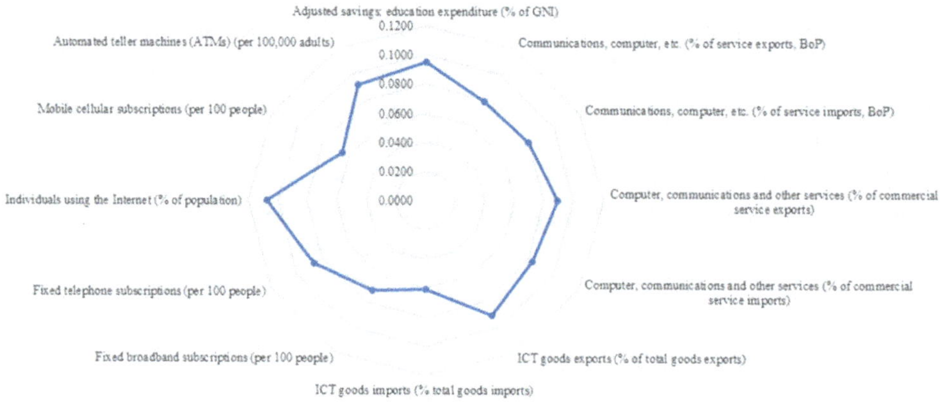

Source: *Indicators are specified as per the existing theoretical and empirical review.*

3.3. Inclusion of Variables

3.3.1. Core DVs and IVs:

The progress of MS can be determined by many variables like total factor productivity (*TFP*), investment, productivity, annual turnover, profits, sale pattern, total revenue and exports at micro level (Kupriyanova et al., 2023; Wen et al., 2024). *DII* and share of MS in GDP are applied as proxy variables for DI and growth of MS in this chapter (Singh, 2024). Accordingly, the above-mentioned variables are applied as core *DVs* and *IVs* in the regression models.

Table 1. List of the variables

Variables	Unit	Code
Digital Inclusion Index	Number	*DII*
Manufacturing value added (% of GDP)	%	*MVA*
Population growth (annual %) (*PG*)	%	*PG*
Population density (people per sq. km of land area)	Number	*PD*
Foreign direct investment net inflows (% of GDP)	%	*FDINI*
Employment to population ratio, 15+, total (%)	%	*EPRT*
Gender inequality index (value)	Number	*GII*
Exports of goods and services (% of GDP)	%	*EGS*

Source: Finalized by authors as per the existing literature.

3.3.2. Control Variables (*CVs*) and *Explanatory Variables* (*EVs*):

Other variables like population growth (*PG*) (Jyoti & Singh, 2023), population density (*PD*), foreign direct investment net inflows (*FDINI*) (Singh & Jyoti, 2023; Singh, 2024), employment to population ratio (*EPRT*) (Jyoti & Singh, 2023) and gender inequality index (*GII*) are considered as *CVs* in regression analysis. Population growth has significant impact on MS (Jyoti & Singh, 2023). Thereupon, population density is creative to increase the physical infrastructure. *FDINI* is an essential driver to create a feasible ecosystem for MS and DI. *EPRT* is a significant driver of MS and DI. Gender equality shows the equal distribution of resources among the male and female. Thus, it has a significant impact on MS and DI. Accordingly, *PG*, *PD*, *EPRT* and *GII* are applied a *CVs* and *EVs* in the regression analysis.

Mediators Variable (MV): Exports of goods and services (*EGS*) is a significant variable to explore the trade performance of a nation. *EGS* shows the progress of trade openness in a country. Trade openness has a crucial impact on MS and DI. Hence, *EGS* is used as a mediator variable in the proposed models. The brief summary of the variables is provided in Table 1. Figure 2 explores the research framework of this chapter.

Figure 2. Research framework

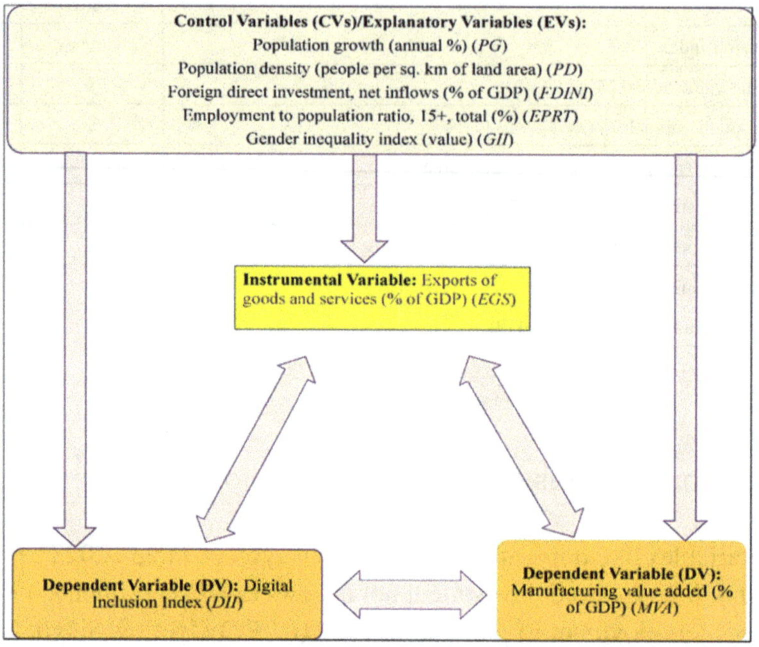

Source: Finalized by authors as per the existing literature.

3.4. Theoretical Framework

Theoretical foundation of industrial development is very old and comprehensive (Singh & Kumar, 2021; Singh & Singh, 2025). For instance, Kupriyanova et al. (2023) applied mathematical and computer models to detect the effect of digitalization on productivity and efficiency of MS in Russia. Singh and Jyoti (2023) applied a log-linear model to examine the impact of *MVA* on digitalization. Wang and Shao (2024) develop a dynamic panel regression model to examine the impact of digital transformation on *TFP* of manufacturing firms in China. Wen et al. (2024) also applied a dynamic panel model to observe the impact of digitalization on *TFP* of the manufacturing industry in China.

3.5. Descriptive and Empirical Analysis

The correlation coefficient matrix and trend analysis are applied for descriptive analysis of the statistical data. Thereupon, log-linear regression models are used to estimate the coefficients of *CV*s and *MV* with DI and MS. These variables are compiled as country-wise panel data during the stated period. It assumes that

DII is a function of *MVA* and stated variables in Table 1. *MVA* is a function of *DII* and other variables. The functional association of *DII* and *MVA* with *EV*s and *MV* are defined as:

$$(DII) = f\{(MVA), (PG), (PD), (FDINI), (EPRT), (GII), (EGS)\} \quad (1)$$

$$(MVA) = f\{(DII), (PG), (PD), (FDINI), (EPRT), (GII), (EGS)\} \quad (2)$$

The description of all the variables is given in Table 1. Following log-linear regression models are applied:

$$ln(DII)_{ct} = \alpha_0 + \alpha_1\, ln(MVA)_{ct} + \alpha_2\, ln(PG)_{ct} + \alpha_3\, ln(PD)_{ct} + \alpha_4\, ln(FDINI)_{ct} + \alpha_5\, ln(EPRT)_{ct} + \alpha_6\, ln(GII)_{ct} + \alpha_7\, ln(EGS)_{ct} + \mu_{ct} \quad (3)$$

$$ln(MVA)_{ct} = \beta_0 + \beta_1\, ln(DII)_{ct} + \beta_2\, ln(PG)_{ct} + \beta_3\, ln(PD)_{ct} + \beta_4\, ln(FDINI)_{ct} + \beta_5\, ln(EPRT)_{ct} + \beta_6\, ln(GII)_{ct} + \beta_7\, ln(EGS)_{ct} + \delta_{ct} \quad (4)$$

Here, *ln* is natural logarithms; α_0 and β_0 are constant coefficient; α_1, ..., α_7 and β_1, ..., β_7 are coefficients of corresponding variables; μ_{ct} and δ_{ct} are error-term; *c* is cross-sectional countries; and *t* is time period in equation (3) and equation (4).

3.6. Robustness Check

The robustness of empirical results is tested through including instrumental variables in the proposed empirical models. One lags of *DV*s and *IV*s are included as instrumental variables in the models. Following models are formulated to check the robustness of the empirical results:

$$ln(DII)_{c,t-1} = \xi_0 + \xi_1\, ln(MVA)_{c,t-1} + \xi_2\, ln(PG)_{c,t-1} + \xi_3\, ln(PD)_{c,t-1} + \xi_4\, ln(FDINI)_{c,t-1} + \xi_5\, ln(EPRT)_{c,t-1} + \xi_6\, ln(GII)_{c,t-1} + \xi_7\, ln(EGS)_{c,t-1} + \phi_{c,t-1} \quad (5)$$

$$ln(MVA)_{c,t-1} = \gamma_0 + \gamma_1\, ln(DII)_{c,t-1} + \gamma_2\, ln(PG)_{c,t-1} + \gamma_3\, ln(PD)_{c,t-1} + \gamma_4\, ln(FDINI)_{c,t-1} + \gamma_5\, ln(EPRT)_{c,t-1} + \gamma_6\, ln(GII)_{c,t-1} + \gamma_7\, ln(EGS)_{c,t-1} + \lambda_{c,t-1} \quad (6)$$

Here, ξ_0 and γ_0 are constant terms; ξ_1, ..., ξ_7 and γ_1,..., γ_7 are coefficients of one lag of respective variables; *t*- 1 is one lag of associated variables; $\phi_{c,t-1}$ and $\lambda_{c,t-1}$ are the error-terms in equation (5) and equation (6). The coefficients of *EV*s, *CV*s and *MV* with *DII* and *MVA* are estimated through *PCSEs* model to avoid the presence of all statistical diagnostics in the heterogenous panel (Singh et al., 2020).

3.7. Causality Test and Use of Statistical Software

Dumitrescu and Hurlin (2012) Granger non-causality test is applied to examine the causal association among the variables (Jyoti & Singh, 2023; Singh & Singh, 2025). SPSS and STATA software are used to create DII, and proposed descriptive and empirical analysis.

4. DESCRIPTIVE RESULTS AND DISCUSSION

4.1. Performance of DI and MS Across countries

Figure 3. Mean values of DII and MVA during 2006 - 2011

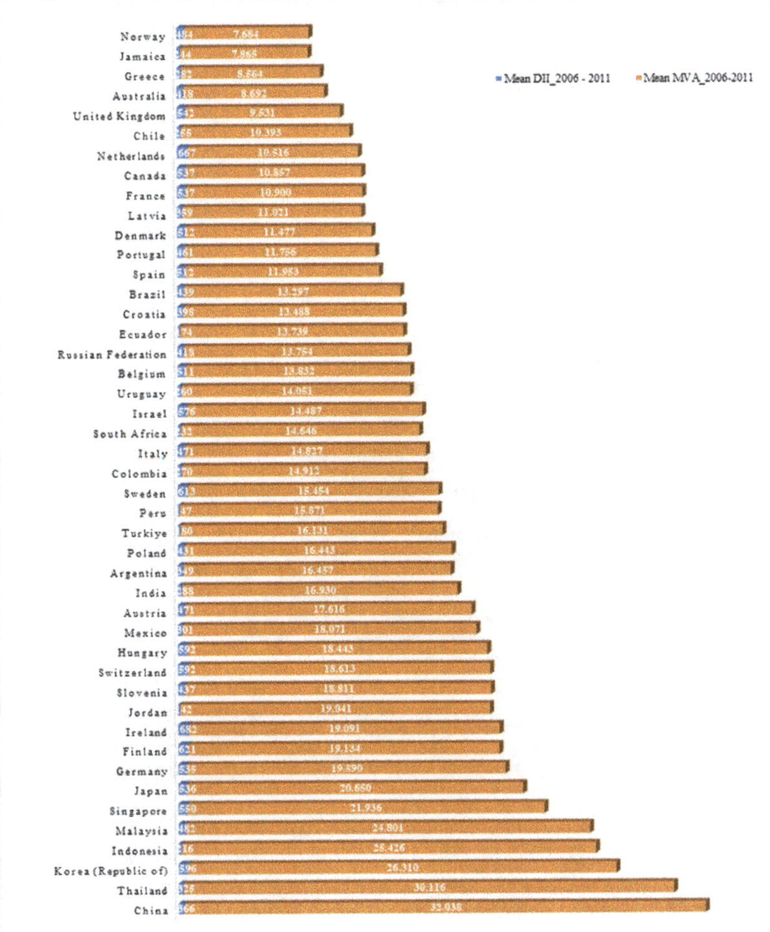

Source: *DII is estimated by authors and statistical values of MVA taken from WDI.*

Figure 4. Mean values of DII and MVA during 2012 - 2017

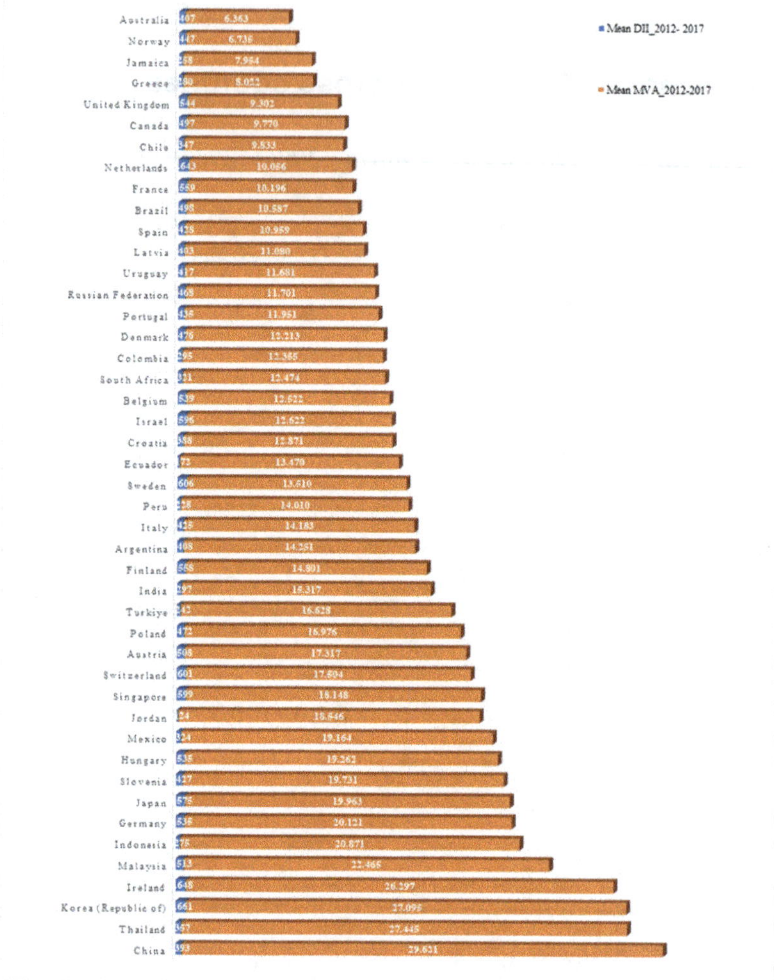

Source: *DII is estimated by authors and statistical values of MVA taken from WDI.*

Figure 5. Mean values of DII and MVA during 2018 - 2022

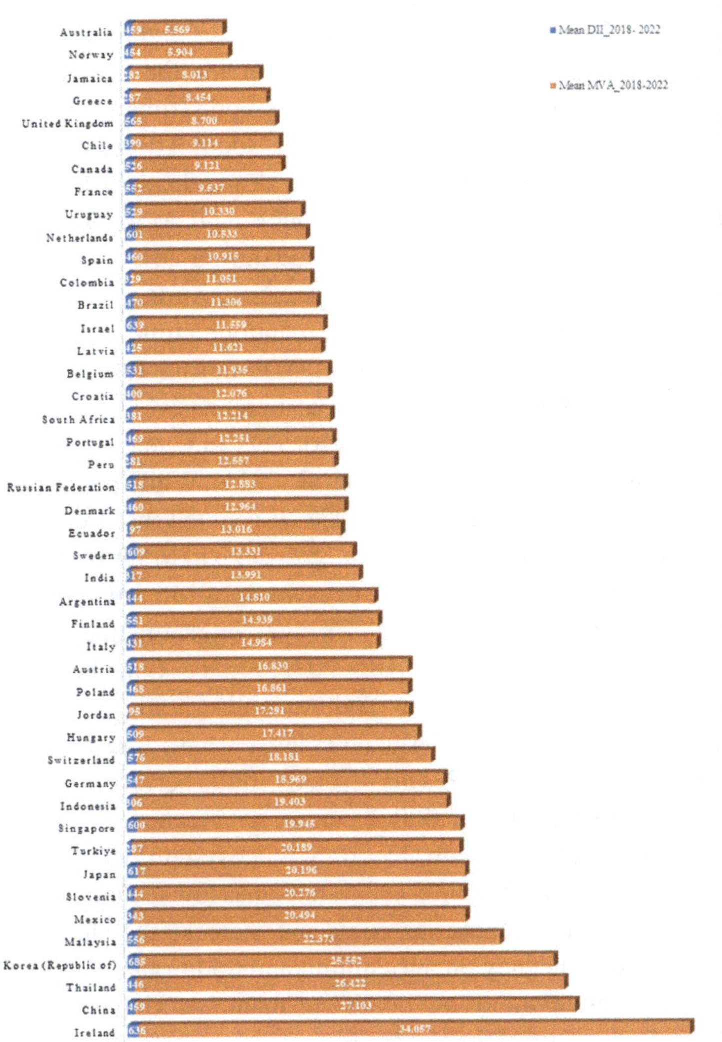

Source: *DII is estimated by authors and statistical values of MVA taken from WDI.*

The performance of DI and MS of 45 countries in three-time spans are given in Figure 3, Figure 4 and Figure 5. The mean values of *DII* and manufacturing value added (*MVA*) in GDP during 2006 – 2011, 2012 – 2017 and 2018 – 2022 are used to explore the relative performance of these countries in DI and MS. There seems to be a high diversity in DI across 45 countries as per the mean values of *DII* in the mentioned durations.

Singh (2024) also estimated the high diversity in digital infrastructure across 34 nations. Belgium, Canda, Singapore, Sweden and the United Kingdom could maintain consistent improvement in DI in the mentioned time span. The performance of Uruguay, Latvia, Mexico, Norway, Slovenia, South Africa, Thailand and Russia in DI could be improved during 2018 – 2022. Argentina, Australia, Chile, China, Colombia and Croatia have positive performance in DI. Moreover, the performance of Brazil, Denmark and Spain in DI decreased during 2018 – 2022. These figures also imply that many countries like Türkiye, Peru, Jordan, Jamaica and Ecuador have the poorest positions in DI during the mentioned time periods.

China has the largest share of MS in its GDP during 2006 – 2011 and 2012 – 2017. These figures show that China has a comprehensive ecosystem of MS in the world (Zhang et al., 2025). Extensive use of DTs in Chinese MS has become its greater position in the world's MS (Chang et al., 2025). While Ireland has the greatest share of MS in its GDP during 2018 – 2022. It is noted that Ireland, China, Thailand and South Korea have the more than 25% share in MVA of their GDP. Hence, these countries have the most effective ecosystem of MS among the 45 countries. Many countries like Mexico, Slovenia, Türkiye, Singapore and Switzerland could improve MS performance during 2018 – 2022. France, Canada, Chile, the United Kingdom, Greece, Jamaica, Norway and Australia should increase the share of MS in their GDP. Moreover, undertaken countries have a significant diversity in contribution of MS in their GDP. This diversity in the performance of MS happens due to many reasons like lack of skilled labour, poor technological development & innovation, poor R&D ecosystem, low digital ecosystem, inefficient government mechanism, inappropriate business ecosystem, and low technological transfer and commercialization. Many countries could not increase the share of MS in their GDP due to high dependency on agricultural and traditional sectors. Furthermore, lack of renewable energy and low quality of manufacturing products are making their poor performance in global MS. Hence, poor performing countries in MS can improve their positions by adopting sustainable policies in this sector. These countries should increase import of highly advanced technologies from developed countries to increase the share of MS in GDP.

4.2. Correlation Matrix

DII has a positive correlation coefficient with *MVA* (r=0.097**), *PD* (r=0.198**), *FDINI* (r =0.274**), *EGS* (r =0.384**) and *EPRT* (r = 0.241**) (Table 2). The estimates reveal that *MVA*, population density, FDI, trade openness and employment to population ratio are positively associated with DI. Singh and Jyoti (2023) also noted a positive correlation of digitalization with *EPRT*, *FDINI* and *MVA*. Laiwang et al. (2025) also noted a positive correlation between digitalization and population density. *MVA* has a positive correlation coefficient with *PD* (r=0.138**), *FDINI* (r=0.072*), *EGS* (r=0.325**), *EPRT* (r=0.194**) and *GII* (r=0.101**). Hence, the estimates imply that population density, foreign direct investment, trade openness, *EPRT* and gender inequality are positively associated with manufacturing value added. Foreign direct investment is positively associated with DI, *MVA* and trade openness. Thereupon, trade openness has positive correlation with *DII*, *MVA* and *PD*. Gender inequality is negatively correlated with *DII*, *PD*, *FDINI*, *EGS* and *EPRT*. Most specifically, *DII* and *GII* have a negative and statistically significant association. Hence, estimated values of *DII* have significant consistency.

Table 2. Correlation among the variables

Variables	*DII*	*MVA*	*PG*	*PD*	*FDINI*	*EGS*	*EPRT*	*GII*
DII	1	0.097**	-0.252**	0.198**	0.274**	0.384**	0.241**	-0.762**
MVA	0.097**	1	0.059	0.138**	0.072*	0.325**	0.194**	0.101**
PG	-0.252**	0.059	1	0.211**	0.039	0.171**	-0.114**	0.288**
PD	0.198**	0.138**	0.211**	1	0.298**	0.710**	0.173**	-0.161**
FDINI	0.274**	0.072*	0.039	0.298**	1	0.497**	0.087**	-0.148**
EGS	0.384**	0.325**	0.171**	0.710**	0.497**	1	0.158**	-0.353**
EPRT	0.241**	0.194**	-0.114**	0.173**	0.087**	0.158**	1	-0.085**
GII	-0.762**	0.101**	0.288**	-0.161**	-0.148**	-0.353**	-0.085**	1

Source: Estimated by authors. **. Correlation is significant at the 0.01 level (1-tailed). *. Correlation is significant at the 0.05 level (1-tailed).

5. EMPIRICAL RESULTS AND DISCUSSION

5.1. Impact of *EV*s, *MV* and *CV*s on *MS* and *DI*

The coefficients of *MVA* with *DII* are appeared positive and statistically significant (Table 3 and Table 4). The estimates suggested that MS help to increase the DI. The positive association of *MVA* with *DI* is consistent with the findings of Singh (2024); Singh and Jyoti (2023). MS sector is an originator of DTs and other digital devices like mobile, computer, and web-based software which create a conducive environment for digitalization. Thus, performance of MS has a spillover influence on digitalization. DI has a positive impact on MS. The statistical value of coefficient reveals that DI is projected to be improved by 0.1103% due to 1% increase in

MVA, while *MVA* is expected to be increased by 0.1469% due to 1% increase in DI. Accordingly, it is reported a positive and bilateral relationship between DI and MS in the selected countries. Furthermore, it is determined that DI would be favourable to increase the sustainability of MS.

Table 3. Impact of MVA, EVs, CVs and MV on DI

Number of obs.		765	**Wald Chi²**			1499.34
Number of groups		45	**Prob > Chi²**			0.000
Obs. per group		17	**Estimated autocorrelations**			0
R-squared		0.5333	**Mean VIF**			1.44
ln(DII)=[DV]	**Coef.**	**Std. Err.**	**z**	**P>|z|**	**[95% Conf. Interval]**	
ln(MVA)	0.1103	0.0189	5.850	0.000	0.0733	0.1473
ln(EPRT)	0.5513	0.0553	9.970	0.000	0.4429	0.6598
ln(EGS)	-0.0268	0.0182	-1.480	0.139	-0.0624	0.0087
ln(FDINI)	0.0260	0.0104	2.510	0.012	0.0057	0.0463
ln(PD)	0.0073	0.0027	2.740	0.006	0.0021	0.0126
ln(PG)	-0.0499	0.0072	-6.920	0.000	-0.0640	-0.0357
ln(GII)	-0.2797	0.0228	-12.270	0.000	-0.3244	-0.2351
Con. Coef.	-3.9152	0.1745	-22.440	0.000	-4.2572	-3.5733

Source: Estimated by authors.

The coefficients of *EPRT* with *DII* and *MVA* are found positive. Thus, the estimates clarified the employment to population ratio has a positive impact on DI and MS. *EPRT* is a significant driver to increase income and purchasing power of people. High purchasing power of people incentivises them to apply DTs in their daily activities. Jyoti and Singh (2023); Singh and Jyoti (2023) also recognised that *EPRT* is favourable to increase digital infrastructure. DI and *MVA* are likely to be increased by 0.5513% and 0.2916%, respectively. Thus, it is noticeable that *EPRT* is useful to increase DI and MS.

Table 4. Impact of DI, EVs, CVs and MV on MVA

Number of obs.		765	**Wald Chi²**			3233.67
Number of groups		45	**Prob > Chi²**			0.000
Obs. per group		17	**Estimated autocorrelations**			0
R-squared		0.319	**Mean VIF**			1.56
ln(MVA)=[DV]	**Coef.**	**Std. Err.**	**z**	**P>|z|**	**[95% Conf. Interval]**	

continued on following page

Table 4. Continued

ln(DII)	0.1469	0.031	4.800	0.000	0.0869	0.2068
ln(EPRT)	0.2916	0.033	8.860	0.000	0.2271	0.3562
ln(EGS)	0.2747	0.022	12.400	0.000	0.2313	0.3181
ln(FDINI)	-0.0637	0.014	-4.700	0.000	-0.0903	-0.0372
ln(PD)	0.0741	0.003	23.160	0.000	0.0678	0.0804
ln(PG)	-0.0170	0.005	-3.250	0.001	-0.0273	-0.0068
ln(GII)	0.2081	0.014	15.150	0.000	0.1812	0.2350
Con. Coef.	0.7437	0.1493	4.980	0.000	0.451084	1.0363

Source: Estimated by authors.

The coefficients of *EGS* with *DII* are detected negative, while its coefficient with *MVA* appears positive. Thus, trade openness has a negative and positive impact on DI and *MVA*, respectively. Thereupon, *FDINI* produces a positive and negative impact on DI and *MVA*, respectively. The estimates imply that trade openness and *FDINI* do not produce a similar implication on DI and *MVA* in the selected countries. Singh (2024); Singh and Jyoti (2023) also suggested that *FDINI* is positive to increase digital infrastructure.

The coefficients of *PD* and *PG* with *DII* and *MVA* are found statistically significant. *PD* and *PG* are used to capture the impact of demographic change on DI and MS in this chapter. The private sector increases the progress of digital services in highly populated densities areas. Hence, a massive populated area helps to develop appropriate digital infrastructure. DI and *MVA* to be declined by 0.0499% and 0.0170%, respectively as population growth increases by 1%. Jyoti and Singh (2023); Singh (2024) also determined that population growth is caused to reduce digital infrastructure. Singh and Jyoti (2023) argued that population growth is causing extensive pressure on resources which may reduce digital infrastructure.

The impact of gender inequality on DI and *MVA* are also found statistically significant. However, the coefficients of *GII* with DI are noted as negative. The estimated value claimed that DI may be decreased by 0.2797% due to 1% increase in gender inequality by 1%. Women have low digital skills, poor accessibility of digital devices, and inefficiency to use DTs. Low economic capacity and purchasing power of women denied them the opportunity to use DTs. Thus, high gender inequality may be caused to reduce digital infrastructure. India is good example where females have lower contribution in digital activities as compared to male. While digitalization is suitable to increase gender equality through creating jobs for females, and reducing wage inequality (Wu et al., 2024). Further, the positive coefficient of *GII* with *MVA*, infer that gender inequality may be positive to increase *MVA*. The statistical value proposes that *MVA* is likely to be increased by 0.2081% due to 1% increase in gender inequality. Gender inequality may be helpful to increase *MVA* due to wage differences between male and females. However, earlier studies claimed that gender equality helps to increase *MVA* in Chile (Wu & Cheng, 2016). High gender equality will be positive to increase firm's growth rapidly (Wu & Cheng, 2016). Further, equal distribution of resources among the male and females will enhance productivity growth of manufacturing firms (Wu & Cheng, 2016). Thus, the positive association of *GII* with *MVA* is not rational due to many reasons in this research.

5.2. Results of Robustness Check

Table 5. Impact of MVA on DI with instrumental variables

Number of obs.		720	**Wald Chi²**			1436.01
Number of groups		45	**Prob > Chi²**			0.000
Obs. per group		16	**Estimated autocorrelations**			8
R-squared		0.5405	**Mean VIF**			1.45
$ln(DII)_{t-1}$=[DV]	**Coef.**	**Std. Err.**	**z**	**P>|z|**	**[95% Conf. Interval]**	
$ln(MVA)_{t-1}$	0.1081	0.020	5.400	0.000	0.0689	0.1473
$ln(EPRT)_{t-1}$	0.5375	0.058	9.330	0.000	0.4246	0.6504
$ln(EGS)_{t-1}$	-0.0245	0.019	-1.270	0.204	-0.0622	0.0133
$ln(FDINI)_{t-1}$	0.0252	0.011	2.350	0.019	0.0042	0.0462
$ln(PD)_{t-1}$	0.0072	0.003	2.590	0.010	0.0018	0.0127
$ln(PG)_{t-1}$	-0.0482	0.007	-6.630	0.000	-0.0625	-0.0340
$ln(GII)_{t-1}$	-0.2893	0.0238	-12.170	0.000	-0.3359	-0.2427
Con. Coef.	-3.8755	0.1799	-21.540	0.000	-4.2282	-3.5228

Source: Estimated by authors.

Table 6. Impact of DI on MVA with instrumental variables

Number of obs.		720	**Wald Chi²**			3014.28
Number of groups		45	**Prob > Chi²**			0.000
Obs. per group		16	**Estimated autocorrelations**			8
R-squared		0.3191	**Mean VIF**			1.58
$ln(MVA)_{t-1}$=[DV]	**Coef.**	**Std. Err.**	**z**	**P>|z|**	**[95% Conf. Interval]**	
$ln(DII)_{t-1}$	0.1431	0.0320	4.470	0.0000	0.0803	0.2059
$ln(EPRT)_{t-1}$	0.3068	0.0319	9.620	0.0000	0.2443	0.3693

continued on following page

Table 6. Continued

$ln(EGS)_{t-1}$	0.2736	0.0228	12.000	0.0000	0.2289	0.3183
$ln(FDINI)_{t-1}$	-0.0642	0.0135	-4.740	0.0000	-0.0908	-0.0377
$ln(PD)_{t-1}$	0.0740	0.0034	22.050	0.0000	0.0674	0.0805
$ln(PG)_{t-1}$	-0.0150	0.0045	-3.380	0.0010	-0.0238	-0.0063
$ln(GII)_{t-1}$	0.2093	0.0146	14.310	0.0000	0.1806	0.2380
Con. Coef.	0.6888	0.1478	4.660	0.000	0.3991	0.9785

Source: Estimated by authors.

The robustness of empirical findings is estimated through using the one lags of all variables in the model. The coefficients of one lag of all *IV*s with one lag of *DII* and *MVA* provide similar results in terms of signs (Table 5 and Table 6). Thus, the estimates infer that empirical results are robust. The coefficients of one lag of *MVA*, *EPRT*, *FDINI* and *PD* with one lag of *DII* are determined positive. Further, one lag of *EGS*, *PG* and *GII* have a negative impact on one lag of *DII*. Thereupon, the coefficients of one lag of *DII*, *EPRT*, *EGS*, *PD* and *GII* provide a positive impact on previous lag of *MVA*.

Table 7. Impact of MVA and DI on Mediator Variable (MV)

Number of obs.		720	**Wald Chi²**			2625.15
Number of groups		45	**Prob > Chi²**			0.000
Obs. per group		16	**Estimated autocorrelations**			8
R-squared		0.1881	**Mean VIF**			1.00
ln(EGS)=[DV]	**Coef.**	**Std. Err.**	**z**	**P>|z|**	**[95% Conf. Interval]**	
ln(MVA)	0.4082	0.0338	12.090	0.000	0.3420	0.4744
ln(DII)	0.4713	0.0365	12.930	0.000	0.3999	0.5428
Con. Coef.	2.9667	0.1217	24.370	0.000	2.7281	3.2053

Source: Estimated by authors.

Table 8. Impact of MVA and DI on Mediator Variable (MV) with instrumental variables

Number of obs.	720	**Wald Chi²**	3294.77
Number of groups	45	**Prob > Chi²**	0.000
Obs. per group	16	**Estimated autocorrelations**	8
R-squared	0.1915	**Mean VIF**	1.00

continued on following page

Table 8. Continued

$ln(EGS)_{t-1}$=[DV]	Coef.	Std. Err.	z	P>|z|	[95% Conf. Interval]	
$ln(MVA)_{t-1}$	0.4175	0.0342	12.220	0.000	0.3505	0.4844
$ln(DII)_{t-1}$	0.4723	0.0379	12.480	0.000	0.3981	0.5465
Con. Coef.	2.9342	0.1223	23.990	0.000	2.6944	3.1739

Source: Estimated by authors.

The validity of *MV* is tested as estimating the impact of *MVA* and *DII* on it (Table 7). Accordingly, one lag of *MV*, *MVA* and *DII* are used to check the robustness of empirical results (Table 8). The estimates proposed that MS and DI help to increase trade openness. Thereupon, *EGS* has a mediator impact on *DI* and *MVA*. Furthermore, the coefficients of one lag of *MVA* and *DII* with the prior lag of *EGS* have a similar sign (Table 8). Thus, regression coefficients of *MVA* and *DII* with *EGS* have reliability.

5.3. Causal Association Among the Variables

Table 9. Dumitrescu and Hurlin (2012) Granger non-causality test results

DV/IVs	*DII=[DV]*			*DV/IVs*	*MVA =[DV]*		
	W-bar	*Z-bar*	*Z-bar tilde*		**W-bar**	**Z-bar**	**Z-bar tilde**
MVA	1.733	3.4768*	1.9158**	*DII*	2.3229	6.2751*	3.9664*
EPRT	2.484	7.0406*	4.5273*	*EPRT*	2.2183	5.7790*	3.6028*
EGS	-	-	-	*EGS*	-	-	-
FDINI	1.709	3.3646*	1.8335*	*FDINI*	1.9623	4.5646*	2.7129*
PD	3.367	11.2254*	7.5939*	*PD*	3.5777	12.2273*	8.3281*
PG	2.737	8.2385*	5.4051*	*PG*	2.7782	8.4346*	5.5488*
GII	2.356	6.4304*	4.0802*	*GII*	3.1323	10.1142*	6.7796*

Source: Estimated by authors.

The causal association among the variables are given in Table 9. The estimates imply that undertaken variables have a causal relationship with DI and *MVA*. DI and *MVA* have a significant dependence on the included variables in the models. Moreover, these variables can be applied to examine the predicted values of DI and *MVA* for the selected countries in future.

6. CONCLUSION

Existing literature implies that DI brings positive and negative implications for the manufacturing and business sector. Findings of most studies argued that DI would be beneficial to increase the sustainability of MS. MS is positive to increase DI. Moreover, DI has a direct and indirect implication on MSMEs and SMEs. SMEs promoting variables like financial accessibility, resource allocation, government regulation, innovation, and marketing have a direct impact on MSMEs and MS (Egere et al., 2022). DTs are helping to increase profits, sales patterns, consumer's satisfaction and revenue generation in SMEs (Mukherjee, 2018; Rosyidiana & Narsa, 2024; Faiz et al., 2024). Entrepreneurs and businessmen are developing new business models using DTs. DTs are increasing the sustainability and social values, digital marketing, and financial performance of SMEs (Vrontis et al., 2022; Sharabati et al., 2024).

The previous review emphasized that DTs and digitalization are highly imperative to increase the growth of MS in the global countries. The developing countries should create a favourable digital infrastructure for improving the performance of SMEs and MSMEs which would be positive to increase the sustainability of MS (Singh et al., 2024). Also, MS is unable to apply GIS technologies to achieve various tasks in many developing nations. MS is also facing many challenges due to lack of managerial skills, low financial availability for small and new entrepreneurs, strict government's tax policies, and inefficient digital infrastructure in developing countries (Singh et al., 2024). Low infrastructure and capital formation are also challenges for further growth and sustainability of MS. Moreover, MS has low trust on domestic technologies due to flexibility in IPRs regime and low potential of technologies commercialization. MS is unable to increase technological upgradation due to lack of advanced technologies. Hence, the above-mentioned factors can be included in policy implementation to increase the sustainability of MS and DI in the global countries.

The trend analysis based on estimated values of *DII* and share of *MVA* in GDP of selected 45 countries during 2006 – 2022, proposed that these nations have a high and significant variation in DI and progress of MS. The correlation coefficient among the *DII* and *MVA* is shown a positive and statistically significant association among these. Thus, it is testified that DI and *MVA* have a positive association. The other variables like *PD*, *FDINI*, *EGS* and *EPRT* are positively associated with DI and MS. Population growth and gender inequality are negatively associated with DI. The regression results indicate that *MVA* and *DI* have a positive and bi-directional causal relationship with each other. Thus, performance of MS is positive to increase DI and vice-versa. DI improves as *MVA*, *EPRT*, *FDINI* and *PD* increase. MS, FDI and population density appear positive to increase DI. Population growth, gender inequality and trade openness shows a negative impact on DI. Thereupon, *MVA* increases as an increase in DI, *EPRT*, *EGS* and *PD*. Therefore, growth of MS would increase as DI, trade openness and population density increase.

7. POLICY IMPLICATIONS

The review and empirical results revealed that DI is beneficial to increase the share of MS in GDP of selected countries. However, DI is providing greater benefits to high and upper-income group countries as these have a greater experience to use DTs and have already created the digital landscape. Lower and low-income groups are unable to get significant benefits from DI. Low digital literacy of people, inefficient digital infrastructure, low spending of government in digital sector, low digital security, and inappropriate cyber security are creating problems to develop efficient digital infrastructure in developing nations. Therefore, appropriate technological diffusion of DTs is essential in the global countries to increase DI. Consequently, promotion of digital manufacturing would enhance sustainability of MS (Mangla et al., 2024). FDI, trade openness, gross capital formation and technological advancement and patenting are supportive to increase the development of the MS. Patenting is highly essential to increase the legal security of invention and new technologies. Accordingly, these indicators would be important to increase the DI and *MVA*.

MS should improve their technological upgradation strategies to increase its growth. MS should increase their association of research organizations and international firms to increase its performance. International trade, commercialization of technologies and knowledge transfer must be improved to increase the share of MS. MS should take a specific policy measure to increase the sustainability of MS. MS should apply GIS technologies to achieve multiple goals like improving supply chain, environmental compliance, transparency in delivery of products, and monitoring of environmental issues. Use of green innovation, green technologies and less carbon emitter technologies, and practices of minimum resource use in the MS would be fruitful to increase its sustainability.

8. INNOVATION AND NOVELTY OF THE CHAPTER

This chapter develops a theoretical association of DI with MS. Benefits and challenges of MSMEs due to digitization and digital transformation are presented in this chapter. It examines the significance of GIS technologies in sustainability of MS as per theoretical literature. Most specifically, it develops *DII* or selected countries in a time series using a *Composite Z-score* method. The statistical values of *DII* provide the relative progress of digitalization across countries. The estimated values of *DII* are beneficial for low-ranking countries to implement policies to improve their performance in DI. Log-linear regression models are formulated to observe the existence of cause-and-effect relationship between DI and MS. Finally, this chapter come with research directions for undertaking further studies.

9. LIMITATIONS AND SCOPE OF FURTHER RESEARCH

In this chapter, *DII is* the integrated composition of 12. The estimated values of *DII* greatly depend on undertaken variables and countries. Thus, ranking of these countries in *DII* may not be as stable as any addition or elimination of variables or countries in their estimations. It automatically changes the performance and ranking of undertaken countries in DI. Thereupon, values of *DII* depend on the adopted method for measuring it. Further, study should include more reliable information about DI and MS through primary data of SMEs and MSMEs. This chapter considered the share of *MVA* in GDP of respective countries for measuring the performance of MS. However, there are many indicators like productivity, efficiency, and TFP which also explore the performance of MS. This chapter does not develop an empirical model to examine the direct impact of DI on sustainability of MS. Hence, existing researchers should finalize the sustainability related indicators in further study. Moreover, researchers can estimate the role of GIS technologies to increase the sustainability of MS in further studies.

REFERENCES

Ahmadi-Gh, Z., & Bello-Pintado, A. (2024). Towards sustainable manufacturing: How does digitalization and development affect sustainability barriers? *Journal of Cleaner Production, 476*(1), 1–13. DOI: 10.1016/j.jclepro.2024.143792

Ahmed, F., Ali, I., Kousar, S., & Ahmed, S. (2022). The environmental impact of industrialization and foreign direct investment: Empirical evidence from Asia-Pacific region. *Environmental Science and Pollution Research International, 29*(20), 29778–29792. DOI: 10.1007/s11356-021-17560-w PMID: 34993824

Almeida, F., & Okon, E. (2025). Impact of digitalization on SMEs performance: The mediating role of IOT. *Digital Transformation and Society, 4*(4), 407–422. DOI: 10.1108/DTS-11-2024-0240

Arif, M. E., & Herawan, T. (2025). Driving performance in SMEs: Exploring entrepreneurial ecosystems, leadership, and dynamic capability in emerging economies. *Journal of the International Council for Small Business, 1*(1), 1–21. DOI: 10.1080/26437015.2025.2531873

Autio, E., Komlósi, E., Szerb, L., Tiszberger, M. G., Park, D., & Jinjarak, Y. (2025). Digital entrepreneurship landscapes in developing Asia: Insights from the global index of digital entrepreneurship systems. *European Journal of Innovation Management, 28*(7), 2845–2872. DOI: 10.1108/EJIM-02-2024-0121

Boffa, E., & Maffei, A. (2024). Investigating the impact of digital transformation on manufacturers' business model: Insights from Swedish industry. *Journal of Open Innovation, 10*(2), 100312. DOI: 10.1016/j.joitmc.2024.100312

Buteau, S. (2021). Roadmap for digital technology to foster India's MSME ecosystem—Opportunities and challenges. *CSI Transaction on ICT, 9*(1), 233–244. DOI: 10.1007/s40012-021-00345-4

Chang, X., Fu, K., & Conteh, M. (2025). How digitalization impacts the high-qulality development of manufacturing industry: Evidence from China. *Sustainability (Basel), 17*(12), 5586. DOI: 10.3390/su17125586

Chen, X., Despeisse, M., & Johnasson, B. (2020). Environment sustainability of digitalization in manufacturing: A review. *Sustainability (Basel), 12*(24), 1–31. DOI: 10.3390/su122410298

Chhotaray, P., Behera, B. C., Moharana, B. R., Muduli, K., & Sephyrin, F. T. R. (2023). Enhancement of manufacturing sector performance with the application of Industrial Internet of Things (IIoT). In *Smart Technologies for Improved Performance of Manufacturing Systems and Services* (pp. 1–19). CRC Press. DOI: 10.1201/9781003346623-1

Chowdhury, E. K. (2025). Drivers of entrepreneurship development in South Asia: Empirical insights and policy recommendations. *Journal of Innovation and Entrepreneurship, 14*(46), 1–30. DOI: 10.1186/s13731-025-00512-w

Cunningham, J. A., Damij, N., Modic, D., & Olan, F. (2023). MSME technology adoption, entrepreneurial mindset and value creation: A configurational approach. *The Journal of Technology Transfer, 48*(1), 1574–1598. DOI: 10.1007/s10961-023-10022-0

Díaz-Pelaez, A., Chura-Quispe, G., Clemente-Almendros, A. A., & Velarde-Molina, J. F. (2024). Analysing the influence of digital strategy and innovation on MSMEs' performance in emerging countries. *Creativity and Innovation Management, 34*(2), 351–363. DOI: 10.1111/caim.12642

Dumitrescu, E. I., & Hurlin, C. (2012). Testing for Granger non-causality in heterogeneous panels. *Economic Modelling, 29*(4), 1450–1460. DOI: 10.1016/j.econmod.2012.02.014

Egere, O. M., Mass, G., & Jones, P. (2022). A critical analysis of the Nigerian entrepreneurial ecosystem on transformational entrepreneurship. *Journal of Small Business Management, 62*(3), 1187–1218. DOI: 10.1080/00472778.2022.2123109

Embia, G., Moharana, B. R., Mohamed, A., Muduli, K., & Muhammad, N. B. (2023). 3D Printing Pathways for Sustainable Manufacturing. In Nayyar, A., Naved, M., & Rameshwar, R. (Eds.), *New Horizons for Industry 4.0 in Modern Business. Contributions to Environmental Sciences & Innovative Business Technology* (pp. 253–272). Springer International Publishing., DOI: 10.1007/978-3-031-20443-2_12

Erumban, A. A., & de Vriers, G. J. (2024). Structural change and poverty reduction in developing economies. *World Development, 181*(1), 106674. DOI: 10.1016/j.worlddev.2024.106674

Faiz, F., Le, V., & Masli, E. K. (2024). Determinants of digital technology adoption in innovative SMEs. *Journal of Innovation and Knowledge, 9*(4), 1–20. DOI: 10.1016/j.jik.2024.100610

Galgal, K. N., Ray, M., Moharana, B. R., Behera, B. C., & Muduli, K. (2023). Quality control in the era of IoT and automation in the context of developing nations. In *Smart Technologies for Improved Performance of Manufacturing Systems and Services* (pp. 39–50). CRC Press. DOI: 10.1201/9781003346623-3

Galvani, S., Carloni, E., Bocconcelli, R., & Pagano, A. (2025). When digitalization meets sustainability: Exploring interactions within a manufacturing firm. *Journal of Business and Industrial Marketing*, *40*(13), 30–46. DOI: 10.1108/JBIM-03-2024-0132

Halkos, G., de Alba, J. M., & Todorov, V. (2021). Analyzing manufacturing sector and selected development challenges: A panel data analysis. *Energy*, *235*(1), 121253. DOI: 10.1016/j.energy.2021.121253

Jyoti, B., & Singh, A. K. (2023). Does digitalization have a causal relationship with economic development? an experience from a country-wise panel data statistical analysis. In R. Sine Nazlı & G. Sari (Eds.), *Handbook of Research on Perspectives on Society and Technology Addiction* (pp. 34-61). IGI Global. DOI: 10.4018/978-1-6684-8397-8.ch003

Karyakarte, Y. (2024). Enhancing manufacturing efficiency with advanced GIS solutions. https://edstechnologies.com/newsletter/2024/june/Manufacturing-GIS.pdf

Kupriyanova, M., Evdokimova, E., Soloviova, I., & Simikova, I. (2023). The impact of digitalization on the manufacturing industry. E3S Web of Conferences, 431, 05030. DOI: 10.1051/e3sconf/202343105030

Laiwang, X., Abbas, A., Kunpeng, A., & We, X. (2025). Quantifying the role of digitalization, social factors and energy transition to achieve sustainability in East Asia. *Humanities & Social Sciences Communications*, *12*(1), 1613. DOI: 10.1057/s41599-025-05893-7

Larios-Francia, R. P., & Ferasso, M. (2023). The relationship between innovation and performance in MSMEs: The case of wearing apparel sector in emerging countries. *Journal of Open Innovation*, *9*(1), 100018. DOI: 10.1016/j.joitmc.2023.100018

Li, W., Xue, D., & Huang, X. (2018). The role of manufacturing in sustainable economic development: A case of Guangzhou, China. *Sustainability (Basel)*, *10*(9), 3039. DOI: 10.3390/su10093039

Maffei, A., Grahn, S., & Nuur, C. (2019). Characterization of the impact of digitalization on the adoption of sustainable business models in manufacturing. *Procedia CIRP*, *81*(1), 765–770. DOI: 10.1016/j.procir.2019.03.191

Mangla, S. K., Luthra, S., Garza-Reyes, J. A., Jose, C., Jabbour, C., & Brem, A. (2024). Sustainability and Industry 4.0: The role of social, environmental and technological factors in the development of digital manufacturing. *Technological Forecasting and Social Change*, *201*(1), 123223. DOI: 10.1016/j.techfore.2024.123223

Mittal, R., Sheetal, , & Soriya, S. (2025). Fostering competitiveness of Indian MSMEs through IT and digitalization. *Information Technology for Development*, *31*(4), 1066–1089. DOI: 10.1080/02681102.2025.2453211

Mu, X., Chen, D., & Zhang, H. (2024). The impact of digital transformation on the sustainable development of enterprises: An analysis from the perspective of digital finance. *Journal of Applied Econometrics*, *28*(1), 1–13. DOI: 10.1080/15140326.2025.2464488

Muduli, K., Luthra, S., Garza-Reyes, J. A., & Huisingh, D. (2023). Application of blockchain technology for addressing reverse logistics challenges: current status and future opportunities. *Supply Chain Forum: An International Journal*, 24(1), 1-6. DOI: 10.1080/16258312.2023.2165279

Mukherjee, S. (2018). Challenges to Indian micro small scale and medium enterprises in the era of globalization. *Journal of Global Entrepreneurship Research*, *8*(28), 1–14. DOI: 10.1186/s40497-018-0115-5

Ochoa, M., & Nonnecke, B. M. (2019). Developing a digital inclusion index at the state level: The case of Mexico. TPRC47: The 47th Research Conference on Communication, Information and Internet Policy 2019. DOI: 10.2139/ssrn.3432615

Prah, K., & Gajšek, B. (2024). Geographical information system modeling for planning internal transportation in a manufacturing plant's outdoor area. *Processes (Basel, Switzerland)*, *12*(1), 1–6. DOI: 10.3390/pr12010180

Qiao, N., & Niu, L. (2024). The impact of the industrial innovation ecosystem on innovation performance – using the equipment manufacturing industry as an example. *Systems*, *12*(12), 1–21. DOI: 10.3390/systems12120578

Qing, C. H. G., Oktaviani, E., & Heripracoyo, S. (2019). The literature review of manufacturing industry 4.0 using mapping design with GIS approach. *2019 International Conference on Information Management and Technology (ICIMTech)*, Jakarta/Bali, Indonesia, 2019, pp. 438-443, DOI: DOI: 10.1109/ICIMTech.2019.8843824

Rosyidiana, R. N., & Narsa, I. M. (2024). Micro, small, and medium-sized enterprises (MSMEs) during the post-pandemic economic recovery period: Digitalization, literation, innovation, and its impact on financial performance. *Cogent Business &amp. Cogent Business & Management*, *11*(1), 1–14. DOI: 10.1080/23311975.2024.2342488

Sankar, M. S. K., Gupta, S., Luthra, S., Jagtap, S., Singh, S., & Kumar, A. (2024). The emergence of digitalization to the manufacturing sector in the sustainability context: A multi-stakeholder perspective analysis. *Journal of Cleaner Production, 468*(1), 142983. DOI: 10.1016/j.jclepro.2024.142983

Shah, S., Menon, S., Ojo, O. O., & Naghi, G. E. (2020). Digitalisation in sustainable manufacturing – a literature review. *2020 IEEE International Conference on Technology Management, Operations and Decisions (ICTMOD)*, Marrakech, Morocco, pp. 1-6. DOI: DOI: 10.1109/ICTMOD49425.2020.9380605

Shang, Z., & Zhang, L. (2022). The sustainable digitalization in the manufacturing industry: A bibliometric analysis and research trend. *Mobile Information Systems, 2022*(1), 1–14. DOI: 10.1155/2022/1451705

Sharabati, A. A. A., Ali, A. A. A., Allahham, M. I., Hussein, A. A., Alhhet, A. F., & Mohammad, A. S. (2024). The impact of digital marketing on the performance of SMEs: An analytical study in light of modern digital transformations. *Sustainability (Basel), 16*(19), 1–25. DOI: 10.3390/su16198667

Singh, A. K. (2024). Causality of digitalization with social sustainability in selected countries. In Siyal, S. (Ed.), *Impact of Digitalization on Education and Social Sustainability* (pp. 39–78). IGI Global., DOI: 10.4018/979-8-3693-1854-6.ch002

Singh, A. K., & Jyoti, B. (2023). Does digitalization have a bi-directional causal relationship with the entrepreneurship ecosystem? evidence from a country-wise panel data analysis. In Yanamandra, R., & Indiran, L. (Eds.), *Handbook of Research on Designing Sustainable Strategies to Develop Entrepreneurial Intention* (pp. 178–172). IGI Global., DOI: 10.4018/978-1-6684-8781-5.ch007

Singh, A. K., & Kumar, S. (2021). Assessing the performance and factors affecting industrial development in Indian states: An empirical analysis. *Journal of Social Economics Research, 8*(2), 135–154. DOI: 10.18488/journal.35.2021.82.135.154

Singh, A. K., Singh, B., & Ashraf, S. N. (2020). Implications of intellectual property protection, and science and technological development in the manufacturing sector in selected economies. *Journal of Advocacy. Research in Education, 7*(1), 16–35. https://oaji.net/articles/2021/1704-1615668997.pdf

Singh, A. K., & Singh, R. (2025). Implications of digital infrastructure on social sustainability in global countries. In Jaswal, A., Mukherjee, D., Gunasekaran, A., & Kandpal, V. (Eds.), *Green Futures. Approaches to Global Sustainability, Markets, and Governance* (pp. 381–403). Springer., DOI: 10.1007/978-981-96-0843-0_19

Singh, A. K., Singh, S., & Gupta, M. (2024). A literature review for the association of social media with micro and small enterprises in India. In Hossain, S. (Ed.), *Cases on Social Media and Entrepreneurship* (pp. 38–62). IGI Global., DOI: 10.4018/979-8-3693-1781-5.ch003

Surya, B., Menne, F., Sabhan, H., Suriani, S., Abubakar, H., & Idris, M. (2021). Economic growth, increasing productivity of SMEs, and open innovation. *Journal of Open Innovation*, *7*(1), 1–37. DOI: 10.3390/joitmc7010020

Tan, Y.-M., Amorós, E., Autio, E., Fu, K., & Park, D. (2025). The impact of digitalization on entrepreneurial innovation: a comparative analysis between developing economies in asia and in other regions. ADB Economics Working Paper Series No. 776, DOI: 10.2139/ssrn.5233001

United Nation Development Programmes (UNDP), Human Development Reports. (2026). [Online]. https://hdr.undp.org/data-center/documentation-and-downloads

Vrontis, D., Chaudhuri, R., & Chatterjee, S. (2022). Adoption of digital technologies by SMEs for sustainability and value creation: Moderating role of entrepreneurial orientation. *Sustainability (Basel)*, *14*(13), 1–19. DOI: 10.3390/su14137949

Wan, X., Kazmi, S. A. A., & Wong, C. Y. (2022). Manufacturing, exports, and sustainable growth: Evidence from developing countries. *Sustainability (Basel)*, *14*(3), 1646. DOI: 10.3390/su14031646

Wang, D., & Shao, X. (2024). Research of the impact of digital transformation on the production efficiency of manufacturing enterprises: Institution-based analysis of the threshold effect. *International Review of Economics & Finance*, *91*(1), 883–897. DOI: 10.1016/j.iref.2024.01.046

Wen, H., Liu, Y., & Liu, Y. (2024). Impact of digitalization on investment and productivity of manufacturing industry: Evidence from China. *SAGE Open*, *14*(3), 21582440241281862. Advance online publication. DOI: 10.1177/21582440241281862

World Development Indicators (WDIs), The World Bank (WB). (2026). [Online]. https://databank.worldbank.org/source/world-development-indicators

Wu, R., & Cheng, X. (2016). Gender equality in the workplace: The effect of gender equality on productivity growth among the Chilean manufacturers. *Journal of Developing Areas*, *50*(1), 257–274. DOI: 10.1353/jda.2016.0001

Wu, Y., Li, L., & Zheng, Y. (2024). The impact of digitization in manufacturing on female employment and gender wage gap. *Journal of Asian Economics*, *95*(1), 101821. DOI: 10.1016/j.asieco.2024.101821

Yang, Q., & Jin, S. (2024). Exploring the impact of digital transformation on manufacturing environment, social responsibility, and corporate governance performance: The moderating role of top management teams. *Sustainability (Basel), 16*(11), 4342. DOI: 10.3390/su16114342

Zen, F., & Faizin, M. A. (2021). An overview of digitalization and internationalization MSMEs: A conceptual paper. *Advances in Economics. Business and Management Research, 193*(1), 195–200.

Zhang, J., van Gorp, D., & Kievit, H. (2023). Digital technology and national entrepreneurship: An ecosystem perspective. *The Journal of Technology Transfer, 48*(1), 1077–1105. DOI: 10.1007/s10961-022-09934-0 PMID: 35602312

Zhang, K., Wang, J., & Wu, Y. (2025). A study of the impact of manufacturing input digitization on firms' organizational resilience: Evidence from China. *Sustainability (Basel), 17*(3), 897. DOI: 10.3390/su17030897

Chapter 3
Technological Innovations for Sustainable Waste-Sorting Systems

Mariem Smaoui
National Engineering School of Sfax (ENIS), Tunisia

Mouna Rekik
http://orcid.org/0000-0003-1869-2394
National Engineering School of Sfax (ENIS), Tunisia

Lotfi Krichen
National Engineering School of Sfax (ENIS), Tunisia

ABSTRACT

Increasingly, traditional waste management systems are struggling to keep up due to rising global waste generation and an associated demand for intelligent automation solutions, which are quickly rising to meet that need. The current chapter details the development and operation of an autonomous smart waste sorting system. This autonomous system uses artificial intelligence along with multisensor technologies to improve the sorting performance of the various materials that it sorts. The machine learning model used in the system supports visual identification of the materials being sorted as well as separate detection of all metals, glass and organic waste by means of unique physical and chemical properties. The operational data for this system is communicated to a centralized digital platform, which provides real-time

DOI: 10.4018/979-8-2600-2383-9.ch003

) 2027, IGI Global Scientific Publishing. Copying or distributing in print or electronic forms without written permission of IGI Global Scientific Publishing is prohibited. Use of this chapter to erative artificial intelligence (AI) technologies is expressly prohibited. The publisher reserves all rights to license its use for generative AI training and machine learning model development.

visibility and management of the system through a user interface. The experimental results of the prototype presents an encouraging avenue for growth of scalable smart waste management solutions.

1-INTRODUCTION

Municipal Solid Waste: A New Global Concern

In the last few years, the management of municipal solid waste has been one of the critical concerns in the world amidst the growing threats of a tremendous increase in the volume of municipal solid waste being generated in the country. In the year 2020, the global generation of municipal solid waste was found to be 2.1 billion tons yearly, which is expected to escalate up to 3.8 billion tons by the year 2050 if immediate solutions are not formulated against this ever-increasing waste generation in the country; therefore, the authorities are in a great need to introduce an effective waste management strategy while considering the sustainability aspects in this context in the country (Kaza et al., 2018; Maalouf et al., 2022; UN Environment Programme, 2024).

The development within the human lifestyles and patterns of consumption in the development countries, related more to the urbanization and industrialization phenomenon, results mainly in the generation of significant amounts of waste per capita within the country. This process is generally related to the financial well-being level of the country, with such a level increasing proportionally to the Gross Domestic Product (GDP) created within the country.

This particular instance may be given for the United States of America with an output of merely 12% or less than 5% of the total world population that creates only 12% or even more than 12% of the total waste generated in the world (UN Environment Programme, 2024).

It is striking to notice, for example, how North America and Central and South Asia produce similar amounts of waste per capita, despite their different economic and demographic dynamics. In middle-income countries where urbanization is fast-growing, waste management has increasingly been a daunting challenge faced by cities with inadequate collection and disposal infrastructures. For example, India faces problems on many fronts: an overburning of landfills, inefficient sorting of waste, and a lack of sufficient enforcement of laws and regulations. These are just some of the major problems that should be tackled urgently by incorporating technological innovations and stronger collaborations with the private sector to better manage, process, and recycle wastes (Singh, 2019, Khemakhem et al., 2022, Worldbank, 2018).

Transitioning to Sustainable Waste Management

Sustainable waste management is an essential concern in dealing with the environmental, social, and economic challenges facing the world. Not only does waste management help mitigate the harmful effects on the environment, including soil, air, or water pollution, but it also guarantees social justice, mainly because it addresses inequality among certain sections of the population that are most impacted (Adeniyi, 2023; Ferronato & Torretta, 2019).

At the core of all these processes, the position of sorting waste cannot be overstated. In other words, sorting waste is the starting point toward the effective recovery of materials and greatly contributes to the success of all approaches to reducing, reusing, and recycling. More particularly, sorting waste has the advantage of recovering a larger quantity of the materials that can be recycled, reducing the quantity of waste disposed of in landfills, and the integration of organic waste into composting or anaerobic digestion processes (Almtireen et al., 2025; Haitao, 2021).

In such a scenario, the salience of the circular economy as a crucial strategic option comes to the forefront. The objectives that are sought to be achieved in the pursuit of this ideal are related to stretching the life cycle of products and materials and supporting responsible resource management. Reducing waste production at the initial stages and optimizing waste materials through effective sorting are some prerequisites to resource management, and the involvement in the fight against climate change in industries with higher greenhouse impacts such as steel, plastic, aluminum, and cement (MacArthur, 2013; Mayouf et al., 2022; Kirchherr et al., 2017).

For example, Material Economics has found that the EU could save up to 296 million tonnes of CO2 each year by 2050 by optimising the use of existing materials, notably through optimising source sorting and available routes of materials.

The key to a successful transformation would depend on collaborations between nations, corporations, and societies. This, in turn, necessitates an increased level of investment in sorting technology, awareness, and innovation.

Speaking of this, it is proper to highlight that the digital revolution, for instance, opens up new possibilities for improvement through the application of artificial intelligence (AI), the Internet of Things (IOT), and automation to greatly improve sorting practices, optimize waste transport operations, and support the decision-making process based on data provided in real time. This would facilitate "smart" waste management that is both efficient and responsive to territorial dynamics and evolution (Material Economics, 2018, Arindam et al., 2022).

Thus, in brief summary, sustainable waste management involves a multi-pronged approach that integrates various concepts in terms of environment, equity, and innovation. Waste sorting is a major driver in terms of waste management practice that forms the foundation of this multi-pronged approach to providing a foundation to a more circular society.

Analysis of Existing Approaches to Waste Sorting and Classification

Generally, the traditional automated sorting system normally utilized in industrial waste management practices heavily relies on the utilization of Programmable Logic Controllers, as well as basic sensors like proximity sensors, limit switches, as well as push buttons. Although such a system has been effective in detecting the presence of items, it has a major drawback in that it lacks the ability to pinpoint the type of waste deposited in a given waste management system, especially in heterogeneous types of waste, where accurate recycling largely depends on accurate waste type identification, i.e., papers, plastic, metallic waste, etc.

Studies from previous works regarding waste management, as well as waste management in the health sector, have revealed that operational as well as technologic limitations have a way of affecting the operational efficiency of an automated system, thus leading to a lack of sustainable operational efficiency in waste management practices, as revealed in a study done by (Swain et al. in 2017).

Besides, irregular shapes of objects in the form of varying sizes and random positions where objects are placed on the conveyor belt add another layer of complexity in achieving efficiency in object recognition. For example, an object can be easily recognized as being present; however, it will not be clear whether the object is made of plastic or metal. Therefore, the separation of plastic and metal materials will not be efficiently achieved (Sathish et al., 2017; Khemakhem et al., 2019). Similar issues have been documented in the general adoption of technology in sustainable operations because of the complexity of the technology involved as well as its non-adaptability in integration with the operations of the organization (Swain et al., 2024).

Conventional sensor-based systems are reported to be less effective in classifying objects that are characterized by random shapes and orientations—conditions that are prevalent in reality—according to a study by Bozma et al. (2002). This addresses the technical limitations of an appropriate waste classification solution that should not only rely on the physical detection of material objects but also their identification based on other properties like color, texture, and surface quality. Emerging in the literature is an important aspect that considerably affects an appropriate technological solution; that is, apart from technical feasibility and potential of solutions like waste classification, organizational feasibility should also be considered in an effective technological solution—this is reported by Swain et al. (2022).

To mitigate the limitations of the traditional sensing technologies, optical sorting technologies have emerged as an alternative option. These technologies involve the use of camera technologies to analyze the visible features of the waste materials using appropriate levels of illumination. Depending on the observed features, the waste materials can be classified according to the type of collection bin to which the materials will be dumped. The use of optical technologies has yielded desirable results in the sorting of lighter materials like paper, cardboard, and soft plastics. However, the performance of optical sorting has faced performance limitations in the handling of heavier materials due to the limitations of the pneumatic machines (Tianchen et al., 2023; Xiaoyu et al., 2020).

In order to solve these challenges, robotic systems have been proposed as solutions. Equipped with different types of cameras and robot arms, these automated and advanced systems are able to see and pick up different items of waste individually. In this regard, different complexities in waste types are able to be sorted or managed (Maria et al., 2021; Zeli et al., 2020).

Of all these innovations, robots with "delta mechanism" fitted with suction cups as grippers have provided impressive results as waste management machines. This is due to their efficiency in identifying and sorting wastes with the help of AI computer software such as "Mask R-CNN." The use of suction cups as grippers is favorable compared to mechanical grippers as long as the objects to be picked and moved are either flat enough to be moved around with such grippers, as objects that have very rough edges may not be moved perfectly with this mechanism (Laski et al., 2015; Konstantia et al., 2022; Hassan et al., 2022; Xinxing et al., 2022).

As a result of integrating AI into sorting systems, significant improvements in recycling efficiency and quality have been realized. For instance, Wilts et al., 2021, proved that AI not only increases speed during sorting but also the purity of the materials recovered. The "ZRR for municipal waste" project clearly elaborates on these benefits with respect to operational performance and safety in achieving the goal of a sustainable circular economy.

Different studies confirm that there is relevance for AI and machine learning in improving waste sorting and automating it. For example, Lubongo et al., (2024) analyzed the usage of AI together with spectroscopic methods and systems of computer vision for sorting plastics, and the outcome was impressive, reaching an accuracy of as high as 99.99% and a sorting capacity of up to 10 tons per hour. Though there are still some issues with plastics of complicated composition, the potential of these systems is outstanding.

They developed an intelligent robot that could identify PET bottles or colored glass containers using computer vision and specialized grippers (Cheng et al., 2024) . Feng et al. (2022) have designed a smart bin with an integrated AI model called GECM-EfficientNet that can automatically classify waste with an accuracy of more than 94% .

However, in spite of the above advancements in the technology used in waste management systems incorporating AI technologies, some limitations still exist in the smarts systems available in the market today that make them unable to recognize more than three to five types of wastes in real-world applications. Additionally, the cost of the systems available today is still on the higher side in spite of the advancements in the technology used in the system for creating an intelligent waste management system that is accurate in its functioning while being cost-effective.

One of the common problems facing the environment and human health is that of waste management. There has been a notable rise in urban areas where there is no monitoring system of garbage bins. For example, cleaning services do not get clear feedback on when the bins are full. In some instances, there has been an incidence of garbage overflowing from bins and spreading on the roads. This is damaging to the health of urban dwellers since it affects their state of living (Arindam et al., 2022).

This can be managed in the context of the smart city if the city employs a monitoring system for the trash bins to be emptied in due time. Already under the current system of garbage disposal, some authorities are already making route optimization techniques for the disposal of waste, similar to the approaches proposed in Shan-Huen et al., (2015), albeit not considering real-time data from the smart bins.

Using the IoT-based modern technologies opens ways to monitor waste more effectively than traditional approaches and to control it. For instance, Kellow et al., 2020, presented a unified solution at both hardware and software levels to enhance waste management.

Many research has been carried out in this area by different researchers, each in a particular aspect. Vera et al., (2013) came up with a method combining bin allocation and garbage truck routing, although separately and without smart bin application. Others developed containers capable of estimating their fill level but without applying the recent technologies, for instance, Internet-connected systems as was done by Vicentini et al., (2009).

Later, Yann et al., (2015) came up with the idea of a "smart bin," with the aim of maintaining cleanliness in the urban environment. Soumyabrata et al., (2023) have suggested a whole IoT-compatible waste management system, with the aim of deploying in the context of smart cities. A system based on wireless sensors was proposed in the study conducted by Ramson et al., (2017).

However, no researcher has extensively investigated the implementation of the Internet of Things in the entire framework of waste management. To bridge the existing implementation gap, Patric et al., (2019) proposed the development of a smart city infra that enables the integration of IoT in waste management. Further, Nidhya et al., (2020) proposed the implementation of an intelligent algorithm, namely ERS.

To mitigate the aforesaid constraints faced by the existing waste management system, as emphasized by various researchers in the literature review section, this study aims to propose a reliable yet efficient waste management model by implementing a combination of AI Technology and Internet of Things Technology.

The proposed system is autonomous, intelligent, and connected. In the proposed system, the classification of the various types of waste, plastic, metal, glass, cardboard, organic waste, is done without human intervention by using high-resolution cameras and AI models trained on large annotated datasets. Specialized sensors, electromagnetic sensors for metals, optical sensors for glass, and gas sensors for organic wastes, provide complementary support to this identification by vision. The identified material follows its path, guided by the actuators to the corresponding bin.

Concurrently, the level sensors attached to the bins monitor the status of their fill in order to avoid overflow. All this information reaches in real time through the IoT network to a platform, accessible by the waste managers, including through a mobile application. This allows for easy supervision, route planning of the collector trips, and overall optimization of the process of managing the produced waste.

In comparison with the above techniques, which were previously presented in the scientific literature, the system we propose includes a series of improvements.

- ✓ •Comprehensive Integration of AI, IoT, and sensors: This is different from other works that are based only on one component of; optical sorters, smart bins, and sensors-based detection techniques and other similar ones, as we are considering all of them as part of a comprehensive and fully-integrated solution.
- ✓ •Real-Time Multi-Material Recognition: The proposed approach extends beyond just plastic, paper, metal objects by employing various other sensors that are capable of detecting various material classes under practical circumstances, especially where objects have irregular shapes and are moving.
- ✓ Dynamic Optimization of Collection Routes: Using real-time information on fill levels provided through IoT technology, our method facilitates the intelligent scheduling of waste collection through the dynamic optimization of routes. Agents make the most out of real-time monitoring through a specially developed mobile app for this purpose, a feature not taken into account in previous studies by Shan-Huen et al., (2015) or Vera et al., (2013).

The rest of the paper is structured as follows. Section 2 introduces the reader to the phenomenon of the intelligent waste sorting system. Section 3 explains the general idea behind the proposed solution, while Section 4 explains the implementation of the proposed solution. Section 5 presents the experimental results, while Section 6 concludes the paper.

2-INTELLIGENT WASTE SORTING SYSTEM DESCRIPTION

The intelligent waste sorting and management system relies on the integration of AI and IOT approaches. It aims at automating every stage of sorting, keeping monitoring on an optimal level, and managing the flow of wastes in real time. This is the innovative way to improve the efficiency of processing while minimizing human intervention as depicted in Figure 1. At the heart of the system architecture is a motorized conveyor belt, which automatically transports the waste from the deposit point to the appropriate collection bins. During this transfer, a vision module will capture images of moving objects. Such images are analyzed with AI algorithms trained by representative databases containing examples of common materials, such as plastic or cardboard. This machine learning allows the system to perfectly identify the nature of waste at once, with immediate decision-making on the proper sorting action. In addition to the visual recognition method, the system employs different sensors dedicated to the detection of different materials. Some can recognize different features of organic wastes. Other sensors can recognize the special light reflection of glass surfaces. Electromagnetic sensors are used to recognize the presence of metal objects; besides, special sensors are used to recognize objects entering the sorting area. Following this identification of material type, motorized means are used to channel each piece of waste to the appropriate compartment in accordance with the decision made by the control system. All of this is done smoothly and in a synchronized fashion based upon updated information. Similarly, the bin management system is automated through this arrangement, as level sensors track the filling up of the different compartments in a timely manner to prevent overflowing. Through the IoT technology provided, the system will remotely transmit all the data it is collecting to the remote platform. The system includes an intuitive mobile application with access to sorting information and real-time bin operations, as well as alerts for abnormalities or bin saturation.

Figure 1. Block diagram of the intelligent waste sorting system

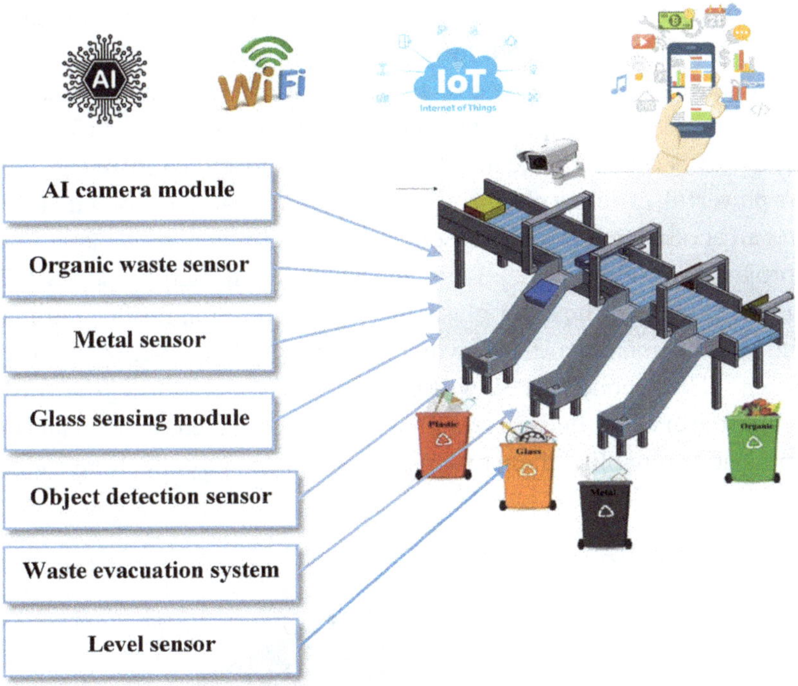

3- Proposed System Design

The intelligent sorting system is composed of an effective combination of selected hardware and software components. In the following section, the specific technical characteristics of all the electronic devices employed in the smart sorting system will be proposed, including the relevant software-related techniques required in order to process data among all the users.

3.1- ESP32-WROOM-32

The microcontroller is the core of the system and is responsible for analyzing and managing all the stages of automated waste sorting. A comparison was done between several boards: ESP32, NodeMCU-ESP8266, and Raspberry Pi Pico W to identify the best solution for our connected project. This analysis identified the ESP32-WROOM-32, presented in Figure 2, as ideal because of the perfect balance among performance, characteristics, and cost. The board is tacked by a dual-core processor capable of running up to 240 MHz, 520 KB of SRAM, and usually 4 MB of flash memory, which enables it to run complex applications in real time. It also provides Wi-Fi and Bluetooth for reliable and versatile wireless connectivity, making it suitable for a wide range of IoT applications such as home automation and industrial monitoring. It provides multiple communication interfaces (UART, SPI, I2C, ADC, PWM) that make its integration with different sensors and actuators easy. It is a powerful, efficient platform for developing smart devices with low-power modes and an economical price (Hussein et al. 2024).

This project uses two ESP32 boards: the first is connected to the ESP32-CAM and several sensors, like those for glass, metal, organic, etc., for the automation of the sorting and the control of the servomotors that guide the waste into the corresponding bins. The second ESP32 will be dedicated to measuring the fill level of the bins with sensors and sends in real time the data collected to the managers via a mobile application, enabling them to supervise and manage collections more easily.

Figure 2. ESP32-WROOM-32

3.2- ESP32-CAM

It is evident that object detection is an innovative computer vision application that allows for object recognition in diverse environments for classification purposes. Over the years, this computer vision application has received an impetus with the advancement of machine learning and AI principles. Looking into the implementation of this computer vision application with respect to the development of the waste sorting system project, the use of AI-enabled camera objects such as Raspberry Pi camera, NVIDIA Nano camera, and ESP32 camera were evaluated based on several factors such as their costs, modes of implementation, power consumption, speed of execution, and functionality with the development platform. Following this parallel comparison, this project's perfect camera option, considering this project's budget-friendly goal of designing a plastic and cardboard waste detection solution using this IoT project using this camera module dubbed "AI-Thinker ESP32-CAM camera module (Module Features)," as shown in Figure 3 below:

Figure 3. AI-Thinker ESP32-CAM camera

As can clearly be noted from Figure 3 above on camera characteristics representation, this camera can connect to a Wi-Fi network and utilize Bluetooth connectivity in addition to offering an OV246 camera module featuring a flash and being capable of delivering a 2-megapixel camera module whose image resolutions can go up to 1600 x 1200 PIXELS in addition to delivering 32 Mbits of camera module flash memory besides 512 KB of internal camera module memory and 4 MB of external camera module memory in addition to a highly efficient ESP32 D0WD camera processor module(Shraddha et al., 2025).

3.3- KY-008 laser module and photoresistor

The KY-008 laser module, alongside the photoresistor component, presented in Figure 4, is utilized to determine the presence of glass material within the sorting system. The component emits laser beam light towards the object, while the other component measures the presence of the laser beam as reflected from the materials. Due to the special properties of glass materials when exposed to laser beam illumination, the device can easily determine the presence of glass materials. These components can easily be connected, operated, and utilize low power consumption levels of 3.3 V to 5 V, thus providing easy operations at low-cost pricing suitable for the sorting of garbage materials (Maria et al., 2022).

Figure 4. KY-008 laser module and photoresistor

3.4- Metal sensor

A comparison was conducted between the inductive proximity sensors CJM18M-8A1 and LJ12A3-4-Z/BY. However, the sensors were found not to be effective in the longer range or were too expensive. To limit the aforementioned issues of the two sensors, an inductive proximity sensor was designed and developed. It can be used to sense metallic objects at longer range as well as be more cost-effective (Rui et al., 2022). This sensor works based on similar technical know-how as any other normal inductive sensor. It detects objects using an electromagentic field created by a copper coil. For its make-up, four principal items are used. They are a copper coil, a diode, a capacitor of 10 nF rating, and a resistor of rating 220 ohms. They are all wired as depicted in Figure 5 to form an R-C circuit. The presence of a metallic object as it approaches the coil interrupts the emitted electromagnetic field. This interruption is picked up by the circuit, which sends a signal to the ESP32. Here, the signal is processed, compared, and an output set accordingly. This model, although basic, proves a viable solution in metal detection over a range, reducing the production costs while creating an optimal solution as per requirement (Texas Instruments Incorporated 2021).

Figure 5. Inductive proximity sensor

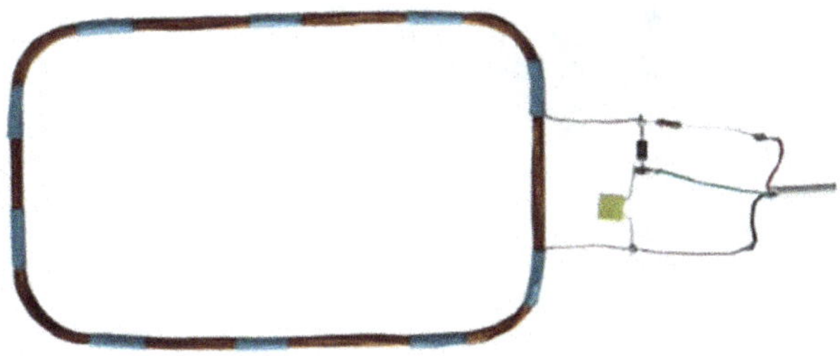

3.5- MQ135 gas sensor

Within the scope of this project, some selected gas sensors are tested in order to define which one is best suited to sensing the gases coming from the organic waste materials. Though the MQ7 sensor, which works best in detecting the carbon monoxide level, as well as the MQ-3 sensor that detects alcohol, ethanol, and smoke, may be effective in a number of given scenarios, these could not sufficiently account for the detection of the usual kinds of gases coming from decomposing materials. However, the best sensor that could be chosen in order to suit best in detecting the gases such as ammonia, benzene, and chemical vapors that are coming from the decomposing materials, as shown in Figure 6, could be the MQ135 sensor. It is also completely compatible with the ESP32, which makes it easier for us to integrate it with our ecosystem. From a technical standpoint, the MQ135 works on 5V, although it can also work on 3.3V with reduced stability. It also works with a voltage of 60 mA, a temperature range of -20 degrees centigrade to 50 degrees centigrade, and an average response time of 30 seconds (Astuti et al., 2021).

Figure 6. MQ135 gas sensor

3.6- LM393 FC-51 IR sensor

Figure 7 illustrates the FC-51 infrared sensor module, which is used in the proposed sensor module for detecting the presence of another object. This sensor module works on the principle of the detection of infrared rays reflected off the surface of an object. This sensor can be used in sorting automation devices because the detection range can be easily modified with the help of the potentiometer used with the sensor module, which can go as high as 30 cm and as low as 2 cm. The voltage requirement of the sensor module is in the range of 3.0 V to 6.0 V. Also, the detection angle of the sensor module is about 35 degrees. Because this sensor module has low power consumption with high reliability, the proposed sensor module can be used in the automated sorting device (Yelbek et al., 2024).

Figure 7. LM393 FC-51 infrared sensor

3.7- SG90 Digital Servo 180°

Next, the servomotor is actuated, which directs the object towards the relevant bin. For the proposed automatic waste collection system, the servomotor employed is the mini servomotor type, also known as the SG-90. This type of servomotor, as shown in Figure 8, was emphasized due to the speed involved, which could make the evacuation task more efficient compared to using a robotic hand. It weighs only 9 grams, thus measuring 23 x 12.2 x 29 mm. It works on 4.8 V, driving the servomotor torque amounting to 1.8kg/cm, operating over 0.1 seconds per 60 degrees, thus revolving quite quickly. Its temperature capability makes it perfect for use, as it works well within 0°C - 55°C, which serves the task well, as stated in the following extract from (Shraddha et al. (2025)).

Figure 8. SG90 digital servo

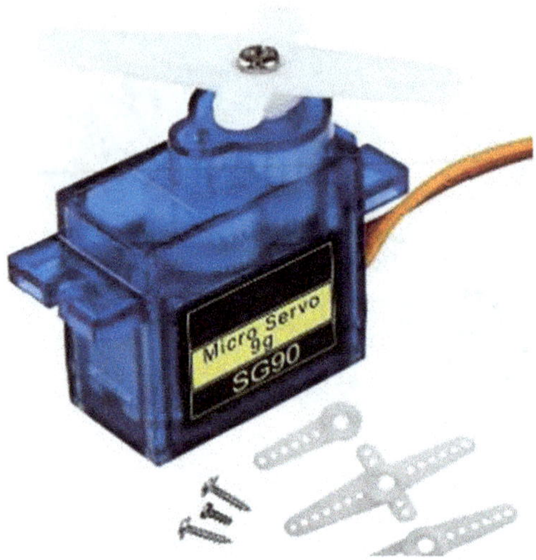

3.8- HC-SR04 sensor

The HC-SR04, as depicted in Figure 9, is an ultrasonic sensor. It is used in our proposed smart sorting system since it has an excellent price/performance ratio and is energy-efficient. The sensor is powered by a 3.3V/5V DC supply. The power consumption of the ultrasonic module is about 15 mA. The ultrasonic wave operates at 40 kHz. It can be used up to 2 cm and 4 meters. The ultrasonic sensor has an accuracy of ±3 mm. Thus, it is ideal for detecting the level of filled garbage bags. It is compatible with ESP32 and other popular microcontrollers (Arindam et al., 2022).

Figure 9. HC-SR04 sensor

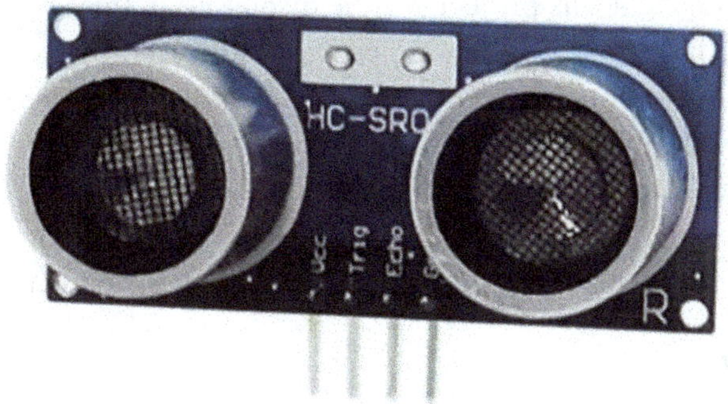

4- SOFTWARE DEVELOPMENT TOOLS

The software section is implemented for intelligent waste detection, such as plastic, cardboard, and so on, using a machine learning model trained on the Edge Impulse platform. It is an advanced solution for dataset gathering and model training and deployment on edge devices. In fact, this platform was found to be easily integrated within edge MLOps workflows that could efficiently train the model on the camera of ESP32-CAM AI Thinker and generate seamless code object detection (Shraddha et al., 2025). Besides that, ESP32 board programming has been designed by using Arduino IDE, which is a development environment known for simplicity and a wide range of compatibility with nearly all kinds of microcontrollers (Arindam et al., 2022). For the user interface, the Smart Waste mobile application is created on the MIT App Inventor platform, which helps users make Android applications easily with minimum programming experience (Anuradha et al., 2023). Communication between the mobile application and the embedded system is done through Firebase, a cloud server selected for reasons including ease of integration, full compatibility with App Inventor, and the advanced level of its features related to a real-time database, authentication, hosting, and cloud functions, ensuring that data synchronization is smooth and secure (Wen-Tsai et al., 2023).

5-IMPLEMENTATION OF THE SYSTEM

The different stages of this project will be discussed in this section. The discussion will begin with AI models' training process, followed by the development of control programs. The specific interconnection of the required electronic components to enable the smooth functioning of this system will be described. The construction process of this conveyor will be explained. The final stages of assembling all these components will be presented in this discussion to show their interrelation in providing smooth operation of the automated sorting device created in the system.

5.1- Camera Model Training

Training of the camera model is one of the most important parts of the project, and it starts by creating a new project in the Edge Impulse online platform. The online platform is specialized in embedded machine learning, which allows efficient management of data gathering, annotation, and training. A user first creates an account and, after logging on, makes a project with a relevant name connected with waste sorting. Then, depending on the needs of this application-which can be image classification or object detection-the appropriate type of project must be selected. The ESP32-CAM is the embedded camera of our system, registered on the platform to ensure smooth and direct image transfer. These images were gathered using the Arduino IDE and uploaded to Edge Impulse. Acquisition parameters (like image resolution and format) are configured to make sure that data are consistent in quality. These collected images need to be manually annotated with bounding boxes to help the model understand where precisely the objects are to be detected. Finally, the images will be resized to 320x320px automatically, as this is a typical size for a lot of models by default. The selected model is MobileNetV2 SSD FPN-Lite, which is known for its lightweight model and embeddability. This object detection model consists of 267 layers and 15 million parameters. This model is highly efficient, thus providing real-time object detection even when computational power is limited. This makes this model suitable for device types such as smartphones. After being created, this MobileNetV2 SSD model is only 63 MB when stored on a file. This makes the model suitable for various small embeddable devices, especially when compared to our device of choice, ESP32-CAM. Various training cycles are done to improve the object detection model's accuracy. Achieved is a confidence rate of 82%. This shows that this model is reliable when detecting waste. Another important process is to export the model as an optimized library in Arduino form, following validation of performance based on thorough testing of its efficacy. This library is incorporated into the ESP32-CAM program to develop the essential element of the automated system for sorting garbage in real-time (Figure 10).

Figure 10. Camera Training Steps

5.2- System Programming

Programming of the Waste Sorting System. The code developed for the ESP32 is divided into several functional blocks, each responsible for one type of automatic sorting of different kinds of material. The first ESP32 module is connected with an ESP32-CAM by UART and with several sensors (KY-008 laser module, photo-resistor, metal sensor, MQ135 gas sensor, LM393 FC-51 infrared sensor, etc.), and all of these elements allow control over automation of the sorting process, including the servomotors that drive the residues to the chosen bin. Figure 11 presents the general functioning of the sorting system by its flowchart. The ESP32-CAM module, together with the material detection sensors of different types, is mounted under a cover located at the inlet of the conveyor. The cover encloses components for protection from interference light and ensures stable detection. When a piece of waste is placed on the moving conveyor and passes underneath this cover, immediately, through a presence sensor, this will be able to detect that something is in place. Then, the conveyor stops automatically to allow the analysis of the material.

The ESP32 camera module will capture the image of the object first. If it recognizes through computer vision that it is plastic waste, it will restart to bring it to either of the corresponding bins. Once it has been identified by the object detection sensor installed in front of that bin that there is indeed an object to be disposed of, it will turn on the corresponding servomotor to enable it to spin 180 degrees to allow it to be ejected into either of those plastics disposal bins. This same procedure is repeated for each of the other materials, such as cardboard, metal, glass, organic, so that sorting occurs in an automated manner, precisely, and in real time.

A specific sensor will detect a specific waste in front of a specific container, thus activating a specific servomotor, such that sorting occurs correctly. When, however, the waste does not correspond to a specific predefined waste, that is, if the waste is unknown, movement along the conveyer will only take place up until a predefined point, where a specific container for rejects will be located. Programming of the Bin fill level detection system. The first step in carrying out this task involves setting up a project using Firebase, which is a cloud-based platform that offers the required tools to carry out efficient handling of real-time data. Upon setting up the project, essential information regarding the database and the API key, among other things, will automatically be generated by Firebase.

Figure 11. Flowchart of the Waste Classification Process

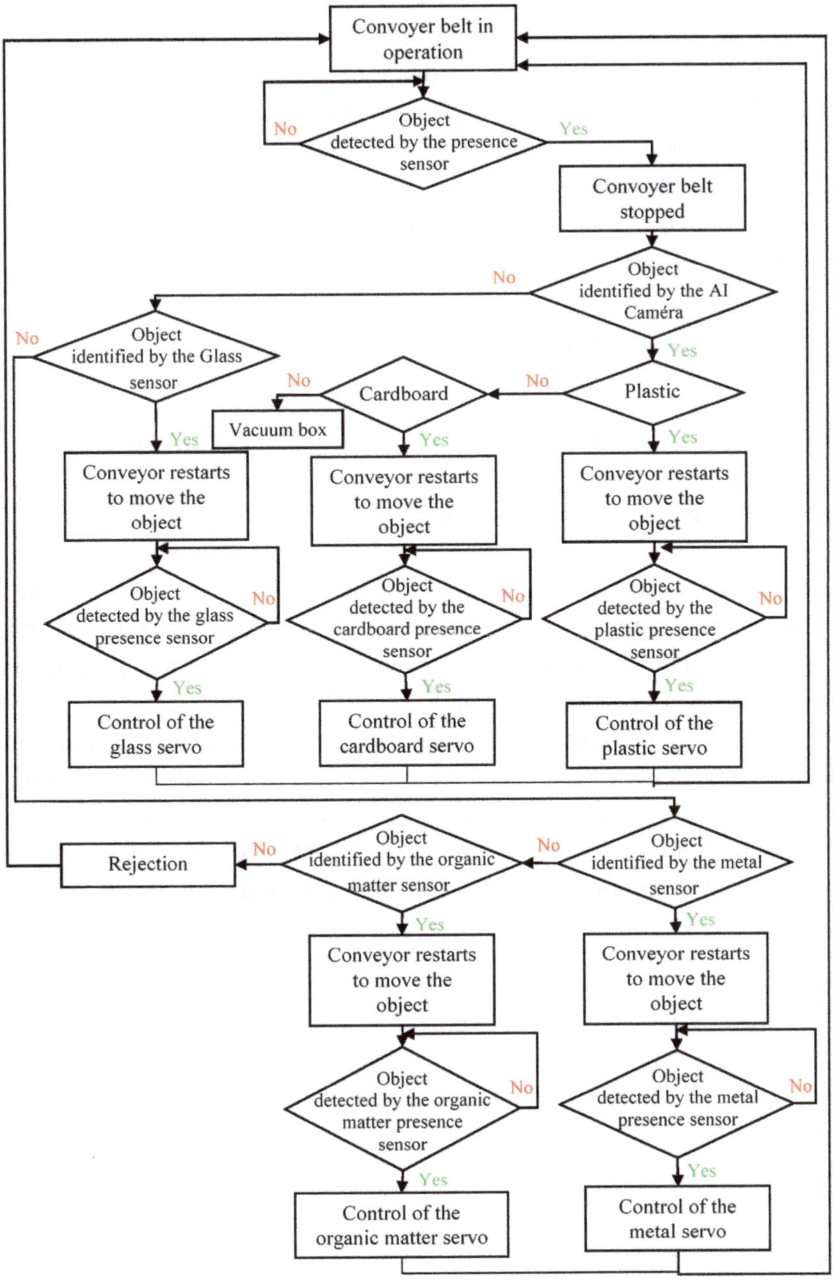

Code was then written that connects to Wi-Fi using the ESP32 circuit board and interacts with the Firebase database. The code reads data from the HC-SR04 sensor that monitors the waste bin fill level in terms of distance measurements.

In this configuration, the maximum distance measured between the sensor and the bottom of the bin is denoted as D_{max} (representing an empty bin, i.e., 0%), while the minimum distance D_{min} (representing a full bin, i.e., 100%) is reached when the waste gets close to the sensor.

Between these two values, the fill percentage is dynamically calculated based on the measured distance using the following formula:

$$\text{Fill Level } (\%) = \left(1 - \frac{D - D_{min}}{D_{max} - D_{min}}\right)$$

- ✓ D is the current distance (in cm) measured by the HC-SR04 sensor,
- ✓ D_{min} corresponds to a full bin,
- ✓ D_{max} corresponds to an empty bin.

This formula also automatically allows the measured distance to be converted to the fill percentage and updates the values in real time. To enable intuitive and remote monitoring of the data, a mobile app has been designed using the MIT app inventor software. It receives the data sent through the Firebase system and demonstrates the fill levels of all bins. Thus, users can easily monitor the data in real-time using this app, making it efficient in waste management by avoiding overflow conditions. The architecture representing the communication system of the electronic circuit, Firebase, and the mobile app is depicted in Figure 12. The steps to program the system in accordance with the detection of fill level in bins are demonstrated in Figure 13.

Figure 12. Communication architecture between the electronic circuit, Firebase, and the mobile application

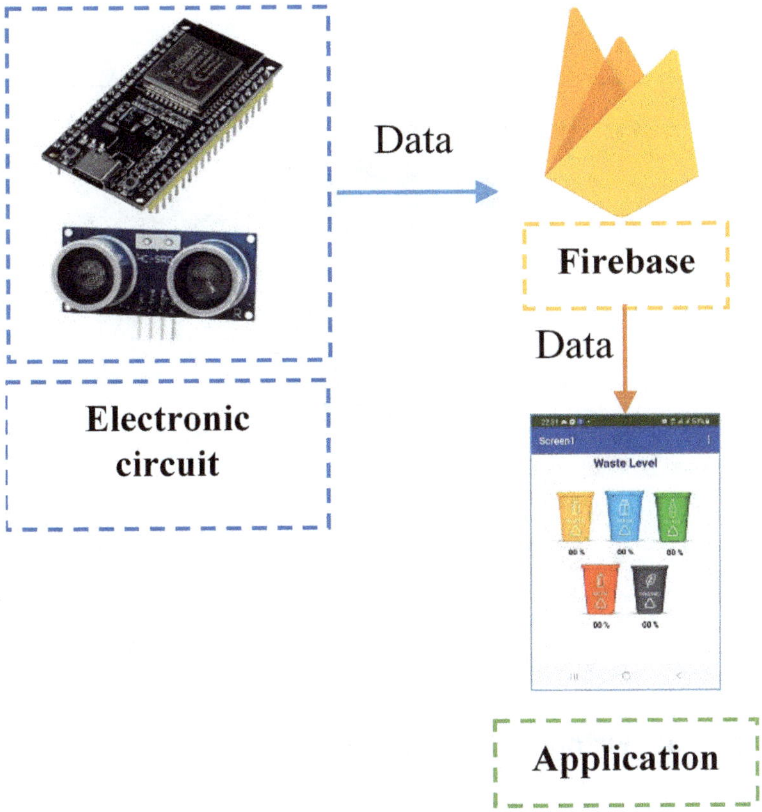

Figure 13. Programming Steps for the Bin Fill Level Detection System

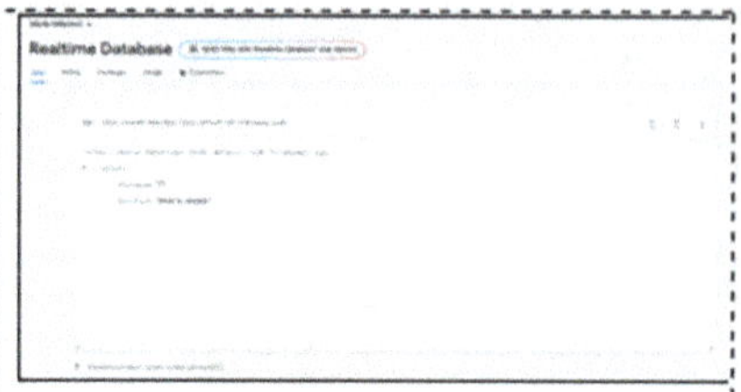

Step 1: Creating the Project in Firebase

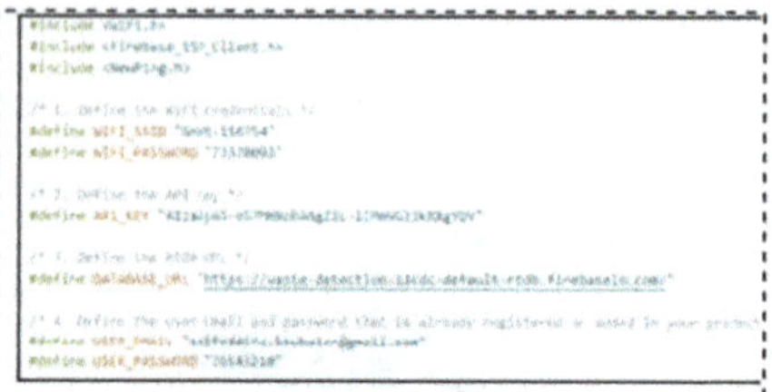

Step 2: Programming the ESP32

Step 3: Creating the "Waste Level" Application

5.3- Electrical Wiring of the System

Figure 14 is the wire diagram that connects the ESP32 Circuit board with the ESP32 Cam camera module. When the figure is combined with Figure 13, the entire diagram with the KY008 sensor, the ESP32 Cam camera module, the various sensors like the photoresistor or light sensor and the metal sensor, the servomotor that directs the waste.

Figure 14. Complete wiring diagram of the waste sorting and bin fill-level detection systems.

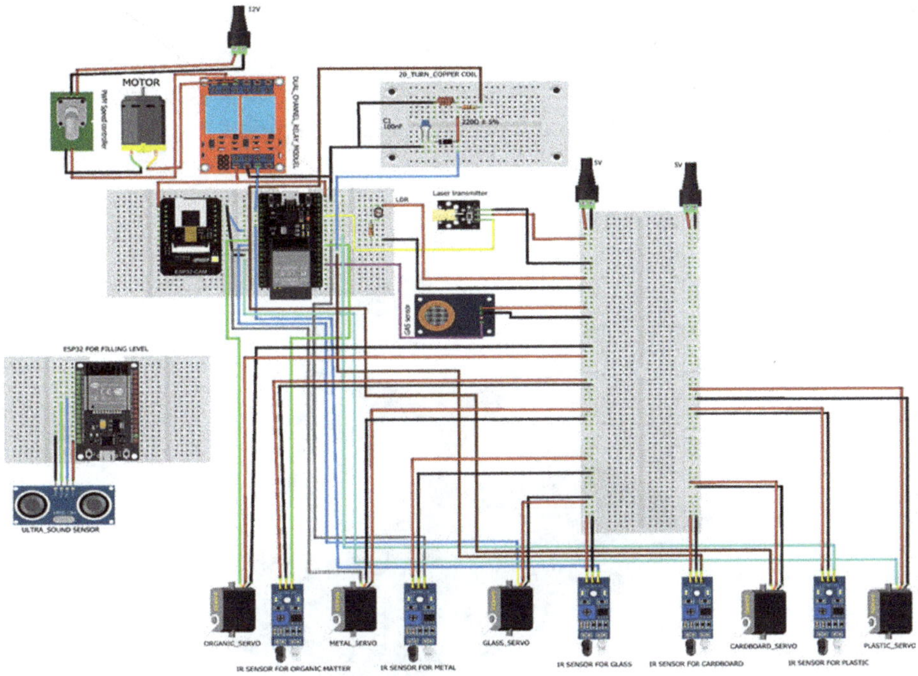

5.4- Design and Construction of the Conveyor

Figure 15 presents a conventional treadmill before any modification. This work has been designed to transform this device into a conveyor belt system for waste sorting, presented in Figure 16. This has involved a sequence of mechanical designs and validation. With respect to the simulation of the conveyor system, this process has involved a number of steps. To begin with, a three-dimensional model of the conveyor system is created through the use of SOLIDWORKS. This model serves to visualize the system as a whole. At the same time, there is a chance to specify dimensions and details of each mechanical assembly. Further, a function analysis is made. This analysis serves to assess components such as weight, movements, and possible shocks on the system. As a result of this assessment, a proper option is selected in terms of material and chassis. Once this was achieved, the next step was to conduct Finite Element Analysis (FEA), where the 3D model was subjected to various loads and boundary conditions, followed by the generation of the mesh to ascertain the various stresses and determine any possible deformation that might occur to the system during its manufacture and use. This led to the enhancement and correction of various weaknesses while ensuring the reliability and efficiency of the system during its manufacture and use. A simulation report was produced to show the mechanical performance and justify various technological decisions based on the design simulations conducted during the design phase.

Figure 15. Conventional treadmill

Figure 16. Conveyor belt system

The data obtained from the chassis simulation process proved and confirmed its robustness and reliability in its structure. Furthermore, in regards to Von Mises stress analysis, it was clear that a lower stress level of 3.432 * 10^6 N/m^2 compared to the material's elasticity limit of 2*10^7 was observed. Moreover, a safety factor of 5.828 was attained, ensuring that not only was elasticity maintained but also durability. Also, from a stability prospect, a smaller displacement of 2.547 mm was obtained, suggesting a high level of rigidity in the structure. Additionally, a minimum level of strain of 1.625 * $10^{(-3)}$ was obtained; thus, there was a high likelihood that the structure would maintain its geometric shapes, a crucial attribute in ensuring system efficiency. Moreover, from a physical persepective, a 2.8815 kg model, with a volume of 0.005763 m^3 and a density of 500 kg/m^3 typical of a material in MDF form, was obtained. Therefore, it was possible to obtain a rough estimate of estimated conveyor weights, enhance efficiency in terms of optimization of costs in terms of materials and ergonomically. As indicated in the discussion, it was clear that a well-designed chassis would result in high levels of stability and durability.

After verifying the robustness and stability of the chassis, the second stage of this phase consists of choosing a motor control system that meets the needs of the previously determined mechanical parameters. Consequently, a brush-type DC motor, 12V, 30W, and 0.2Nm torque, as shown in Figure 17, had already demonstrated effectiveness in home use, including a sufficient power supply to operate the conveyor belt without compromising the current electronic components and power supply. The advantage of a simple control circuit without the use of a gear motor or other control circuit diagrams represents a major advantage, especially considering microcontroller control through a simple circuit referred to as a Direct Drive Circuit. However, a control method referred to as a pulse width modulation regulator, as shown in Figure 18, allows us to control the speed of rotation without significant power loss and maintain strong torque even during slow rotation, unlike other control methods, while allowing us to control it through a microcontroller, thus allowing smart control and management.

Figure 17. 12V DC brushed motor

Figure 18. PWM voltage regulator

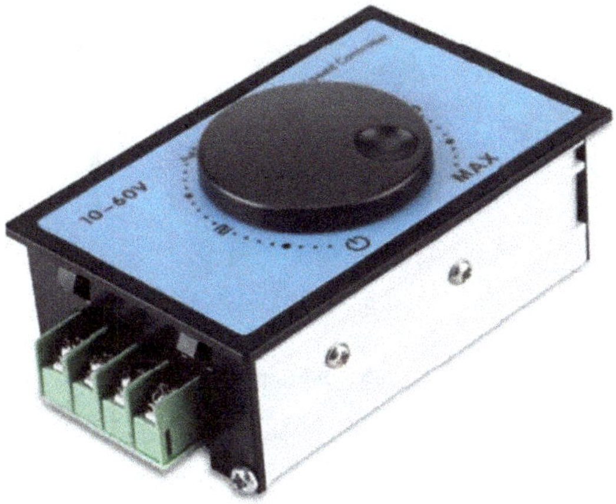

After the design and simulation phase, the assembly phase becomes viable to take further steps towards the completion of the conveyor belt. Essentially, the system under question consists of different components specifically designed to work in concert to achieve the desired goals. Firstly, the frame acts as the primary component since it defines the basic stability required in the system. Secondarily, the placement of the table becomes instrumental in enhancing the integration of the frame to ensure the required level of robustness. Thirdly, the integration of the support component becomes viable to ensure the proper installation of the belt to achieve the desired movement. Essentially, the integration of the pulleys becomes viable to facilitate the movement required to convey the waste. On the same note, the installation of the electric motor becomes plausible to achieve the dynamic movement by integrating the speed regulators. A protective cover is installed to guard sensitive devices against light, electromagnetic disturbance, and mechanical harm, thus ensuring system reliability. Next, the electrical enclosure is fastened to the structure so that it serves as a housing and protective casing for the electronic devices such as control panels, relays, and motors from dust, atmospheric fluctuations, and shocks. Figure 19 illustrates the conveyor system components. The actual assembly of the conveyor was carried out in Figure 20 to physically test the prototype of the paper.

Figure 19. Conveyor Components

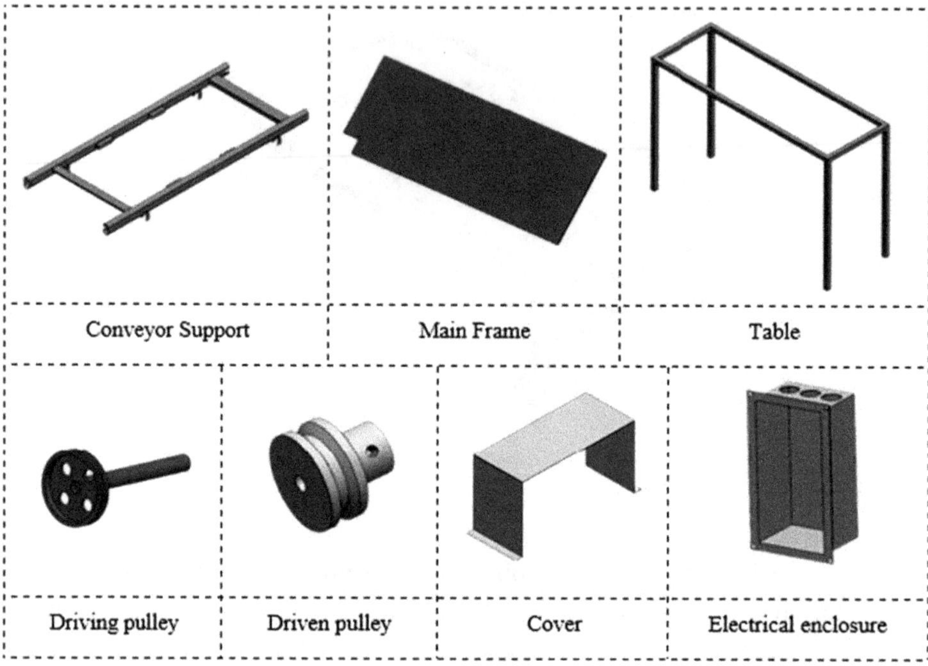

Figure 20. Some steps for assembling the conveyor

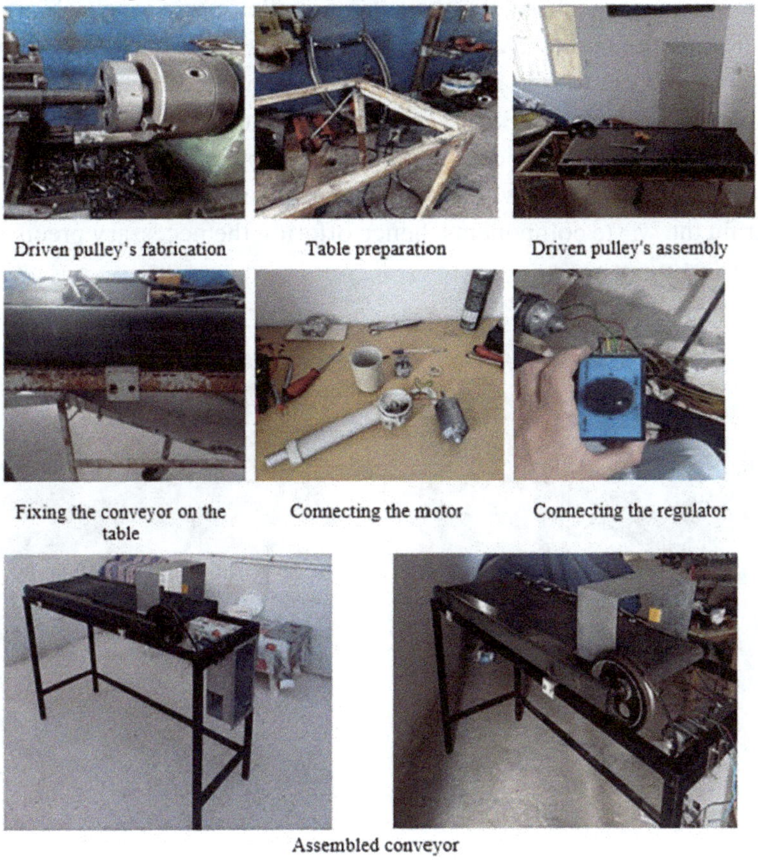

5.5- Complete assembly of the system

The full assembly of the system will be shown in the next section for the entire system including the conveyor belt as well as the electronic part of the system. This step is part of the physical realization of the entire system as an automated sorting system. It effectively demonstrates the realization of the system design as outlined by the various design steps as shown in the coming section.

Figures 21 and 22 reveal different perspectives of the fully assembled prototype, emphasizing the comprehensive integration of mechanical and electronic components. Figure 21 shows the prototype with views from both sides, containing components that have been wired to the conveyor system, as well as the system ready for the experimental phase. Figure 22 (a) shows an enlarged image of the electric enclosure side, containing all control modules, power supply units, and significant system connections that have been made in this aspect. Figure 22 (b) shows the inside of the protective cover, in which all the connected sensor modules have been arranged in order to guarantee accuracy in the detection process during waste sorting. Meanwhile, the ESP32 camera is positioned at the top, while all other sensors are positioned on the sides. Figure 22 (c) shows the metal sensor placed at the base of the conveyor belt, which plays a significant role in identifying metallic objects within the waste components, hence offering the necessary precision in the entire waste sorting process.

Figure 21. View of the prototype from both sides

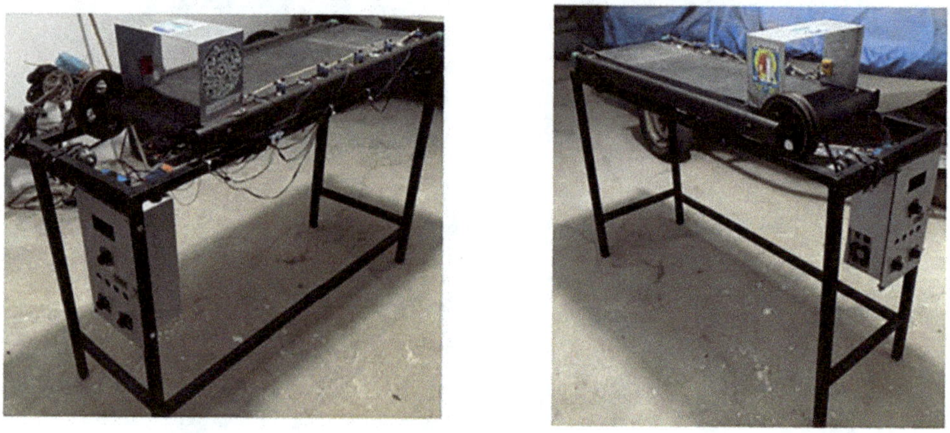

Figure 22. Views of the electrical enclosure and the integrated sensors

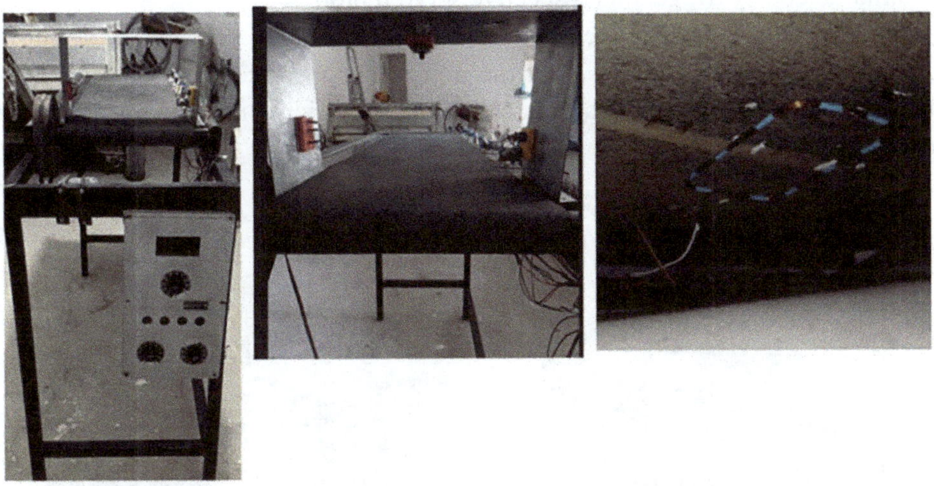

(a) View of the prototype from the electrical enclosure side.

(b) View of the different sensors connected inside the cover

(c) Metal sensor connected beneath the conveyor belt.

6- RESULTS AND DISCUSSION

To verify the appropriate function of an intelligent waste sorting system, an experiment was done to assess the reliability and robustness of the intelligent waste sorting method. The first experiment was carried out to test the ability of the intelligent waste sorting method in detecting a piece of waste made of plastic as depicted in Figure 23.

To begin with, the object made of plastic is placed at the point where the object enters the conveyor. During the movement along the belt, it enters the area where it is detected using a sensor placed underneath the protection cover. Such detection leads to the stopping of the conveyor belt. Here begins the analysis phase. It is the point where the object is exposed to the camera esp32 placed above the belt. Here, the object is recognized by the embedded intelligent device capable of performing appropriate recognition regarding the specific type of the object. Here, the object enters the category of plastic objects identified as such in state 1.

Once the object is categorized, the conveyor turns on again automatically to move the item to the specified sorting zone. The system then sends a signal from the software to light up the object detection sensor in that lane of plastic sorting, and simultaneously triggers its respective servo motor.

When the object crosses the sensor placed near the bin, it is able to detect its passage (state 2). Immediately after the object has overtaken the sensor, a 180° rotation is made by the servomotor (states 3 and 4), which sends the object into the bin directly under it (state 5).

Figure 23. Automatic identification process of a plastic waste item

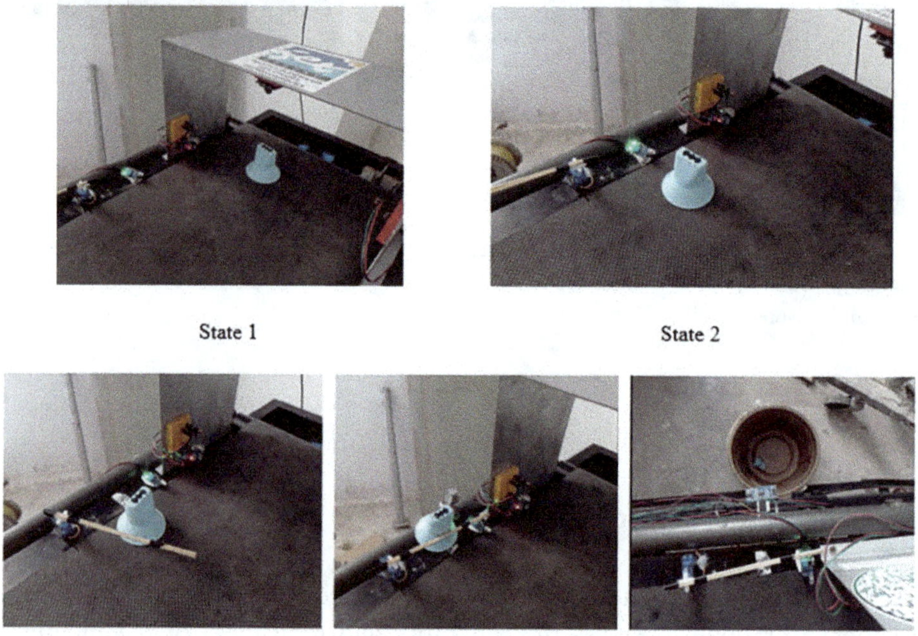

State 1 State 2

State 3 State 4 State 5

A different kind of waste was used to verify the functioning of the prototype as presented in Figure 24. In the case of the prototype verification tests, the object has been a waste item made of cardboard. The waste item has been identified with the ESP32-CAM .

Once the object rests on the conveyor belt, it starts to travel toward the analysis area by itself. A presence sensor placed underneath the object's protective cover recognizes the object's presence. This action causes the conveyor to stop at once, initiating the identification stage.

Next, the image of the waste is then taken by the ESP32-CAM located above the conveyor belt. It then identifies whether it is made of plastic or cardboard based on the information it has from the artificial intelligence unit. After a few seconds, it precisely identifies the waste material, which in this case was made of cardboard.

Once this process is finished, the conveyor will restart, moving the item to the designated sorting area. Once this item reaches the designated collection bin, another presence sensor will detect the event, allowing the servomotor assigned to this particular bin to be activated, completing a swift rotation that will direct the item to the cardboard waste collection bin situated directly beneath.

Figure 24. Automatic identification process of a cardboard waste item

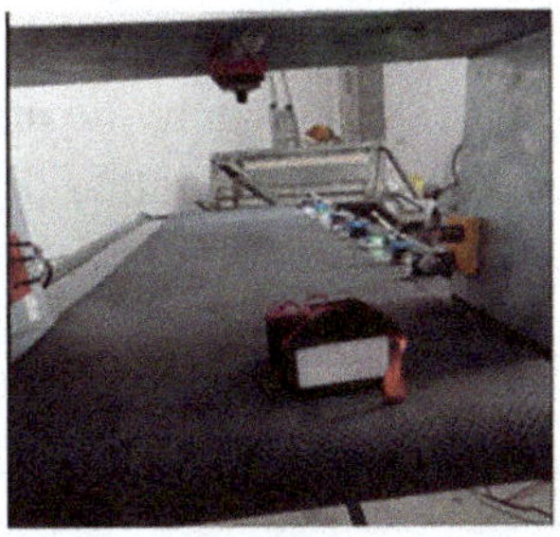

State 1

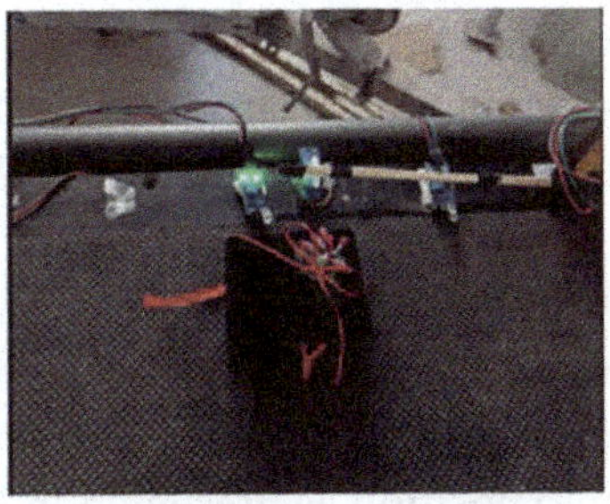

State 2

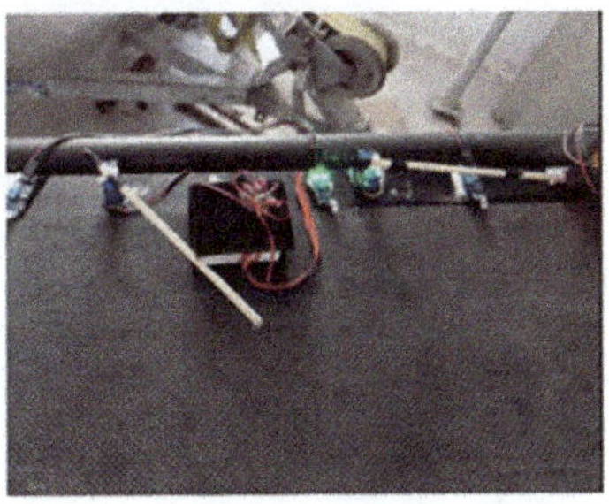

State 3

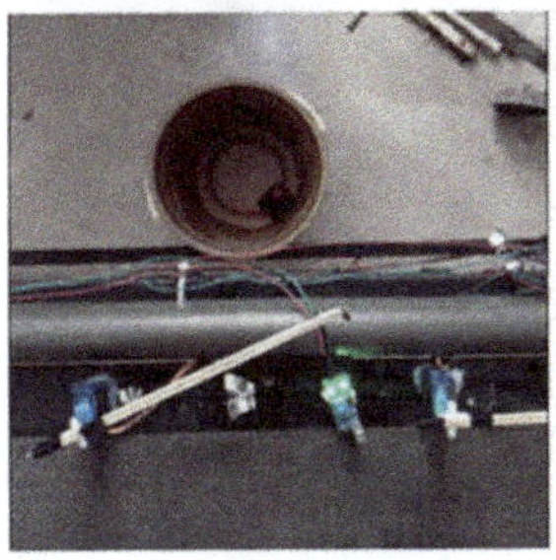

State 4

Once the test is done with this program, the two wastes identified by the camera as different types of solid waste—plastic and cardboard—were further examined. The tests were done to verify the capabilities of the ESP32 camera module to recognize different types of wastes with the artificial intelligence capabilities built within.

For a more detailed exploration of the efficiency and reliability of each sensor engaged in the process of material classification, like that of glass, organic matter, and metals, a further piece of experimentation was carried out. This was in a bid to confirm the capacity of each sensor in recognizing and recognizing metal in a precise and exact manner.

The object is placed at the entrance of the conveyor belt. The moment this object touches the presence sensor, the conveyor belt will stop immediately. It is at this point that the ESP32-CAM captures a photo of this object. The major challenge, however, lies in the fact that this object is neither cardboard nor plastic.

While simultaneously, a metal sensor that has previously been installed under the conveyor belt detects that there is a presence of a metallic material. Further, the reading from the photo resistor shows that it is maximal, which implies that there is blocking of the laser beam; hence, if the object were not opaque, it should not be glass. However, as to whether or not gas detection, whose main aim is to detect the presence of an organic matter, shows any emission, it will still be disregarded because metal will be a priority.

By eliminating the variables from the data which don't match, the nature of the object is thus deduced to be metallic. This information is transmitted digitally, activating a command on the ESP32 board. This command enables both the presence sensor and the associated servomotor. Following this identification, there is the resumption of motion on the conveyor, thus facilitating the motion of the object towards the compartment area. This motion is detected again before the servomotor directs it into position as presented in Figure 25.

Figure 25. Automatic identification process of a metal waste item

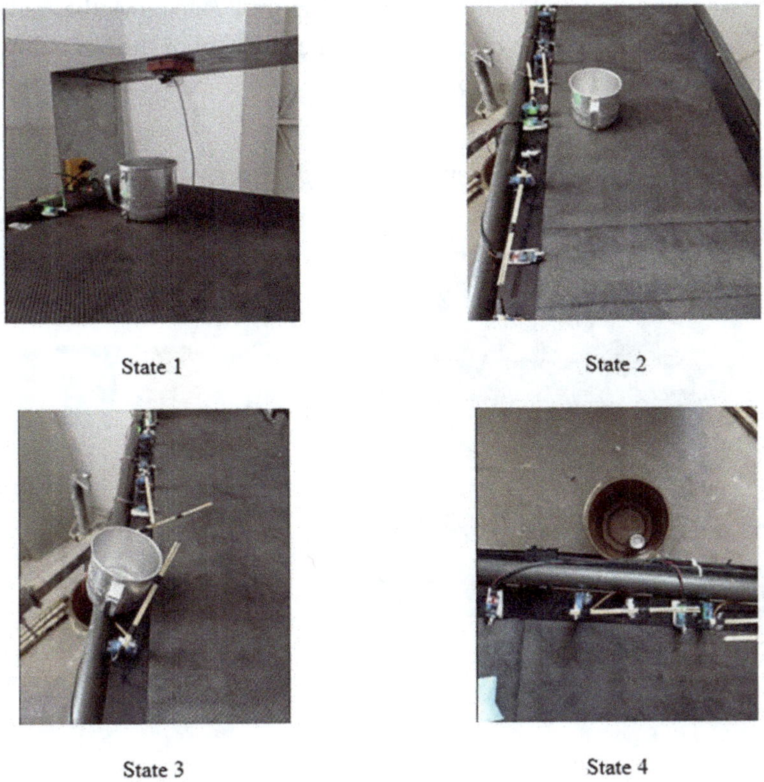

In order to verify the efficiency of the bin fill level detection system based on the bin filled with waste items, a specific experiment is conducted with the bin filled with plastic waste. At the beginning of the experiment, the distance between the ultrasonic HC-SR04 sensor and the bottom of the bin is noted as 75 cm. When the bin is not filled with any waste materials or can be considered an empty bin, the fill level is 0%. When the waste materials are added gradually to the bin, the distance between the ultrasonic sensor and the bottom of the bin is reduced proportionally. When the waste materials fill the bin completely and can be considered a full bin, the distance between the ultrasonic sensor and the bottom.

During all tests, the percentage of filling was dynamically calculated based on a predefined equation and sent in real-time to the Firebase platform. Two kinds of scenarios occurred during this process. The first scenario is illustrated in Figure 26, in which it was observed that the bin was completely empty. It was instantly sent to the Firebase platform through the mobile application. As far as the second scenario is concerned, it is seen that the bin is half filled. As plastic waste continued to be deposited into the bin, the distance reduced. At a particular time, when the fill level of the bin touched 72%, this value is sensed, computed using the ESP32 microprocessor, and updated to the database. Figure 27 displays the real physical position of the bin along with a live display of the fill level of the bin being updated to 72% through a mobile application.

Figure 26. Empty bin and 0% fill level in the application

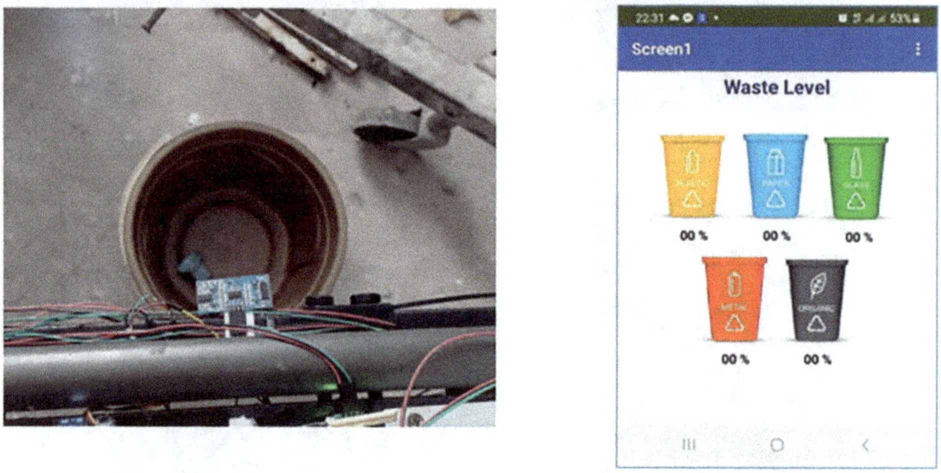

Figure 27. Full bin with a 72% fill level in the application

7- CONCLUSION

This chapter has presented an explanation of the architecture and main elements that characterize the intelligent system for sorting and managing different kinds of waste-plastic, metal, glass, cardboard, and organic-based on the advanced integration of AI with IoT. The major technological choices have been highlighted, which are essentially based on the use of supervised learning algorithms for the exact and automated recognition of materials, while the respective hardware implementation consists of ESP32 modules and a set of specialized sensors. The design, development, and validation of the prototype proved the technical and functional feasibility of the project. The evidence of this phase shows good integration of hardware and software parts and significant effectiveness in the automatic sorting that took place in real conditions. Consequently, this prototype constitutes the very solid basis for any future industrialization or scaling up of the system for larger applications. Beyond the purely technical aspects, this system clearly illustrates the synergy between AI and IoT in combining operational performance, environmental sustainability, and substantial reduction in human intervention. Economic and ecological advantages are relevant: minimization of human errors, optimization of operating costs, improvement in quality and speed of sorting, and ample prospect for large-scale automation.

ACKNOWLEDGEMENTS

This chapter was developed within the framework of the PEJC research project (Code: PEJC2024-D3P01).

REFERENCES

Adeniyi, L. A. (2023). Assessment of the effects of location of disposal-sites on waste disposal practices of urban residents in Nigeria. *GeoJournal, 88*(4), 3889–3903. DOI: 10.1007/s10708-023-10835-4

Almtireen, N., Reddy, V., Sutton, M., Nedvidek, A., Karn, C., Ryalat, M., Elmoaqet, H., & Rawashdeh, N. (2025). PLC-Controlled Intelligent Conveyor System with AI-Enhanced Vision for Efficient Waste Sorting. *Applied Sciences (Basel, Switzerland), 15*(3), 1550. DOI: 10.3390/app15031550

Anuradha, P., Vasanth, K., Renuka, G., & Rao, A. R. (2023). IoT based enabling home automation system for individuals with diverse disabilities. e-Prime - Advances in Electrical Engineering. *Electronics and Energy, 6*, 100366.

Arindam, R., Apurba, M., Jungmin, K., & Ilkyeong, M. (2022). IoT-based smart bin allocation and vehicle routing in solid waste management: A case study in South Korea. *Computers & Industrial Engineering, 171*, 108457. DOI: 10.1016/j.cie.2022.108457

Astuti, S. D., Tamimi, M. H., Pradhana, A. A. S., Alamsyah, K. A., Purnobasuki, H., Khasanah, M., Susilo, Y., Triyana, K., Kashif, M., & Syahro, A. (2021). Gas sensor array to classify the chicken meat with E. coli contaminant by using random forest and support vector machine. *Biosensors & Bioelectronics: X, 9*, 100083. DOI: 10.1016/j.biosx.2021.100083

Bozma, H. I., & Yalçın, H. (2002). Visual processing and classification of items on a moving conveyor: A selective perception approach. *Robotics and Computer-integrated Manufacturing, 18*(2), 125–133. DOI: 10.1016/S0736-5845(01)00035-7

Cheng, T., Kojima, D., Hu, H., Onoda, H., & Pandyaswargo, A. H. (2024). Optimizing Waste Sorting for Sustainability: An AI-Powered Robotic Solution for Beverage Container Recycling. *Sustainability (Basel), 16*(23), 10155. DOI: 10.3390/su162310155

Feng, Z., Yang, J., Chen, L., Chen, Z., & Li, L. (2022). An intelligent waste-sorting and recycling device based on improved EfficientNet. *International Journal of Environmental Research and Public Health, 19*(23), 15987. DOI: 10.3390/ijerph192315987 PMID: 36498058

Ferronato, N., & Torretta, V. (2019). Waste mismanagement in developing countries: A review of global issues. *International Journal of Environmental Research and Public Health, 16*(6), 1060. DOI: 10.3390/ijerph16061060 PMID: 30909625

Haitao, Ch. (2021). Optimization of an Intelligent Sorting and Recycling System for Solid Waste Based on Image Recognition Technology. *Advances in Mathematical Physics, ID, 4094684*, 12.

Hassan, G., Chemori, A., El Rafei, M., Gouttefarde, M., Francis, C., Hervé, P.E., Sallé, D. (2022). Ultra-Local Model-Based Intelligent Robust Control of PKMs: Theory and Simulations. IFAC-PapersOnLine, 55, 12, 2022, 634-640.

Hussein, N.M., Mohialden, Y.M., Salman S.A. (2024). Impact of IoT-Based Environmental Monitoring on Lab Safety and Sustainability. Babylonian Journal of Internet of Things, 16-26.

Kaza, S., Yao, L., Bhada-Tata, P., & Van Woerden, F. (2018). *What a Waste 2.0: A Global Snapshot of Solid Waste Management to 2050.* World Bank Publications. DOI: 10.1596/978-1-4648-1329-0

Kellow, P., Joel, J. P. C. R., Ousmane, D., Ashok, K. D., Victor, H. C. A., & Sergei, A. K. (2020). A Smart Waste Management Solution Geared towards Citizens. *Sensors (Basel), 20*(8), 2380. DOI: 10.3390/s20082380 PMID: 32331464

Khemakhem, S., Rekik, M., & Krichen, L. (2019). Impact of Electric Vehicles integration on residential demand response system to peak load minimizing in smart grid. *19th International Conference on Sciences and Techniques of Automatic Control and Computer Engineering (STA).* DOI: 10.1109/STA.2019.8717304

Khemakhem, S., Rekik, M., & Krichen, L. (2022). A Smart Home Power Supervision Strategy Based on Optimal Plug-in Electric Vehicles Integration into Smart City Context for Peak Loads Profiles Flattening. *Arabian Journal for Science and Engineering, 47*(3), 3383–3396. DOI: 10.1007/s13369-021-06315-x

Kirchherr, J., Reike, D., & Hekkert, M. (2017). Conceptualizing the circular economy: An analysis of 114 definitions. *Resources, Conservation and Recycling, 127*, 221–232. DOI: 10.1016/j.resconrec.2017.09.005

Konstantia, M., Freiderikos, R., George, L., Stavros, O., Maria, P., Michalis, Z., & Michail, M. (2022). Intelligent robotic system for urban waste recycling. IEEE International Conference on Imaging Systems and Techniques, Proceedings, ISBN (Electronic) 9781665481021.

Laski, P. A., Takosoglu, J. E., & Blasiak, S. (2015). Design of a 3-DOF tripod electro-pneumatic parallel manipulator. *Robotics and Autonomous Systems*, *72*, 59–70. DOI: 10.1016/j.robot.2015.04.009

Lubongo, C., Bin Daej, M. A., & Alexandridis, P. (2024). Recent developments in technology for sorting plastic for recycling: The emergence of artificial intelligence and the rise of the robots. *Recycling*, *9*(4), 59. DOI: 10.3390/recycling9040059

Maalouf, A., & Mavropoulos, A. (2022). Re-assessing global municipal solid waste generation. *Waste Management & Research*, *41*(4), 936–947. DOI: 10.1177/0734242X221074116 PMID: 35075952

MacArthur, E. (2013). Towards the circular economy. *Journal of Industrial Ecology*, *2*, 23–44.

Maria, K., Fredy, R., George, P., Nikitas, M., & Michail, M. (2021). Robotic Waste Sorting Technology: Toward a Vision-Based Categorization System for the Industrial Robotic Separation of Recyclable Waste. *IEEE Robotics & Automation Magazine*, *28*(2), 50–60. DOI: 10.1109/MRA.2021.3066040

Maria, V., Priyanka, P., Oluwaseyi, F., Katherine, H. I., Liang, Z., Hossain, S., & Sheikh, I. A. (2022). Development of a Noninvasive Blood Glucose Monitoring System Prototype: Pilot Study. *JMIR Formative Research*, *6*, 8. PMID: 36018623

Material Economics. (2018). *The Circular Economy—A Powerful Force for Climate Mitigation*. Material Economics Sverige AB.

Mayouf, O. M., Rekik, M., & Krichen, L. (2022). Designing a hybrid system using solar cells and batteries to power a clinic in Tajarhi desert village in the Libya-Niger border off-grid. 2022 IEEE 21st International Conference on Sciences and Techniques of Automatic Control and Computer Engineering (STA).

Nidhya, R., Manish, K., Renjith, V. R., & Deepak, V. (2020). Enhanced Route Selection (ERS) algorithm for IoT enabled smart waste management system. *Environmental Technology & Innovation*, *20*, 101116. DOI: 10.1016/j.eti.2020.101116

Patric, M., Diogo, M., Eduardo, D., Jonatan, C., Rodrigo, C., Juergen, R., Edison, P., & Rafael, K. (2019). An IoT-based smart cities infrastructure architecture applied to a waste management scenario. *Ad Hoc Networks*, *87*(1), 200–208.

Ramson, S. R. J., & Moni, D. J. (2017). Wireless sensor networks based smart bin. *Computers & Electrical Engineering*, *64*, 337–353. DOI: 10.1016/j.compeleceng.2016.11.030

Rui, H., Aaron, U., Dian, J., Jiang, Z., & Jae-Won, C. (2022). Inductive proximity sensors within a ceramic package manufactured by material extrusion of binder-coated zirconia. *Sensors and Actuators. A, Physical*, *338*, 113497. DOI: 10.1016/j.sna.2022.113497

Sathish, P. G., Subrata, H., & Atul, T. (2017). A review on automated sorting of source-separated municipal solid waste for recycling. *Waste Management (New York, N.Y.)*, *60*, 56–74. DOI: 10.1016/j.wasman.2016.09.015 PMID: 27663707

Shan-Huen, H., & Pei-Chun, L. (2015). Vehicle routing–scheduling for municipal waste collection system under the "Keep Trash off the Ground" policy. *Omega*, *55*, 24–37. DOI: 10.1016/j.omega.2015.02.004

Shraddha, M. K., Sameer, U., Amogh, R. M., Shreyas, S. A., & Meenaxi, M. R. (2025). Grading and Classification of Tomatoes Using ESP32-CAM and Impulse Cloud based IoT Platform for Sustainability. Seventh International Conference on Recent Trends in Image Processing and Pattern Recognition (RTIP2R-2024). *Procedia Computer Science*, *260*, 938–946. DOI: 10.1016/j.procs.2025.03.277

Singh, M. (2019). Solid waste management in urban India: Imperatives for improvement. *Journal of Contemporary Issues in Business and Government*, *25*(1), 87–92.

Soumyabrata, S., Rituparna C. (2023). IoT based smart waste management system in aspect of COVID-19. Journal of Open Innovation: Technology, Market, and Complexity, 9, 2, 100048.

Swain, S., Muduli, K., Biswal, J. N., Tripathy, S., & Panda, T. K. (2017). Evaluation of barriers of health care waste management in India—a gray relational analysis approach. *Proceedings of the 5th International Conference on Frontiers in Intelligent Computing: Theory and Applications (FICTA 2016)*, Volume 1, 181–188. Springer, Singapore. DOI: 10.1007/978-981-10-3153-3_18

Swain, S., Muduli, K., Kommula, V. P., & Sahoo, K. K. (2022). Innovations in internet of medical things, artificial intelligence, and readiness of the healthcare sector towards health 4.0 adoption. *International Journal of Social Ecology and Sustainable Development*, *13*(1), 1–14. DOI: 10.4018/IJSESD.292078

Swain, S., Muduli, K., Kumar, A., & Luthra, S. (2024). Analysis of barriers of mHealth adoption in the context of sustainable operational practices in health care supply chains. *International Journal of Industrial Engineering and Operations Management*, *6*(2), 85–116. DOI: 10.1108/IJIEOM-12-2022-0067

Texas Instruments Incorporated. (2021). Sensor Design for Inductive Sensing Applications Using LDC, SNOA930C – March 2015 – Revised May 2021.

Tianchen, J., Huaiying, F., Rencheng, Z., Jianhong, Y., Lulu, F., Yangyang, H., & Zhengxing, C. (2023). Low-value recyclable waste identification based on NIR feature analysis and RGB-NIR fusion. *Infrared Physics & Technology*, *131*, 104693. DOI: 10.1016/j.infrared.2023.104693

UN Environment Programme. International SWA. (2024). Global Waste Management Outlook 2024—Beyond an Age of Waste: Turning Rubbish into a Resource. Technical Report, Available online: https://wedocs.unep.org/20.500.11822/44939

Vera, C. H., Karl, F. D., Richard, F. H., & Daniele, V. (2013). Models and Algorithms for the Integrated Planning of Bin Allocation and Vehicle Routing in Solid Waste Management. *Transportation Science*, *48*, 1.

Vicentini, F., Giusti, A., Rovetta, A., Fan, X., He, Q., Zhu, M., & Liu, B. (2009). Sensorized waste collection container for content estimation and collection optimization. *Waste Management (New York, N.Y.)*, *29*(5), 1467–1472. DOI: 10.1016/j.wasman.2008.10.017 PMID: 19103480

Wen-Tsai, S., Indra, G. T. I., & Sung-Jung, H. (2023). An IoT-Based Aquaculture Monitoring System Using Firebase. *Computers, Materials & Continua*, *76*(2), 2179–2200. DOI: 10.32604/cmc.2023.041022

Wilts, H., Garcia, B. R., Garlito, R. G., Gómez, L. S., & Prieto, E. G. (2021). Artificial intelligence in the sorting of municipal waste as an enabler of the circular economy. *Resources*, *10*(4), 28. DOI: 10.3390/resources10040028

Worldbank. (2018). What a Waste 2.0: A Global Snapshot of Solid Waste Management to 2050.

Xiaoyu, W., Jia, L., Linpeng, Y., & Zhenming, X. (2020). Auto-sorting commonly recovered plastics from waste household appliances and electronics using near-infrared spectroscopy. *Journal of Cleaner Production*, *246*, 118732. DOI: 10.1016/j.jclepro.2019.118732

Xinxing, C., Huaiyang, H., Yuxuan, L., Jiqing, L., & Ming, L. (2022). Robot for automatic waste sorting on construction sites. *Automation in Construction*, *141*, 104387. DOI: 10.1016/j.autcon.2022.104387

Yann, G., Arnab, S., & Paul, C. (2015). A Smart Waste Management with Self-Describing Complex Objects. *International Journal On Advances in Intelligent Systems*, *8*, 1–16.

Yelbek, U., Alexandr, N., Timoth, M., Zhanbolat, S., Sungat, A., Alizhan, K., Assel, T.M., Aigul, K.K. (2024). Advancing sanitary surveillance: Innovating a live-feed sewer monitoring framework for effective water level and chamber cover detections. Heliyon, 10, 6, e27395.

Zeli, W., Heng, L., & Xintao, Y. (2020). Vision-based robotic system for on-site construction and demolition waste sorting and recycling. *Journal of Building Engineering*, *32*, 101769. DOI: 10.1016/j.jobe.2020.101769

Chapter 4
Control and Optimization of Grid-Connected Wind Energy Systems Using MPPT and Hybrid Controllers

P. H. Gurunandan
http://orcid.org/0009-0006-9713-9181
Shri Siddhartha Institute of Technology, India

G. N. Shilpa
http://orcid.org/0000-0002-0354-9547
Sri Siddhartha Institute of Technology, India

C. Nataraja
http://orcid.org/0009-0000-8444-5506
Sri Siddhartha Institute of Technology, India

C. R. Mohan Kumar
Sri Siddhartha Institute of Technology, India

M. Praveen
Sri Siddhartha Institute of Technology, India

D. K. L. N. Simha
Sri Siddhartha Institute of Technology, India

ABSTRACT

Integrating wind energy into the power grid offers significant benefits in reducing reliance on fossil fuels, but challenges such as intermittent generation impact system stability and power quality. This chapter explores advanced control techniques for optimizing grid-connected wind energy systems, focusing on Maximum Power Point Tracking (MPPT) and hybrid controllers. Incremental Conductance (IC) MPPT is examined for its effectiveness in maximizing energy capture under varying wind

DOI: 10.4018/979-8-2600-2383-9.ch004

© 2027, IGI Global Scientific Publishing. Copying or distributing in print or electronic forms without written permission of IGI Global Scientific Publishing is prohibited. Use of this chapter to generative artificial intelligence (AI) technologies is expressly prohibited. The publisher reserves all rights to license its use for generative AI training and machine learning model development.

conditions. Hybrid control strategies, combining Proportional-Integral (PI) control, fuzzy logic, and model predictive control (MPC), are discussed for their robustness in addressing grid disturbances. The chapter also covers energy storage integration with wind turbines, improving grid reliability, and fault detection mechanisms to maintain system integrity. The results highlight the role of real-time control, power quality management, and grid compliance in enhancing the efficiency and reliability of renewable energy integration.

INTRODUCTION

The transition to renewable energy sources is increasingly critical in combating climate change and achieving sustainable energy solutions (Menzri et al., 2024). Wind power is considered one of the most promising and scalable renewable energy sources with significant potential to meet global electricity demand (Menzri et al., 2024). However, incorporating wind energy into existing power grids presents challenges due to wind speed variability and the technical constraints of converting wind energy into usable power (Maaruf et al., 2022). As a result, the integration and effective performance of wind turbines in grid connections have become a major research focus (Maaruf et al., 2022). The effectiveness of such systems relies not only on the efficient harnessing of wind energy but also on grid stability and power quality (Mohamed et al., 2021). Advanced control and optimization strategies are essential to address these challenges and ensure smooth integration into the power grid (Hamid et al., 2022).

Maximum Power Point Tracking (MPPT) is a key technique for improving the operation of wind turbines (Al-Ani et al., 2024). It allows turbines to adjust their operating parameters in real-time to extract the maximum power from the wind (Kesavan et al., 2024). Since wind speed is unpredictable, MPPT algorithms ensure that the turbine operates at its maximum power point with minimal energy loss (Borni et al., 2025). One of the most widely used MPPT techniques in wind energy systems is Incremental Conductance (IC), due to its responsiveness to rapidly changing wind conditions (Elmorshedy et al., 2022). In grid-connected systems, MPPT must be integrated with advanced control systems to regulate the produced power and ensure compliance with grid requirements (Rekioua et al., 2024).

Hybrid control strategies have also made significant progress in improving the stability and efficiency of wind energy systems (Babu et al., 2024). Hybrid controllers combine traditional methods like Proportional-Integral (PI) control with fuzzy logic or Model Predictive Control (MPC) to enhance system flexibility and performance under varying wind and grid conditions (Nasef et al., 2022). This integration helps to handle nonlinearities and uncertainties, allowing the system to remain stable during grid disturbances or fluctuations in wind conditions (Gupta et al., 2025). By combining fuzzy logic with conventional control methods, these strategies help maximize power output while maintaining grid stability (Gulzar et al., 2023).

Energy storage systems are vital for reducing the intermittency of wind energy (Zaghba et al., 2024). Since wind energy generation is not constant, storage systems allow for the capture of surplus energy during high wind periods for later use when wind speeds drop (Sharma et al., 2023). By integrating battery storage with wind turbines, consistent power generation and grid export become feasible (Pandey et al., 2024). However, issues such as balancing charge/discharge cycles, efficient storage management, and reducing operational costs remain (Naima et al., 2025). Hybrid control measures are needed to optimize the interaction between the wind turbine and energy storage systems to ensure efficient energy storage and utilization (Sarathkumar et al., 2021).

The reliability and safety of wind energy systems are closely linked to fault detection and control (Gbadega et al., 2022). Wind turbines are complex machines prone to various faults, including mechanical, electrical, and communication failures (Kumar et al., 2024). Early detection of these faults is crucial to prevent damage to turbines and other system components (Rao et al., 2022). Model-based or data-driven fault detection techniques continuously monitor system performance to identify anomalies indicative of potential faults (Abdelwahab et al., 2024). Control strategies must quickly respond to fault detection to minimize the impact on the grid and maintain safe, efficient operation (Tuli et al., 2022). These fault detection and control mechanisms play a critical role in extending the lifespan of wind energy systems and ensuring their long-term viability as a renewable energy source (Vikkurty et al., 2022).

This chapter provides a comparative and analytical evaluation of MPPT techniques and hybrid control strategies for grid-connected wind energy systems. Unlike existing studies, it integrates control design, energy storage coordination, and fault detection within a unified framework supported by simulation-based validation.

CRITICAL REVIEW OF EXISTING MPPT AND CONTROL METHODS

A wide range of Maximum Power Point Tracking (MPPT) techniques and control strategies have been proposed to enhance the efficiency and stability of grid-connected wind energy systems. Traditional methods such as Perturb and Observe (P&O) and Incremental Conductance (IC) are widely used due to their simplicity and ease of implementation. However, these techniques often face limitations under rapidly varying wind conditions, leading to reduced tracking accuracy and oscillations around the maximum power point. To overcome these challenges, advanced optimization-based methods such as Particle Swarm Optimization (PSO) and Genetic Algorithms (GA), as well as artificial intelligence (AI)-based approaches, have been developed to improve tracking performance and adaptability.

Table 1. Comparative Analysis of MPPT Techniques

MPPT Method	Tracking Efficiency	Convergence Speed	Computational Complexity	Best Use Case
Perturb & Observe (P&O)	Moderate	Slow–Moderate	Low	Small-scale systems with stable wind
Incremental Conductance (IC)	High	Moderate–Fast	Medium	Grid-connected systems with variable wind
Genetic Algorithm (GA)	Very High	Moderate	High	Large-scale wind farms
Particle Swarm Optimization (PSO)	Very High	Fast	High	Hybrid renewable energy systems
AI-Based Methods (ANN, RL)	Very High	Fast	Very High	Smart grids and adaptive systems

CRITICAL DISCUSSION AND RESEARCH GAP

Among the various MPPT techniques, Incremental Conductance (IC) offers a strong balance between accuracy and implementation complexity, making it suitable for grid-connected wind energy systems. However, it still suffers from performance degradation under highly turbulent wind conditions. Optimization-based methods such as PSO and GA provide improved global search capability and higher tracking efficiency, but their computational complexity limits real-time implementation in embedded systems.

Artificial intelligence-based approaches, including neural networks and reinforcement learning, demonstrate superior adaptability and performance in nonlinear and uncertain environments. Despite these advantages, their dependency on large training datasets, high computational requirements, and lack of interpretability restrict their widespread practical deployment.

A key research gap identified from the literature is the lack of integrated comparative frameworks that evaluate MPPT techniques alongside hybrid control strategies and energy storage coordination under realistic grid conditions. Additionally, many studies focus on individual techniques without considering their interaction with grid stability, fault tolerance, and power quality requirements. Therefore, there is a need for hybrid and adaptive control approaches that combine the strengths of conventional, optimization-based, and intelligent methods to achieve robust, real-time performance in grid-connected wind energy systems.

RECENT RESEARCH ON CONTROL STRATEGIES IN WIND ENERGY SYSTEMS

Recent research has focused on improving wind turbine efficiency and grid stability through advanced control algorithms. Artificial intelligence based MPPT techniques have gained attention due to their ability to handle nonlinear system behavior and rapidly varying wind conditions. For example, neural-network-based MPPT methods have demonstrated improved tracking efficiency compared with traditional P&O algorithms.

Machine learning approaches such as reinforcement learning and deep learning are also being explored for adaptive wind turbine control. These methods enable turbines to learn optimal operating strategies based on historical wind data and system responses.

Modern grid-connected wind systems increasingly provide advanced grid-support functions such as frequency regulation, reactive power compensation, and voltage support. The integration of wind energy with smart grids and digital monitoring platforms has also enabled improved system coordination and predictive maintenance.

MATHEMATICAL MODELING OF WIND ENERGY CONVERSION SYSTEMS

Wind turbine power generation is governed by aerodynamic principles. The mechanical power extracted from wind can be expressed as:

Wind power equation:

$$P = \frac{1}{2}\rho A V^3 C_p$$

Where:

- P = extracted wind power
- ρ = air density
- A = swept area of turbine blades
- V = wind speed
- C_p = power coefficient

The power coefficient depends on the tip-speed ratio (λ) defined as:
Tip-speed ratio:

$$\lambda = \frac{\omega R}{V}$$

Where:

- ω = rotor angular speed
- R = turbine blade radius

Maximum power extraction occurs when the turbine operates at an optimal tip-speed ratio. Therefore, control strategies such as Maximum Power Point Tracking (MPPT) are required to maintain optimal operating conditions under varying wind speeds.

OBJECTIVES

- To investigate and examine the application of the Maximum Power Point Tracking (MPPT) methods to optimize the power recovery of wind energy systems, to guarantee the maximum efficiency at different wind conditions.
- The study aims at investigating the effectiveness of the hybrid control methods of using traditional Proportional-Integral (PI) control with fuzzy logic and Model Predictive Control (MPC) in improving the stability and performance of grid-connected wind energy systems.
- To explore the concept of energy storage systems and wind turbines and especially the difficulties and advantages of balancing the energy storage and generation to ensure a steady power supply and enhance grid reliability.
- To assess the impact of advanced fault detection and control systems in grid-connected wind energy systems, with a focus on early fault identification and mitigation to ensure the system's long-term operational integrity and safety.
- To evaluate the harmonization of wind energy output with grid standards, including voltage regulation, frequency control, and power quality management, to ensure reliable and efficient integration of wind power into the existing grid infrastructure.

MAXIMUM POWER POINT TRACKING (MPPT) IN WIND ENERGY SYSTEMS

Principles and Importance of MPPT in Wind Energy Systems

Maximum Power Point Tracking (MPPT) ensures that wind turbines operate at optimal conditions to maximize energy extraction under varying wind speeds. Since the power output of a wind turbine is highly dependent on wind velocity, MPPT algorithms continuously adjust system parameters such as rotor speed or voltage to maintain operation at the maximum power point. This not only improves energy efficiency but also enhances grid stability by reducing power fluctuations..

MPPT is not just important to wind energy systems in terms of energy efficiency. It is a key factor in grid stability and reliability because it will help to make the wind turbine work under the best possible parameters, eliminating losses of power or variability. The power output in a turbine may fluctuate considerably with the speed of the wind, and this may cause a lot of instability in the system unless handled appropriately. MPPT will also make sure that these swings are kept to a minimum by automatically changing the states of operation of the turbine. This does not only increase the energy output but also the performance of the entire system, such that it takes up the necessary energy needs.

MPPT not only maximizes energy extraction but also improves overall system stability in grid-connected wind energy systems. By continuously adjusting turbine operating parameters, MPPT algorithms minimize power fluctuations and maintain efficient operation under variable wind conditions.

Figure 1. Principles and Importance of MPPT in Wind Energy Systems

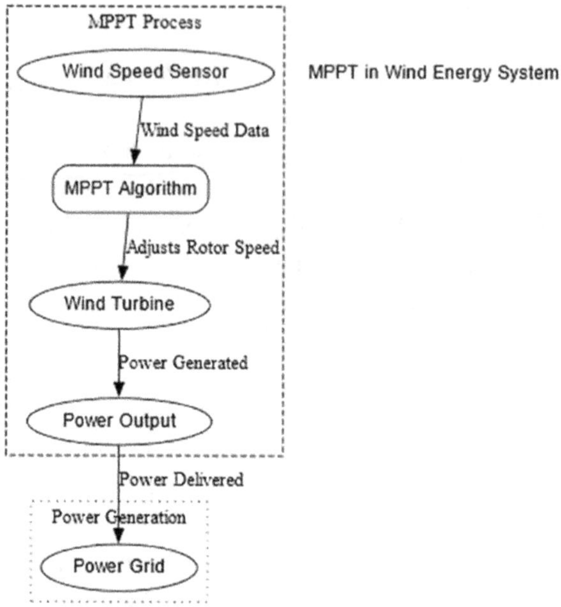

Common MPPT Algorithms: P&O, IC, and Others

Maximum Power Point Tracking (MPPT) is an important method of optimization of wind energy system performance, over that of a wind turbine, which operates at its optimal power point (MPP) under varying wind conditions. The Perturb and Observe (P&O) and Incremental Conductance (IC) are among the most general MPPT algorithms. The P&O algorithm is performed by periodically perturbing the operating point of the operating system and monitoring the change in power output. In case the power becomes larger, the perturbation direction is persisted; in case the power becomes smaller, the perturbation direction is reversed. This basic and commonly used algorithm is good in low turbulence, low-varying wind speed, and is prone to oscillations about the maximum power point under rapidly varying wind speeds.

The Incremental Conductance (IC) algorithm is an enhancement of the P&O algorithm, which finds the optimal operating point through the derivative of the power versus the voltage. This technique determines the conductance increment, which is compared to the conductance of the system at a particular moment. At a point where the two are equal, the system is available at the maximum power point. The IC approach is also better at tracking the MPP when the wind has changed quickly than the P&O method because it can detect the direction of change better and cope with the changes better. Additional methods of MPPT, like the Genetic Algorithm (GA) or Particle Swarm Optimization (PSO), have been put forward to improve the workings of MPPT systems, particularly in more advanced situations of large-scale wind farms or multi-turbine systems. These sophisticated techniques have global search features that may overcome the weakness of local optimization experienced in P&O and IC.

Incremental Conductance (IC) MPPT for Wind Power Optimization

The Incremental Conductance (IC) algorithmic model is a popular algorithm of Maximum Power Point Tracking (MPPT) in wind energy systems because it is capable of offering rapid and precise tracking of the optimal power point with varying wind velocities. The principle of the IC method is that the value of the derivative of the power with regard to the wind speed is zero at the maximum power point (MPP). The IC method dynamically changes the operating point of the wind turbine to maximize the power extraction by constantly comparing the instantaneous value of the power derivative with the instantaneous conductance. This technique is specifically useful in the rapidly changing wind conditions and works better than the traditional techniques such as Perturb and Observe (P&O), particularly in turbulent wind conditions.

Within wind power systems, use of IC-based MPPT increases the efficacy of the power generation process, as it ensures that the wind turbine is used at its functional optimum whenever the wind changes in pace. The algorithm compares the voltage-current-power relationship to decide whether to adjust the operating point up or down, which gives the algorithm a real-time regulation process necessary to maximize the captured energy. With changing wind conditions, the IC method constantly reviews the optimum operating voltage and current, making the turbine draw as much power as possible without the oscillations that are likely to be evident in other methods. This feature renders the IC approach especially appropriate to the larger grid-based wind energy systems in which steady and effective source production is essential.

Figure 2. Incremental Conductance (IC) MPPT for Wind Power Optimization

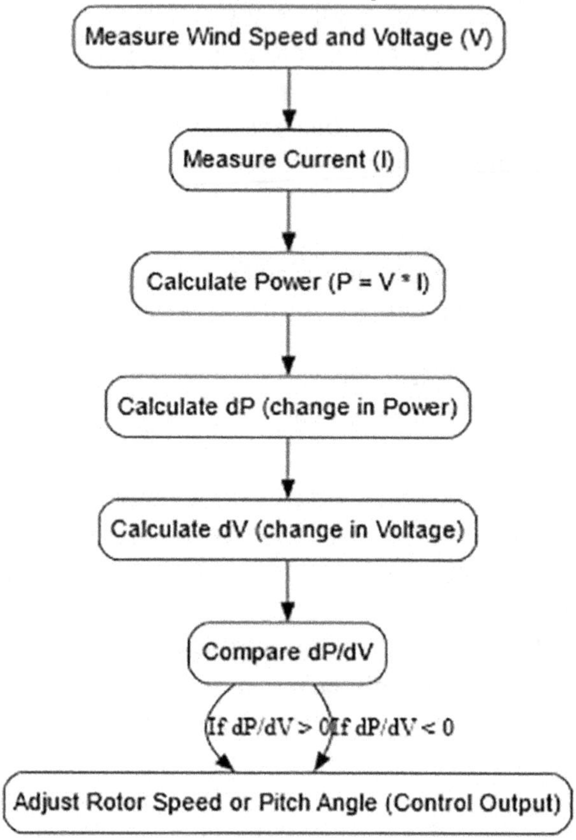

The Incremental Conductance (IC) algorithm determines the maximum power point by evaluating the slope of the power–voltage curve. At the maximum power point, the derivative of power with respect to voltage is zero:

$$\frac{dP}{dV} = 0$$

Since $P = VI$, this condition can be expressed as:

$$\frac{dI}{dV} = -\frac{I}{V}$$

Based on this principle, the IC algorithm operates as follows:

Step 1: Measure the instantaneous voltage V and current I of the wind energy system.

Step 2: Calculate incremental changes ΔV and ΔI.
Step 3: Compute the ratio $\frac{\Delta I}{\Delta V}$ and compare it with $-\frac{I}{V}$.

Step 4:

- If $\frac{\Delta I}{\Delta V} = -\frac{I}{V}$, the system is operating at the maximum power point.
- If $\frac{\Delta I}{\Delta V} > -\frac{I}{V}$, increase the operating voltage.
- If $\frac{\Delta I}{\Delta V} < -\frac{I}{V}$, decrease the operating voltage.

Step 5: Update the control signal and repeat the process continuously.

This method enables accurate tracking of the maximum power point under rapidly varying wind conditions and reduces oscillations compared to conventional P&O techniques

Real-Time MPPT Techniques for Grid-Connected Wind Systems

Figure 3. Real-Time MPPT Techniques for Grid-Connected Wind Systems

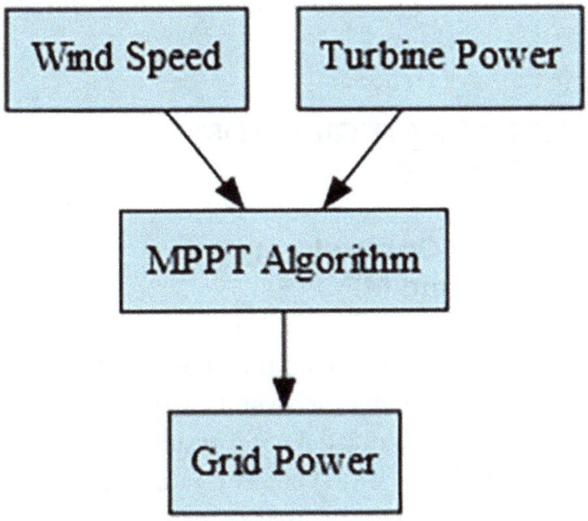

Performance of MPPT algorithms is typically evaluated using quantitative metrics such as:

- MPPT Tracking Efficiency (%) – ratio of extracted power to theoretical maximum power.
- Settling Time (s) – time required for the system to reach steady operation after wind variation.
- Oscillation Magnitude – variation around the optimal operating point.

Higher tracking efficiency and lower settling time indicate better algorithm performance.

Table 2. Comparison of Common MPPT Algorithms Used in Wind Energy Systems

MPPT Method	Advantages	Limitations	Application
Perturb & Observe (P&O)	Simple implementation	Oscillations near MPP	Small wind systems
Incremental Conductance (IC)	Accurate under changing wind	Higher computational effort	Grid-connected turbines
Genetic Algorithm	Global optimization capability	Complex implementation	Large wind farms
Particle Swarm Optimization	Fast convergence	Parameter tuning required	Hybrid energy systems

The comparison illustrates that while traditional algorithms such as P&O offer simple implementation, advanced optimization methods such as PSO and genetic algorithms provide improved convergence and performance in large-scale wind energy systems.

HYBRID CONTROL STRATEGIES FOR WIND ENERGY SYSTEMS

Introduction to Hybrid Controllers: Combining PI, Fuzzy, and MPC

Hybrid controllers are a result of integrating the strengths of the various control strategies with the aim of improving the overall performance of the wind energy systems. When applied to grid-connected wind systems based on wind energy, the application of Proportional-Integral (PI) control, Fuzzy Logic (Fuzzy), and Model Predictive Control (MPC) facilitates the use of more robust and adaptive control to deal with the complex dynamics of wind power generation and integration into the grid. PI controllers are established as being simple and effective in balancing the variables like voltage and frequency and have therefore been used as a good option in the regulation of stable operations under steady-state conditions. Fuzzy logic, however, comes with the facility to deal with non-linearities as well as uncertainties and provides flexibility in dealing with variable and unpredictable conditions like changing wind speeds and grid disturbances. MPC offers a sophisticated optimization methodology since the future system behavior is predicted and the control action is altered so that the system will work optimally over time, considering constraints and system dynamics.

When these control strategies are combined into a hybrid approach, each of the techniques can balance the weakness of the other to achieve greater overall performance and flexibility. A hybrid control system will allow the PI controller to take care of simple control functions like voltage and current control, whereas the fuzzy logic controller can be implemented to deal with nonlinearities and disturbances that will occur due to variations of environmental and operational conditions. The performance of the system may then be optimized using MPC to predict and correct future conditions so that the system is able to perform optimally in real-time. Such PI, fuzzy, and MPC control in hybrid controllers is specifically useful with grid-connected wind energy systems, in which stability, power quality, and the ability to meet dynamic fluctuations in wind conditions and grid demand are crucial.

Although hybrid controllers significantly improve system stability and adaptability, they also introduce certain limitations. The integration of multiple control techniques increases controller complexity and requires careful parameter tuning. Additionally, real-time implementation may demand higher computational resources, particularly when predictive algorithms such as Model Predictive Control (MPC) are included. These factors must be considered when implementing hybrid controllers in large-scale wind energy systems.

Design of Hybrid PI-Fuzzy Controllers for Enhanced Performance

The design of the hybrid PI-Fuzzy controllers of the wind energy system aims at integrating the benefits of the Proportional-Integral (PI) control and Fuzzy Logic Control (FLC) to improve the performance of the system under different working conditions. PI controllers have become popular in wind energy systems since they are very simple and helpful in terms of controlling power output, voltage, and frequency. Nevertheless, they tend to perform poorly in case of nonlinearities or uncertainties within the system, e.g., when the speed of wind suddenly changes or the grids are perturbed. The incorporation of fuzzy logic, conversely, introduces the ability to deal with more complicated uncertainties and imprecise data and is more adaptable and resilient to dynamic settings. The hybrid PI-Fuzzy controller, which is the result of integrating these two types of control, has the ability to offer the stability and precision of PI control as well as the flexibility and adjustability of FLC.

Figure 4. Design of Hybrid PI-Fuzzy Controllers for Enhanced Performance

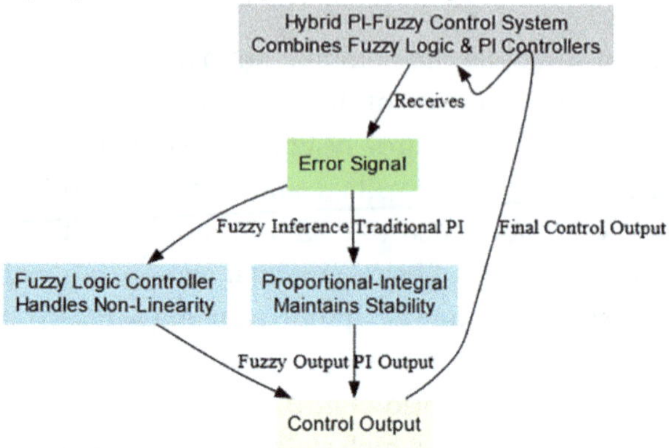

The hybrid PI-fuzzy controller is this controller, which aims at optimizing the operation of the grid-connected wind energy systems by adjusting the PI and fuzzy components in real time. The PI controller usually deals with the low-frequency aspects of the system, like reduction of steady-state error and keeping the voltage constant. The fuzzy controller is a supplement to this in that it handles the high-frequency dynamics, like responding to a quick change in wind speed and grid voltage. It is a two-fold strategy that will be effective in providing the system with rapid resilience to unexpected changes and simultaneously keeping the efficiency in the generation of power and compliance with the grid. The hybrid controller is therefore of great benefit to the traditional PI control in that it has a superior performance in terms of stability of the system, fault tolerance, and energy conservation, particularly when non-ideal or variable conditions exist.

Figure 5. Adaptive Hybrid Controllers for Dynamic Wind Conditions

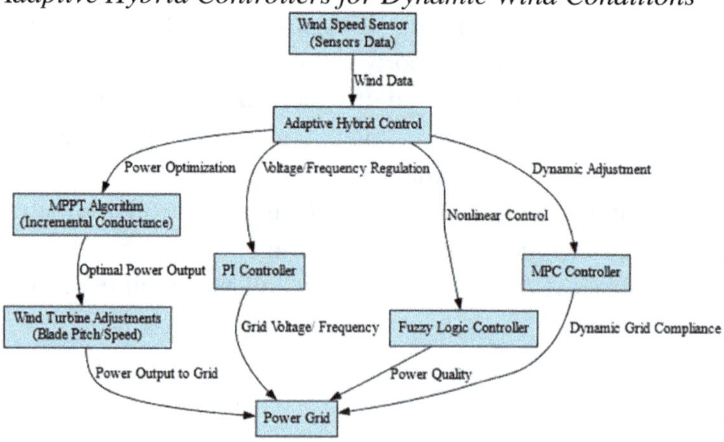

Hybrid control strategies combine PI, fuzzy logic, and Model Predictive Control (MPC) to improve the performance of wind energy systems under dynamic conditions. The PI controller ensures steady-state accuracy, while fuzzy logic handles nonlinearities and uncertainties, and MPC optimizes system behavior based on future predictions.

The conventional PI controller is expressed as:

$$u(t)=Kpe(t)+Ki\int e(t)dt$$

In hybrid configurations, fuzzy logic dynamically adjusts controller gains to improve transient response and system stability. These controllers enable effective voltage regulation, frequency control, and disturbance rejection, making them suitable for grid-connected wind systems operating under variable conditions.

POWER QUALITY AND GRID COMPLIANCE IN WIND ENERGY SYSTEMS

Power Quality Issues in Grid-Connected Wind Systems: Harmonics, Flicker, and Voltage Sags

Grid-connected wind energy systems introduce power quality issues such as harmonics, voltage sags, and flicker due to fluctuating wind conditions and power electronic converters. These issues can affect system stability and equipment performance.

To address these challenges, control techniques such as active and reactive power regulation, filtering methods, and voltage control strategies are implemented. These approaches ensure compliance with grid standards and maintain stable operation under dynamic conditions.

Figure 6. Power Quality Issues in Grid-Connected Wind Systems: Harmonics, Flicker, and Voltage Sags

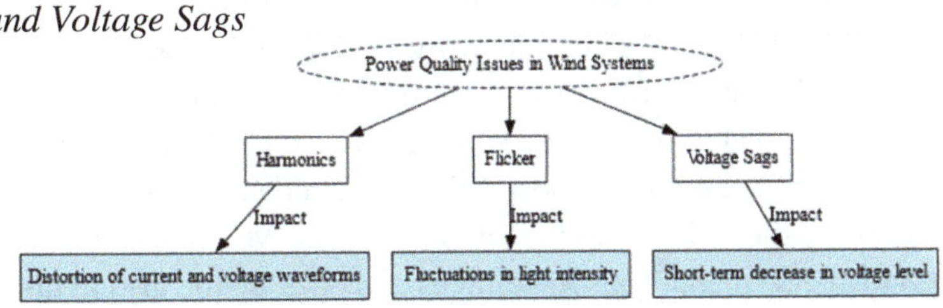

As a way of curbing such power quality problems, there is a need to have sophisticated control measures that ensure the integrity of the power supplied to the grid. Harmonic distortion is dealt with by placing filters in the system that can either cancel out or absorb the unwanted frequency components present in the system, which may be a passive or active filter. Flicker control, in its turn, consists of stabilizing the power output of a wind turbine by modulating the working conditions to minimize sudden increases and reductions of the voltage and current. Voltage sags are normally mitigated by incorporating dynamic reactive power control systems and voltage regulation systems that can react to any abrupt decrease in power output by injecting reactive power into the system. These solutions, together with good monitoring and real-time control, can assist in making sure that the wind energy systems connected to the grid are in line with the required grid codes and that the quality of the power is up to the standard. The quality of power is important to harmonize the introduction of wind energy into the grid and increase the stability of renewable energy sources and eliminate interruptions in the supply of electricity.

Control Techniques for Ensuring Grid Compliance and Stability

An essential part of the integration of wind energy systems into the current power system is grid compliance and stability. The preservation of the level of voltage and frequency, along with the ability to make sure that the power supplied to the grid is of the necessary quality, is necessary to eliminate disturbances and provide the grid with reliability. To overcome these difficulties, the control methods are used to keep the active and reactive power generated by wind turbines within the prescribed range so that the system will not exceed the established boundaries. Such methods are voltage regulation, frequency control, and reactive power compensation, all of which are essential in avoiding problems like fluctuation of voltage, harmonic distortion, grid instability, and much more. The control strategies are required to allow wind turbines to be flexible to fast variations in the wind conditions and meet the strict grid codes and requirements.

The methods of the control used to guarantee the compliance and stability of the grid usually include the usage of sophisticated controllers that combine several functions of the power control. These controllers constantly check grid parameters and control the production of the wind system correspondingly, which allows the system to react to the grid in real-time. As an illustration, grid-side converters have been used to regulate the amount of power that the wind energy system sends to the grid and to stabilize the voltage and frequency levels. Moreover, there is reactive power control to avoid fluctuations in voltages and back up the grid in high demand times. With such control mechanisms, wind energy systems can be made to run in harmony with the grid system, which ensures that the energy provided is reliable, stable, and of quality, and thus grid compliance and operational stability are ensured.

Voltage and Frequency Control in Wind Energy Systems

The control of voltage and frequency is very essential in ensuring the stability and reliability of the grid-connected wind energy systems. Wind turbines, being variable, may also cause variability in the voltage and frequency when they are connected to the grid, causing power quality problems such as voltage swells and depressed voltage sags and frequency variations. To handle these issues, wind energy systems should be provided with control systems that make the output voltage and frequency remain within reasonable limits set out by grid codes. Wind system Voltage control is normally done by means of grid-side converters that control the DC link voltage and ensure the necessary power factor. This will make the produced power compatible with the voltage standards of the grid and therefore will not cause any damage to the wind energy system and the grid to which this is connected.

Figure 7. Voltage and Frequency Control in Wind Energy Systems

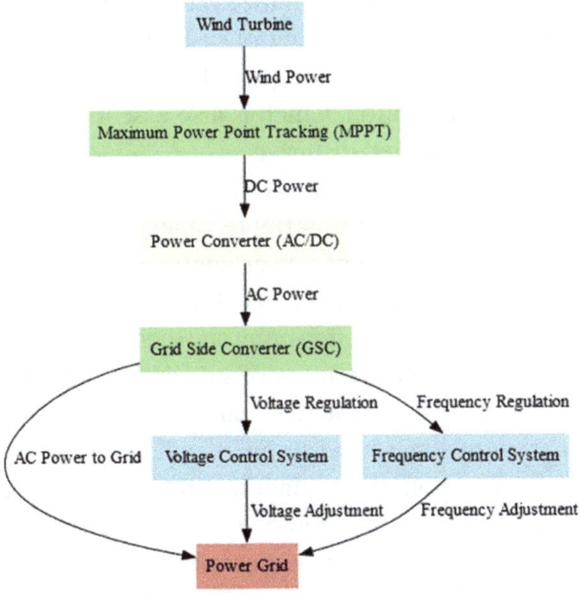

The frequency control of the wind energy systems is also essential since wind turbines should be used in step with the frequency of the grid. The changes in speed of wind may result in variation of the rotational speed of the turbine, which directly influences the frequency of produced power. To counter this, wind energy systems have control measures, which adjust the operating point of the turbine with respect to the change in the grid frequency. These planning methods frequently have feedback loops, which constantly observe grid situations and adapt the power output of the turbine. More specific and dynamic frequency control is also possible by the use of more sophisticated hybrid controllers, which provide the wind energy systems with the capability to react to changes in the frequency of the grid as rapidly and effectively as possible. This will allow wind energy systems to play a role in grid stability, especially in areas where an enormous amount of renewable energy has been integrated into the grid, where the frequency and voltage need to be maintained at constant levels in order to ensure the grid operates adequately.

Role of Reactive Power Control in Power Quality Management

The reactive power regulation is very important in the quality maintenance of power and compliance of the grid in the wind power systems. Reactive power, as compared to active power, is not part of the energy output but is necessary as a regulator of the voltage level and as a facilitator of the efficient operation of the grid. The wind turbines, especially the grid-connected ones, must have the ability to regulate the reactive power in order to stabilize the voltage profile. The systems of wind energy that have a reactive power control system can deliver dynamic voltage support, help to compensate for the variation of power factor, and minimize the probability of the voltage fluctuation, which is also essential when the wind speed is alternating or during grid-related disturbances.

In grid-connected systems, the system has the capability of regulating reactive power so that the system can respond swiftly to changes in the grid voltage, which would otherwise cause the problem of power quality (sags or swells in voltages). By modulating the reactive power output, wind turbines assist in maintaining the required level of voltages at the connection point and therefore make them compatible with grid standards and minimize chances of grid-induced issues like harmonics or over-voltage excursions. This operation is particularly required in offshore wind farms or the installation of large wind turbines, where the distance between the turbines and the grid makes voltage regulation a more complex operation. The efficient control of reactive power also improves the overall stability and reliability of the wind energy system as well as makes sure that the energy system meets the high standards of quality of power established by the grid operators.

Harmonization of Wind Energy Output with Grid Standards

Balancing wind energy system output with grid specifications is of importance in making sure that there is smooth integration and power grid stability. Wind turbines, especially those in grid-connected systems, produce power that could be irregular, like harmonic distortion, variable voltage, and flickering frequency. These variations come about because of the intermittency of wind and the dynamic nature of wind turbine. The grid operators have also set certain standards and codes, including voltages, frequency range, and harmonic distortion levels that wind energy systems should follow so that they do not negatively impact the local grid and the overall power grid in general. The harmonization needs to be effective and this demands accurate control methods that control the quality of power as well as addressing grid needs.

Figure 8. Harmonization of Wind Energy Output with Grid Standards

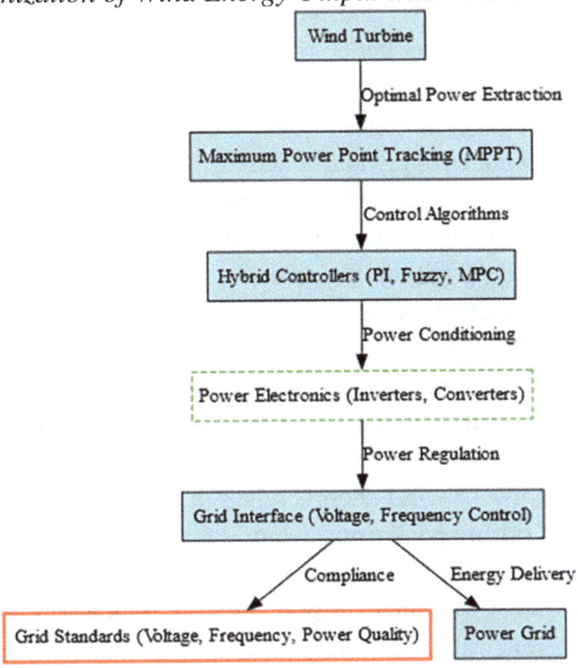

To meet grid criteria, wind energy systems are required to have sophisticated control methods that stabilize voltage and frequency besides reducing harmonic distortion. The use of power electronic devices like inverters and converters, as well as advanced control algorithms, enables real-time control of the output of the turbine to ensure that it does not breach the grid codes. These control systems keep checking the quality of power entering the grid and make corrections, as much as possible, to reduce the fluctuations and have the power within the acceptable limits. The harmonization process does not only take control of the flow power but also controls the reactive power to aid in maintaining the voltage regulation. Through these superior control systems, a grid-connected wind energy system can be used to achieve grid stability without sacrificing power quality, which in the long term leads to the realization of reliable and efficient power delivery.

ENERGY STORAGE INTEGRATION WITH WIND ENERGY SYSTEMS

Importance of Energy Storage for Wind Energy Stability

Storage of energy is important in improving the stability of wind energy systems, especially when coupled with the grid. The natural fluctuation of wind means that it is difficult to synchronize power generation and grid demand in real time, resulting in a fluctuation of electricity supply. A good solution to this problem is battery- or supercapacitor-based energy storage systems that store any excess generated energy when the wind is blowing strongly and release it when wind speeds fall or demand suddenly increases. This storage and on-demand ability to transmit energy assists in evening out the power output of the wind turbines to provide a more steady and dependable supply of electricity to the grid.

A combination of energy storage and wind energy systems also has a major advantage in grid stability and reliability. Through offering a buffer to the short-term variability in wind power, the energy storage helps to decrease the chances of an imbalance in the grid, which would otherwise cause a voltage sag, frequency variability, or expensive curtailment of wind-generated power. Also, energy storage helps in the effective use of wind energy, where it can be stored during high wind energy and it is used during peak times when it is needed most. This load leveling and power smoothing capability make energy storage a key factor in the full utilization of the contribution of wind energy to the grid, which increases its role in the overall transition to renewable energy systems.

Control Strategies for Hybrid Energy Storage Systems in Wind Farms

The combination of energy storage systems (ESS) and wind energy systems is also becoming more and more necessary to relieve the intermittency and variability of wind power generation. The hybrid energy storage systems, which are directly a combination of several types of storage technologies, including batteries, supercapacitors, and flywheels, are highly beneficial in wind farms in balancing supply and demand. In order to successfully integrate such systems of storage, there should be control strategies that would result in the stable functioning of the wind turbines as well as the storage systems. These control methods are aimed at controlling the charging and discharging of the storage equipment in real-time depending on the weather conditions, grid requirements, and needs of the system. These systems are also able to level the swings in power generation through dynamically changing the energy storage levels and provide an uninterrupted supply of energy and curb grid disruptions.

Figure 9. Control Strategies for Hybrid Energy Storage Systems in Wind Farms

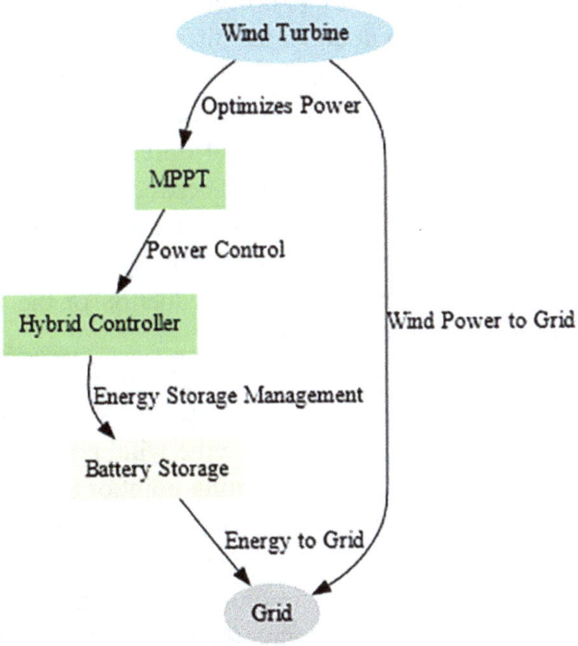

The overall aim of the control strategies applied in the hybrid energy storage systems is to optimize the relationship between the wind turbines and the energy storage units with the aim of maximizing the power output as well as the efficiency. The control system has to make decisions on when to save the surplus energy produced in times of high wind and when to dump some of that energy in times of low wind or at peak grid demand. High-level algorithms are used to forecast energy demands and production trends, and therefore the system can make proactive decisions concerning the management of energy storage. These control measures must also make sure that the energy storage operating systems work within their safe operating limits and that they also meet the needed grid support services, including voltage regulation and frequency control. These control strategies can improve the overall stability, efficiency, and reliability of wind farms so that they can more readily integrate with the electrical grid through the coordination of the operation of wind turbines and hybrid energy storage systems.

MPPT and Energy Storage Coordination for Peak Power Management

The combination of wind energy systems and energy storage is a necessary measure in dealing with interim nature of wind power generation. The concept of Maximum Power Point Tracking (MPPT) is crucial in ensuring that wind turbines work at their optimum power output under the different conditions of the wind, yet the changing nature of the wind still causes periods of overproduction and under-production. Batteries or supercapacitors can be used as energy storage systems to take in excess energy as the wind speeds up and store it to be used in times when the wind slows down. This synchronized control between MPPT and energy storage is useful in smoothing the power supply to provide a more consistent and reliable supply of energy into the grid. The behavior of these systems is essential in terms of the control of peak power demand and the reduction of the fluctuations that otherwise could lead to instability of the grid.

The coordination of effective MPPT and energy storage is performed by using dynamic control algorithms, which control the wind power production and the state of charge of the energy storage system. In case the wind power is higher than the current grid demand, MPPT will set the operating point of the turbine to optimize the power generation, with the excess power stored in the energy storage system. In low wind conditions, the stored energy is released by the energy storage system in order to sustain the necessary power output. This coordination would provide power generation that is consistent, allowing the grid to stay stable and lessening the use of conventional backup sources of power generation. The system will be able to fulfill the changing demand without compromising efficiency and reliability by maximizing the capture and use of stored energy.

Impact of Storage on Grid Stability and Power Quality

The combination of energy storage systems and wind energy systems is an important factor that will help in improving grid stability and power quality. The output of wind energy is always intermittent, whereby the speed of wind varies, hence causing variation in power production. Such fluctuations may lead to voltage sags and frequency instability, among other power quality problems, when wind energy enters the grid directly. The system can also store excess energy by adding an energy storage device like batteries or supercapacitors and then release it when the wind is less, and it captures the energy in storage during spurts of high wind speed to maintain a constant and reliable stream of power. This controllability of the fluctuations in power mitigates the threat of instability in the grid and makes the energy supplied to the grid consistent and within the acceptable limits of quality of grid operators.

Figure 10. Impact of Storage on Grid Stability and Power Quality

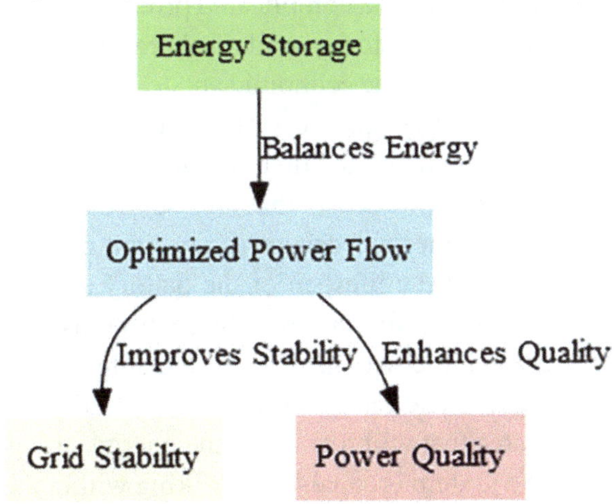

Energy storage particularly has a great influence on the stability of the grid in a highly renewable energy-based system. Storage systems serve as buffers that can help take in energy when there is surplus production to ensure that the grid does not get overloaded or see a rise in voltage. Simultaneously, they are capable of releasing accumulated energy as the level of wind production decreases to ensure that the necessary frequencies and voltage levels are sustained on the grid. This capability of balancing supply and demand in real time increases the overall reliability of the grid, as there is less necessity for traditional fossil-based power backups. Regarding the quality of power, storage can solve some problems like harmonic distortion and flicker as well as voltage by making sure that power variances are reduced as much as possible, thus contributing to the conformity of power quality to set standards.

Challenges in Integrating Battery Storage with Wind Energy Systems

The combination of wind energy systems and battery storage has a number of challenges that need to be overcome to reduce the impediments of an efficient and dependable power supply. This is because controlling the change in wind power generation as well as the energy demand is one of the main challenges. Wind power is also intermittent in nature, and the amount of energy generated depends on the speed of the wind, and battery backup systems should have the capacity to store surplus energy in times of high wind as well as release it in times of low wind. The storage capacity of the energy is greatly determined by the capacity of the battery, the charge/discharge cycles, and the response time, which has to be well synchronized with the output of the wind turbine. The energy stored in the battery and the power received in the wind turbine is a complicated process that needs sophisticated control algorithms to optimize the storage and discharge process.

Besides these operation issues, the economic and technical challenges also exist in terms of integrating the battery storage systems and wind energy. Large-scale battery storage systems are expensive and therefore may not be economically viable in many projects due to high capital costs that are incurred. The other issue is the wear and tear caused to the batteries because of constant charge/discharge cycles that may result in reduction of the lifespan of the battery as well as escalation of maintenance cost. Besides, the assimilation of these systems into the current grid infrastructure must be compatible with grid codes and able to manage the variation in the voltage and frequency. The integration process should not only require sophisticated control measures that will maximize the energy storage and release but should also ensure that the system is capable of working within the grid regulations and can give the required support to the grid when needed.

ADVANCED CONTROL TECHNIQUES FOR GRID-CONNECTED WIND SYSTEMS

Model Predictive Control (MPC) for Wind Energy Systems

Model Predictive Control (MPC) has potentially become an advanced control method in grid-connected wind energy systems because it is able to optimize system performance in real-time through the prediction of future behavior. To forecast upcoming conditions of the wind energy system, MPC uses a mathematical model depicting the system to forecast the power output, voltage, and frequency. It is such predictions that enable the controller to make the best decisions regarding the control inputs, that is, the rotor speed or pitch angle, to maximize power production and meet the grid demands. The extraordinary benefit of MPC is that it can solve multivariable control problems and constraints (e.g., grid stability, power quality, and loss reduction) but takes into account dynamic and uncertain characteristics of both wind conditions and grid parameters.

Figure 11. Model Predictive Control (MPC) for Wind Energy Systems

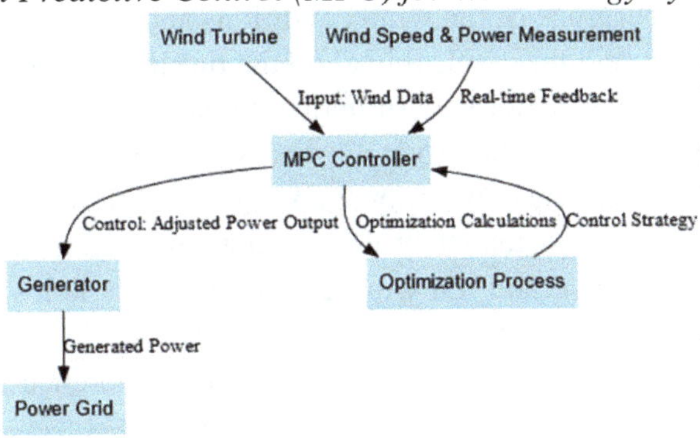

The MPC use in wind energy systems is especially useful in the context of the usage of the variable renewable energy sources in the grid. The continual revision of predictions by real-time measurements makes MPC not only operate the wind turbine at the most efficient level but also take into consideration the constraints of the system, including power factor correction and voltage regulation. This helps the system to react ahead to the weather variations and grid unrest, reducing the chances of the grid running wild. Also, MPC has the potential to enhance the total efficiency of the system by maximizing the energy flow, minimizing the wear and tear of the mechanical parts of the turbine, and increasing the life cycle of the system.

Limitations of Model Predictive Control

Despite its advantages, MPC has several challenges. The technique requires an accurate system model and involves solving optimization problems at every control interval. This increases computational requirements and may limit real-time implementation in low-cost embedded controllers used in wind turbines.

Adaptive Control Methods for Variable Wind Speeds

Adaptive control approaches are important in the optimization of grid-connected wind energy systems, more especially in the case of the variable wind speed. Fluctuations in wind speed are a major issue for the stability and performance of wind turbines since they contribute to changes in the power production, which may influence the functionality of the turbine and also its connection with the grid. The adaptive control methods are aimed at altering the control parameters in real time depending on the variability of the wind so that the turbine works at its optimum power point irrespective of the variations in wind speed. These control systems adjust the operating setpoints of the turbine, e.g., rotor speed and pitch angle, dynamically in reaction to the occurrence of the wind conditions and thereby maximize energy harvesting and reduce the mechanical load on the turbine components.

Adaptive control has a major benefit in that it responds to the uncertainties and nonlinearities present during the production of wind power. The adaptive control systems are in contrast to the conventional fixed control strategies, which are predefined in their values, but instead they are constantly revised to match the feedback of the turbine performance and the wind conditions. This helps them to operate optimally even when the wind is changing or unpredictable at a very high rate. With tools like fuzzy logic, neural networks, or model reference adaptive control, such systems are able to use past data and modify their responses so as to enhance performance over time.

Neural Networks and Machine Learning in Wind System Control

Neural networks and machine learning have been demonstrated as very promising in controlling grid-connected wind energy systems to enhance the performance and efficiency of the systems. Conventional control techniques tended to be based on predetermined mathematical models, which may be constrained by the fact that they are unable to adjust to dynamic and uncertain situations. Neural networks, in contrast, can learn intricate data patterns; thus, they are appropriate when real-time control of the wind energy systems is to be considered. These models can predict the optimal control actions and vary the system parameters by training neural networks on past data in wind turbines, which contained wind speed, turbine performance, and power output. This ability to adapt to learning enables relative enhancement of control strategies, especially in systems where wind conditions and system dynamics are very volatile.

Figure 12. Neural Networks and Machine Learning in Wind System Control

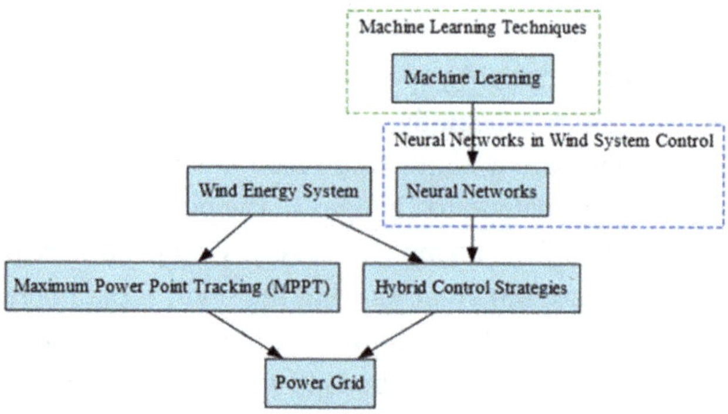

Other benefits of machine learning algorithms are found in control over wind systems, such as reinforcement learning and deep learning. Reinforcement learning also allows systems to maximize control actions by trial-and-error experience with the environment, which is an autonomous method of making wind turbine operation more efficient. Deep learning algorithms are able to learn the large amounts of sensor data in order to identify latent patterns and relationships, which enable the detection of faults and anomalies in real time. With a combination of machine learning methods, wind energy systems will be able to be more reactive to the change of wind characteristics and the demands of the grid so that the turbine will be able to work at the best efficiency and to ensure the quality of power meets the standards. The innovations in the methods of control provide new opportunities to the new generation of intelligent wind energy systems, which are able to autonomously optimize their performance and add the contributions to the more stable and sustainable power system.

Neural-network-based control methods also face several challenges. These models require large datasets for training and validation. In addition, neural networks often behave as black-box models, which reduces interpretability and makes it difficult to guarantee stability under unseen operating conditions. Consequently, hybrid approaches combining machine learning with traditional control methods are often recommended.

Sliding Mode Control (SMC) for Wind Energy Optimization

Sliding Mode Control (SMC) has become one of the useful methods of maximizing the operation of the grid-connected wind energy system because of its robustness and capacity to manage uncertainties and nonlinearities in the system. SMC works by pushing the system trajectory to be sliding on a given surface, called the sliding surface, that is established to push the system to the desired state. This approach is especially useful when it comes to wind power plants, the environment of which can be very erratic and volatile. With the use of SMC, the system can even be stabilized in case of changes in wind speeds, grid disturbances, and other uncertainties. The most important merit of SMC is that it allows effective control actions with low sensitivity to model errors, which is why it is preferred in the systems where it is not easy to model wind conditions and turbine dynamics.

The usage of SMC can greatly enhance the power flow in wind energy systems, control of voltage, and stability of frequencies. By way of example, in the control of the pitch angle of the wind turbine or rotor speed, SMC will make sure that the turbine will work with the maximum efficiency even when the wind speed changes. The natural strength of SMC enables the wind energy system to take up any external disturbances, like sudden changes in wind speed or power faults in the grid, and return the power output of the system to a stable and confident value. SMC can provide a more flexible as well as fault-tolerant control approach than standard approaches such as PI or PID controllers by avoiding the need to have a precise system model to control. Consequently, the addition of SMC to grid-connected wind systems makes them more efficient and better able to function in real time, as well as adds to the stability and reliability of the power grid as a whole.

One major drawback of SMC is the chattering phenomenon, which refers to high-frequency oscillations caused by switching control actions. These oscillations may increase mechanical stress on turbine components and require smoothing techniques to mitigate their effects.

Fault Detection and Control in Wind Energy Conversion Systems

The fault detection and control in the wind energy conversion systems are vital in keeping the reliability of the systems and avoiding damage to the wind turbine and other related parts. Breakdowns related to wind turbines include a number of fault conditions, including mechanical, electrical, and communication, which can have a drastic impact on the functionality and security of the system. Efficient fault detectors are used to check the well-being of the system in real-time by incessantly examining the parameters, including vibration, temperature, current, and voltage. Through sophisticated algorithms, one can detect the presence of anomalies in the behavior of the system at a relatively early stage before the faults get out of control and thus preventive actions can be taken. Such techniques do not only assist in the detection of faults but also in diagnosing their particular causes, which is necessary in reducing downtime as well as maintenance costs.

After the fault has been identified, it is also necessary to initiate measures of control that could help to limit the effects of the fault on the rest of the system. The fault-tolerant control strategies enable the system to be able to sustain itself at a smaller capacity or under a safe state until the problem is fixed. Such types of control methods may include modifying the operating conditions of the turbine, like cutting in the wind speed or altering the pitch angle to stabilize the turbine and avoid additional destruction. With electrical faults, e.g., a change in grid voltage or a short circuit, the state-of-the-art controllers may isolate the faulty parts and reconfigure the system to cause as little power generation loss as possible.

Modern fault detection approaches increasingly use machine learning techniques such as support vector machines, neural networks, and deep learning models. These techniques analyze sensor data from turbines to identify abnormal operating conditions and predict potential failures before they occur. Predictive maintenance strategies based on these methods can significantly reduce maintenance costs and improve the reliability of wind farms.

Future Research Trends in Wind Energy Control Systems

Recent research in wind energy systems is moving toward advanced digital technologies and intelligent grid integration. Emerging topics include digital twin models for predictive maintenance, cybersecurity frameworks for wind farms, and grid-forming converters capable of supporting grid voltage and frequency independently. In addition, coordinated wind farm control strategies are being developed to mitigate wake effects and maximize total farm output. These innovations are expected to significantly enhance the reliability and efficiency of future renewable energy systems.

Simulation and Performance Analysis

To evaluate the effectiveness of different MPPT techniques and control strategies, simulation studies are performed using MATLAB/Simulink. A variable-speed wind turbine model integrated with MPPT controllers and grid interface components is considered for analysis.

Simulation Setup

The simulation model includes:

- Variable wind speed profile
- Wind turbine generator model
- MPPT controller (P&O, IC, PSO)
- Grid-connected inverter system
- Energy storage interface (optional)

Performance Evaluation

The performance of MPPT techniques is evaluated using key metrics such as tracking efficiency, settling time, and oscillation around the maximum power point.

Results and Discussion

Simulation results indicate that:

- The P&O method exhibits oscillations around the maximum power point under rapidly changing wind conditions.
- The IC method improves tracking accuracy and reduces steady-state oscillations.
- The PSO-based method provides faster convergence and higher global optimization capability, particularly in complex operating conditions.

Typical system responses include:
• Wind speed vs. output power curves, showing dynamic adaptation of MPPT algorithms

- Power tracking characteristics, demonstrating convergence toward the optimal operating point
- Voltage and current waveforms, indicating improved stability with advanced methods

Comparative Performance of MPPT Techniques

Method	Tracking Efficiency (%)	Settling Time (s)	Oscillation Level	Overall Performance
P&O	85–92	High	High	Moderate
IC	92–96	Medium	Low	Good
PSO	95–98	Low	Very Low	Excellent

The results clearly demonstrate that advanced optimization-based techniques outperform conventional methods in terms of efficiency and dynamic response, making them suitable for modern grid-connected wind energy systems.

Hybrid Control Section

Mathematical Representation of Hybrid PI-Fuzzy Control

The conventional PI controller is defined as:

$$u(t) = K_p e(t) + K_i \int e(t)\, dt$$

where:

- $e(t)$ is the error signal

- K_p, K_i are proportional and integral gains

In hybrid PI-Fuzzy control, the fuzzy logic system dynamically adjusts K_p and K_i based on system conditions such as error magnitude and rate of change.

Working Principle

- The PI controller ensures steady-state accuracy
- The fuzzy controller:
  - Handles nonlinearities
  - Adapts controller gains in real time
  - Improves transient response

This hybrid structure enhances system robustness, reduces overshoot, and improves performance under fluctuating wind and grid conditions.

MACHINE LEARNING SECTION

Comparison of Control Approaches

Traditional control methods such as PI and IC rely on predefined mathematical models and are effective under stable conditions but struggle with nonlinear and uncertain environments. Artificial Neural Networks (ANN) provide strong pattern recognition capabilities and can approximate complex nonlinear relationships, while Reinforcement Learning (RL) enables systems to learn optimal control strategies through interaction with the environment.

Compared to traditional methods, AI-based approaches offer superior adaptability and predictive capability. However, they require large datasets, higher computational resources, and careful tuning. Therefore, hybrid approaches combining AI with classical control methods are increasingly preferred in modern wind energy systems.

Energy Storage

Control Logic for Energy Storage Integration

The coordination between wind energy generation and energy storage systems can be described using a simple control logic:

- If $P_{wind} > P_{load} \rightarrow$ excess energy is stored (charging mode)

- If $P_{wind} < P_{load} \rightarrow$ stored energy is released (discharging mode)

This strategy ensures:

- Power smoothing
- Reduced grid fluctuations
- Improved reliability

Advanced control systems further optimize this process using predictive algorithms and state-of-charge management techniques.

CONCLUSION

This chapter presented key control strategies for improving the performance of grid-connected wind energy systems. MPPT techniques such as IC and PSO enhance energy extraction efficiency, while hybrid control strategies improve system stability under varying conditions. Energy storage integration further supports grid reliability by mitigating power fluctuations.

Advanced methods including machine learning and predictive control offer promising directions for future research. These approaches will contribute to the development of intelligent, stable, and efficient renewable energy systems.

Future research should focus on integrating artificial intelligence, advanced predictive control strategies, and large-scale energy storage technologies to enhance the reliability of grid-connected wind energy systems. These developments will support the transition toward sustainable and resilient power systems.

REFERENCES

Abdelwahab, S. A. M., El-Rifaie, A. M., Hegazy, H. Y., Tolba, M. A., Mohamed, W. I., & Mohamed, M. (2024). Optimal control and optimization of grid-connected PV and wind turbine hybrid systems using electric eel foraging optimization algorithms. *Sensors (Basel), 24*(7), 2354. DOI: 10.3390/s24072354 PMID: 38610565

Al-Ani, M. A., Zdiri, M. A., Salem, F. B., & Derbel, N. (2024). Optimized grid-connected hybrid renewable energy power generation: A comprehensive analysis of photovoltaic, wind, and fuel cell systems. Engineering, Technology &. *Applied Scientific Research, 14*(3), 13929–13936.

Babu, T. M., Chenchireddy, K., Kumar, K. K., Nehal, V., Srihitha, S., & Vikas, M. R. (2024). Intelligent control strategies for grid-connected photovoltaic wind hybrid energy systems using ANFIS. *International Journal of Advances in Applied Sciences, 13*(3), 497–506. DOI: 10.11591/ijaas.v13.i3.pp497-506

Borni, A., Bessous, N., Zaghba, L., Bouchakour, A., Agmas, M. S., Ali, E., Ghoneim, S. S. M., & Hoballah, A. (2025). Enhancing grid connected wind energy conversion systems through fuzzy logic control optimization with PSO and GA techniques. *Scientific Reports, 15*(1), 27678. DOI: 10.1038/s41598-025-12593-4 PMID: 40730602

Elmorshedy, M. F., Habib, H. U. R., Ali, M. M., Sathik, M. J., & Almakhles, D. J. (2022). Improved performance of hybrid pv and wind generating system connected to the grid using finite-set model predictive control. *IEEE Access : Practical Innovations, Open Solutions, 10*, 110344–110361. DOI: 10.1109/ACCESS.2022.3214996

Gbadega, P. A., & Balogun, O. A. (2022). Modeling and control of grid-connected solar-wind hybrid micro-grid system with multiple-input ćuk DC-dc converter for household & high power applications. *International Journal of Engineering Research in Africa, 58*, 191–224. DOI: 10.4028/www.scientific.net/JERA.58.191

Gulzar, M. M., Iqbal, A., Sibtain, D., & Khalid, M. (2023). An innovative converter-less solar PV control strategy for a grid connected hybrid PV/wind/fuel-cell system coupled with battery energy storage. *IEEE Access : Practical Innovations, Open Solutions, 11*, 23245–23259. DOI: 10.1109/ACCESS.2023.3252891

Gupta, M., Tiwari, P. M., Viral, R. K., Shrivastava, A., Zneid, B. A., & Hunko, I. (2025). Grid-connected PV inverter system control optimization using Grey Wolf optimized PID controller. *Scientific Reports, 15*(1), 28869. DOI: 10.1038/s41598-025-10617-7 PMID: 40774975

Hamid, B., Hussain, I., Iqbal, S. J., Singh, B., Das, S., & Kumar, N. (2022). Optimal MPPT and BES control for grid-tied DFIG-based wind energy conversion system. *IEEE Transactions on Industry Applications*, *58*(6), 7966–7977. DOI: 10.1109/TIA.2022.3202757

Kesavan, P. K., Subramaniam, U., Almakhles, D. J., & Selvam, S. (2024). Modelling and coordinated control of grid connected photovoltaic, wind turbine driven PMSG, and energy storage device for a hybrid DC/AC microgrid. *Protection and Control of Modern Power Systems*, *9*(1), 154–167. DOI: 10.23919/PCMP.2023.000272

Kumar, K., Lakshmi Devi, V., Dhanamjayulu, C., Kotb, H., & ELrashidi, A. (2024). Evaluation and deployment of a unified MPPT controller for hybrid Luo converter in combined PV and wind energy systems. *Scientific Reports*, *14*(1), 3248. DOI: 10.1038/s41598-024-53605-z PMID: 38332236

Kumar, M., Singh, A. K., & Lal, S. (2024). Design And Simulation Of Grid-Connected Solar Wind Hybrid Power System With MPPT Techniques. *Nanotechnology Perceptions*, *20*(S15), 4049–4064.

Maaruf, M., Khan, K., & Khalid, M. (2022). Robust control for optimized islanded and grid-connected operation of solar/wind/battery hybrid energy. *Sustainability (Basel)*, *14*(9), 5673. DOI: 10.3390/su14095673

Menzri, F., Boutabba, T., Benlaloui, I., Chrifi-Alaoui, L., Alkuhayli, A., Khaled, U., & Mahmoud, M. M. (2024). Applications of novel combined controllers for optimizing grid-connected hybrid renewable energy systems. *Sustainability (Basel)*, *16*(16), 6825. DOI: 10.3390/su16166825

Naima, B., Belkacem, B., Ahmed, T., Benbouhenni, H., Riyadh, B., Samira, H., Sarra, Z., Elbarbary, Z. M. S., & Mohammed, S. A. (2025). Enhancing MPPT optimization with hybrid predictive control and adaptive P&O for better efficiency and power quality in PV systems. *Scientific Reports*, *15*(1), 24559. DOI: 10.1038/s41598-025-10335-0 PMID: 40629005

Nasef, S. A., Hassan, A. A., ElMadany, H. T., Zahran, M. B., El-Shaer, M. K., & Abdelaziz, A. Y. (2022). Optimal power management and control of hybrid photovoltaic-battery for grid-connected doubly-fed induction generator based wind energy conversion system. *International Journal of Renewable Energy Research*, *12*(1), 408–421.

Pandey, P. K., Kumar, R., Gupta, V., Kanungo, A., & Diwania, S. (2024). An innovative hybrid controller-based combined grid-connected hybrid renewable energy system. *Electrical Engineering*, *106*(6), 7055–7072. DOI: 10.1007/s00202-024-02363-2

Rao, T. E., Sundaram, E., Chenniappan, S., Almakhles, D., Subramaniam, U., & Bhaskar, M. S. (2022). Performance improvement of grid interfaced hybrid system using distributed power flow controller optimization techniques. *IEEE Access : Practical Innovations, Open Solutions, 10*, 12742–12752. DOI: 10.1109/ACCESS.2022.3146412

Rekioua, D., Mezzai, N., Mokrani, Z., Oubelaid, A., Kakouche, K., Logerais, P. O., Alshareef, M., Bajaj, M., Tuka, M. B., & Ghoneim, S. S. (2024). Effective optimal control of a wind turbine system with hybrid energy storage and hybrid MPPT approach. *Scientific Reports*, *14*(1), 30013. DOI: 10.1038/s41598-024-78847-9 PMID: 39622862

Rekioua, D., Mokrani, Z., Kakouche, K., Oubelaid, A., Rekioua, T., Alhazmi, M., Ali, E., Bajaj, M., Dost Mohammadi, S. A., & Ghoneim, S. S. (2024). Coordinated power management strategy for reliable hybridization of multi-source systems using hybrid MPPT algorithms. *Scientific Reports*, *14*(1), 10267. DOI: 10.1038/s41598-024-60116-4 PMID: 38704399

Sharma, S., Chauhan, B. K., & Saxena, N. K. (2023). Grid connected fuzzy logic control-based MPPT techniques for hybrid photovoltaic wind with battery system. *International Journal of Power and Energy Conversion*, *14*(2-3), 280–309. DOI: 10.1504/IJPEC.2023.134873

Sharma, S., Chauhan, B. K., & Saxena, N. K. (2023). Artificial neural network grid-connected MPPT-based techniques for hybrid PV-WIND with battery energy storage system. Journal of The Institution of Engineers (India): Series B, 104(6), 1217-1226.

Zaghba, L., Borni, A., Benbitour, M. K., Fezzani, A., Alwabli, A., Bajaj, M., Dost Mohammadi, S. A., & Ghoneim, S. S. (2024). Enhancing grid-connected photovoltaic system performance with novel hybrid MPPT technique in variable atmospheric conditions. *Scientific Reports*, *14*(1), 8205. DOI: 10.1038/s41598-024-59024-4 PMID: 38589473

Chapter 5

Mapping Trend Analysis and Research Fronts in AI-Driven Innovations in Indian Agriculture and Allied Sectors: A Comprehensive Scientometric Evaluation

Sumit Kumar
http://orcid.org/0009-0002-2866-715X
Babasaheb Bhimrao Ambedkar University, India

Surya Prakash Patel
http://orcid.org/0009-0002-7825-5313
Babasaheb Bhimrao Ambedkar University, India

Km Sunita
http://orcid.org/0009-0001-3066-4022
Babasaheb Bhimrao Ambedkar University, India

Sharad Kumar Sonkar
Babasaheb Bhimrao Ambedkar University, India

ABSTRACT

Artificial intelligence in Indian agriculture and related industries has become a driving force of innovation, providing solutions for crop tracking, yield prediction, resource

DOI: 10.4018/979-8-2600-2383-9.ch005

© 2027, IGI Global Scientific Publishing. Copying or distributing in print or electronic forms without written permission of IGI Global Scientific Publishing is prohibited. Use of this chapter to erative artificial intelligence (AI) technologies is expressly prohibited. The publisher reserves all rights to license its use for generative AI training and machine learning model development.

efficiency, and sustainability improvement. The massive growth of research resulted in a complex and uneven literature, underscoring the need to conduct systematic assessments of knowledge networks and research patterns. The chapter conducts a thorough scientometric evaluation of the innovations in AI-based agriculture in India based on the bibliographic information published in Scopus. Growth of publications, effects of citation, productivity of authors and institutions, collaboration patterns, and theme development patterns are examined to establish the prevailing research topics and interdisciplinary connections. The results indicate this assessment gives a systematic overview of the intellectual landscape, defines gaps in the existing research, and describes possible future opportunities to explore this area, aiding evidence-based policy-making and strategic innovation in sustainable agriculture.

INTRODUCTION

Agriculture an Indian economic pillar since it supports the livelihood of a large percentage of the population, and it also leads to food security and economic growth in the country (Saxena et al., 2023; Jayadatta, 2024). The industry marked by an uneven distribution of cropping, small and big farms, and dissimilar weather conditions, which in general makes it a problem of maintaining productivity and efficient use of resources (Lokeshwari et al., 2024). The increased development of the agricultural system due to the rapid population growth, the inconsistency of the climate, and the growth in the requirements of quality food have led to the emergence of the necessity of innovative methods to maximize production and reduce the harm to the environment (Mishra et al., 2023). The conventional farming methods that are mostly based on human observations and experience are not enough to handle these complexities on a large-scale basis. The adoption of modern technologies has become fundamental in solving the problems that involve irrigation management, nutrient optimization, pest and disease management, and crop surveillance (Raman et al., 2024; Goel et al., 2024). New digital tools offer the ability to collect, process, and analyze masses of heterogeneous data in time and make decisions and implement timely interventions (Boutracheh et al., 2024a; Jena et al., 2024). In this regard, the idea of applying Artificial Intelligence (AI) to agriculture offers a chance to optimize productivity, sustainability, and resilience across agricultural systems and change traditional operations into data-driven and intelligent systems (Akhavan et al., 2022; Mishra et al., 2023).

The artificial intelligence provides computational procedures that can reproduce patterns in effects of complicated datasets, develop predictive models, and present actionable information to make decisions in agriculture (Subasi et al., 2024; Naeem et al., 2024). The machine learning, deep learning, and hybrid AI methods have gained popularity in a short time period because of their ability to process nonlinear, high-dimensional, and dynamic agricultural measurements (Wu, 2023; Fernberg, 2024). It can be applied in crop yield prediction, detecting disease and pests, assessing soil nutrients, water management, and climate prediction (Saxena et al., 2023; Lokeshwari et al., 2024). Artificial intelligence has helped in health diagnostics, optimization of productivity, and tracking of resources in allied industries such as livestock monitoring and fisheries and forestry monitoring (Arora et al., 2024; Boutracheh et al., 2024a). The usage of complementary technologies, including remote sensors, Internet of Things (IoT) devices, and cloud computing, provides the opportunity to monitor continuously, perform real-time analytics, and make predictions of intervention (Raman et al., 2024; Skubis et al., 2024). Through these technological innovations, precision agriculture systems enable sensor data to combine with geospatial data and historical trends and are more efficient in their work, contribute to sustainable resource exploitation, and minimize the environmental impact (Mishra et al., 2023; Jena et al., 2024).

The academic performance of the AI-based agricultural investigations has increased manifold in the last ten years, which indicates the growing attention to digital transformation (Jayadatta, 2024; Lokeshwari et al., 2024). The bibliometric data show that the field dominated by publications in crop-centric areas, with other areas being livestock, fisheries, forestry, and agri-supply chain management (Raman et al., 2024; Naeem et al., 2024). Academic institutions, national research organizations, and technology-driven enterprises have contributed to the prominence of Indian research, collaborating to find solutions to the local and regional problems (Boutracheh et al., 2024a; Saxena et al., 2023). The joint work of engineers, data scientists, agronomists, and environmental professionals has helped to create new AI applications (Mishra et al., 2023; Jayadatta, 2024). Nevertheless, the literature still disjointed, and little synthesis of knowledge structures, thematic development, and new research fronts takes place. Scientometric analysis gives a systematic method to measure the trends in publications, track the patterns of collaboration, and demonstrate the direction of the development of research in various agricultural sectors (Raman et al., 2024; Wu, 2023).

The visualization of research fronts in the field of AI-based agricultural research helps discover the new priority and interdisciplinary connection (Subasi et al., 2024; Naeem et al., 2024). Thematical clustering and keyword co-occurrence analysis demonstrate that there a transition to deep learning, hybrid intelligence, and explainable AI models, as the agricultural problems are becoming more complex (Boutracheh et al., 2024b; Jena et al., 2024). The practical application and potential innovation of AI in climate-smart agriculture, precision irrigation, and automated detection of diseases have been proven (Saxena et al., 2023; Lokeshwari et al., 2024). Thematic evolution also reveals that the agricultural system interconnected and has to consider allied sectors (dairy, fisheries, forestry, and agri-supply chains) in research, which increasingly being integrated (Mishra et al., 2023; Raman et al., 2024). These research trends have been changing, and it important to understand the trends to be able to align technological interventions to meet sustainability targets, respond to regional diversity, and provide equitable benefits to diverse farming communities.

The intersection of AI, data analytics, and agricultural sciences a radical solution to sustainable food systems and robust farming procedures (Subasi et al., 2024; Naeem et al., 2024). The systematic analysis of the research trends and knowledge structures makes it possible to identify the unexplored areas, the high-impact contributions, and innovative collaboration networks (Boutracheh et al., 2024b; Saxena et al., 2023). Bibliometric analysis insights are being used to help direct future efforts in precision agriculture and application in allied sectors and to help solve problems interdisciplinarily, offering a roadmap to future academic and real-world improvements (Lokeshwari et al., 2024; Wu, 2023). Bibliographic datasets that are indexed in Scopus, along with scientometric tools, provide a formal system of measuring the development of publications, citation patterns, clusters of research, and new fronts in the field of Indian agriculture (Raman et al., 2024; Mishra et al., 2023). By bringing about a wholesome comprehension of the intellectual and thematic landscape, the stakeholders can streamline interventions, maximize resource distribution, and enhance the pace of AI-driven solutions adoption that will support sustainable and technologically advanced agricultural systems that would help India to address current and future food security and economic goals in the country (Jayadatta, 2024; Akhavan et al., 2022).

Although artificial intelligence applications in agriculture have expanded significantly, there is still limited understanding of the global research landscape, collaboration networks, and thematic evolution of this field. Most existing studies focus on specific AI applications such as crop monitoring or yield prediction, while fewer studies examine the overall knowledge structure and research development patterns. Furthermore, limited scientometric studies analyze how AI research in agriculture has evolved within the Indian context and how interdisciplinary collaborations contribute to technological innovation. This chapter therefore aims to provide a comprehensive scientometric analysis of AI-based agricultural research to identify research trends, influential contributions, and emerging thematic areas.

Despite the rapid growth of AI applications in agriculture, there remains a lack of comprehensive scientometric studies that systematically map the intellectual structure, collaboration networks, and thematic evolution of AI-driven agricultural research in the Indian context. Existing studies primarily focus on specific applications such as crop prediction or disease detection, with limited attention to the broader research landscape.

To address this gap, this chapter makes the following contributions:

1. Provides a large-scale scientometric analysis of AI-driven agricultural research using Scopus-indexed publications.
2. Identifies key research trends, influential authors, institutions, and collaboration networks.
3. Maps thematic clusters and emerging research fronts using keyword co-occurrence and citation analysis.
4. Highlights research gaps and future directions for AI adoption in agriculture and allied sectors.

This structured analysis offers a holistic understanding of the evolution of AI in Indian agriculture and supports strategic research and policy decisions.

1.1 Related Studies on Artificial Intelligence in Agriculture

Existing literature on artificial intelligence in agriculture demonstrates significant advancements in predictive modeling, image-based disease detection, and precision farming. For instance, machine learning models such as Random Forest and Support Vector Machines have been widely applied for crop yield prediction and soil analysis, while deep learning approaches—particularly convolutional neural networks—have shown high accuracy in plant disease detection using image datasets.

Several studies have emphasized the integration of AI with IoT-based sensing systems to enable real-time monitoring and decision-making. These approaches support precision agriculture by optimizing irrigation, fertilizer application, and pest control strategies. Furthermore, remote sensing technologies combined with AI have been extensively used for large-scale crop monitoring and environmental assessment.

A critical limitation of existing studies is their fragmented nature, focusing on isolated applications rather than providing a holistic understanding of research development. Limited attention has been given to analyzing collaboration networks, thematic evolution, and knowledge structures within AI-driven agricultural research—particularly in the Indian context.

There is a clear need for a comprehensive scientometric analysis that integrates bibliometric indicators, network analysis, and thematic mapping to provide a structured overview of research trends and emerging directions in this domain..

2. CONCEPTUAL FOUNDATIONS OF AI IN INDIAN AGRICULTURE

2.1 Evolution of Artificial Intelligence in Agricultural Systems

The development of artificial intelligence systems in agricultural systems connected to a stepwise shift from experience-based farming to a more data-based and algorithm-assisted practice of Indian agriculture. Initial computational work done on rule-based expert system-based programs to encode agronomic knowledge; such systems were to select crops, schedule fertilizer, and provide advisory services related to pests. With an increase in digital record keeping and meteorological databases, statistical learning methods became popular in yield estimation and risk assessment that were related to weather. The growth of remote sensing platforms, sensor-based field surveillance, and mobile-based data collection made the use of machine learning methods that understood patterns in massive agricultural datasets faster. These advances signified a transition towards intelligent systems with the ability to learn adaptively and model predictively and provide decision support based on the context and in various agro-climatic conditions.

Figure 1. Evolution of Artificial Intelligence in Agricultural Systems

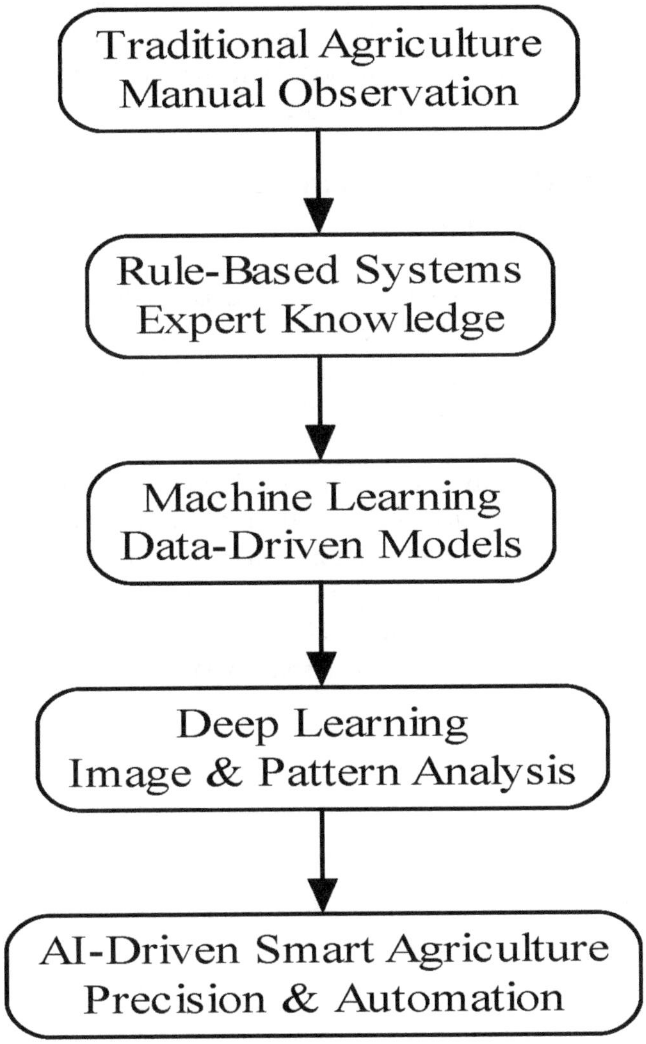

Further changes in the agricultural intelligence system occurred with progress in deep learning systems and high-performance computing that facilitated automated extraction of features of a satellite image, hyperspectral data, or real-time sensor feeds. Convolutional and recurrent learning models were used to solve complex problems like crop disease detection, detection of crop growth stage, optimization of irrigation, and evaluation of livestock behaviors. Connection to the Internet of Things environments and cloud-based analytics allowed maintaining the flow of data and scalable implementation of smart agricultural solutions. In the Indian context, this development corresponded with the national digital agriculture policies and the increased focus on precision farming and innovation based on sustainability. The further development of artificial intelligence within agricultural systems will therefore be a paradigm shift that supports the modern considerations and the up-and-coming bodies of knowledge in AI-based agriculture and other related fields.

2.2 Role of Data-Driven Intelligence in Modern Farming Practices

Intelligence based on data has become a critical part of the contemporary agricultural production and changing the historical decision-making into the predictive and optimized management concepts. The combination of the advanced computational techniques and real-time collection of data will allow us to properly monitor the health and the condition of the soil, the water consumption, and the variations in climatic conditions. Predictive models and machine learning algorithms are used to analyze past and present data to generate the patterns that will guide the planting times, irrigation, fertilization plans, and pest management decisions. This kind of strategy enables engagement in the operations in a resource-efficient way, and at the same time, the quality and the crop yield of the products are increased, aligning the aim of agriculture with the goals of sustainable use and economic feasibility.

Figure 2. Role of Data-Driven Intelligence in Modern Farming Practices

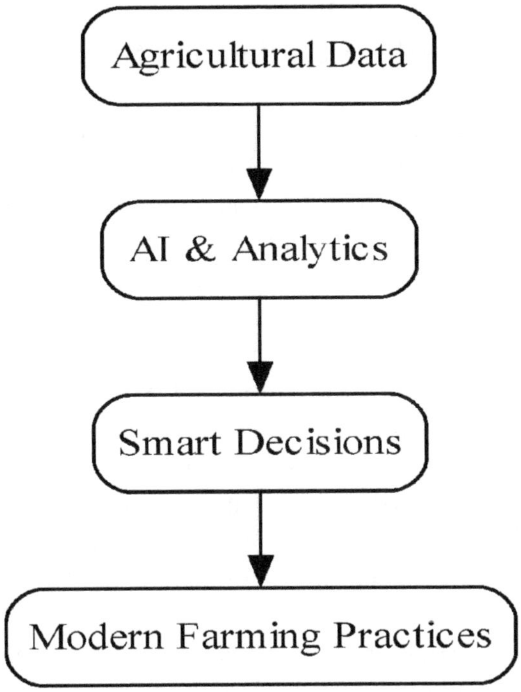

Data-driven intelligence not only applied in managing individual farms but also in larger agricultural systems, such as in optimizing supply chains, making market predictions, and addressing risks. Sensor networks, satellite images, and mobile-based platforms produce massive amounts of data that capture spatial and time variation in different agro-climatic regions. Analytical models use such datasets to predict possible stressors, optimize logistics, and aid in the strategic decision-making in the farms and the region. Higher data visualization tools also support the interpretation of complicated data points, which allows the stakeholders to make effective decisions that limit operational risks and environmental effects.

Besides enhancing productivity, evidence-based intelligence will enhance resilience to the climate variability and resource challenges. The predictive information based on big data helps in the early identification of pest infestations, disease outbreaks, and water stress to implement timely measures. This because historical trends can be integrated with real-time monitoring to create adaptive practices of management in response to dynamic conditions. The intersection of data analytics with new technologies, including Internet of Things (IoT) devices, cloud computing, and artificial intelligence platforms, creates a knowledge-based agricultural ecosystem with the ability to maintain high performance and environmental sustainability and economic stability in the farming environment.

2.3 Integration of AI with Precision and Smart Agriculture

The combination of artificial intelligence and precision agriculture a paradigm shift in the operation of crop production systems. Precision agriculture focuses on monitoring and management of crops, soil, and environmental conditions at a specific location in order to maximize the use of the resources and to improve productivity. Artificial intelligence part of this strategy, as it allows analyzing data in real-time, predicting, and making decisions dynamically. Machine learning algorithms are run on large amounts of multispectral imagery, sensor data, and past agronomic data to detect spatial and temporal variations in the health of crops, soil moisture, and nutrient content. Deep learning systems can be used to automatically identify diseases, pest infestations, and stress, giving viable input that can be used to inform specific actions. Intelligent predictive irrigation, fertilization, and crop scheduling systems can be used to allocate inputs efficiently, avoiding te and minimizing operational expenses and enhancing the quality and reliability of yields in variable farm environment conditions.

Smart agriculture an evolution of precision farming that brings all three elements of connectivity, automation, and decision-support systems into the agricultural ecosystem. The AI serves as the central analytic engine of the smart agriculture systems, combining information streams of the Internet of Things devices, drones, autonomous-controlled machines, and satellite images. The specialized AI systems are able to do round-the-clock monitoring and anomaly detection, which allows the adaptive control of irrigation networks, nutrient provisioning, and pest management measures. Predictive analytics are used to predict the impact of climate and crop performance, which can be used in resilient and sustainable farming practices.

2.4 AI Ecosystem for Indian Agriculture and Allied Sectors

The AI has been a key element of the Indian agricultural technology environment that incorporates several data acquisition, processing, and decision support layers. The ecosystem will include a set of sensing apparatus, including satellite imagery, IoT-based field sensors, drones, and mobile data gathering systems that will deliver real-time data on the health of the soil, the development of crops, the presence of water, and the environment around. Computational models can use this heterogeneous information to come up with insights that inform precision farming practices, optimize the use of inputs, and predict the possible threats to crop yield and quality. Open systems and platforms allow smooth exchange of information between farms and cloud-based analytical systems, forming a continuous feedback mechanism, which contributes to resource efficiency and productivity.

The application of AI in related industries expands the ecosystem to the crop production. Image and sensor data are used by livestock monitoring systems to measure the health of animals, monitor their movement patterns, and optimize the feed schedule. Predictive modeling used in fisheries management in stock assessment, monitoring water quality, and detection of disease outbreaks. Agroforestry and forestry projects use AI-based spatial analysis to provide biodiversity, deforestation, and climate change adaptation solutions. Gathering AI algorithms with mobile applications and cloud infrastructure can help to optimize supply chains and create agri-fintech solutions, which can optimize market access and post-harvest losses and help smallholder farmers manage risks. These technologies working in unison create an integrated digital platform, which has various elements employed to enhance sustainable and efficient farming.

Figure 3. AI Ecosystem for Indian Agriculture and Allied Sectors

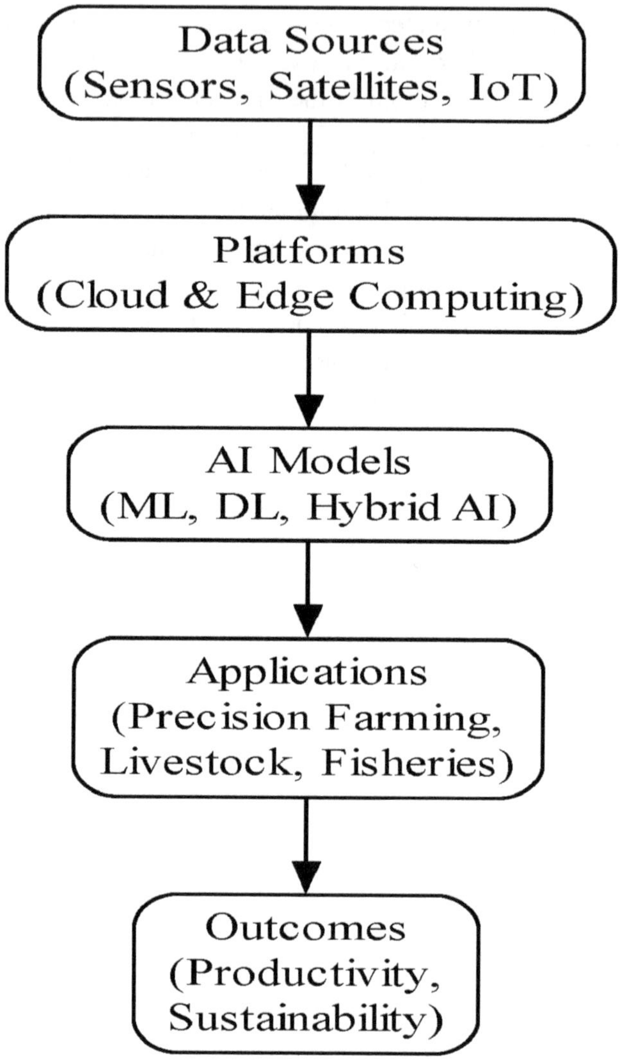

Indian agriculture has been transformed by the interplay of policy environments, research institutions, technology vendors, and primary users to evolve a new AI ecosystem. The promotion of AI-based solutions in various agro-climatic regions has been growing rapidly due to the national efforts put in place to promote digital agriculture, smart farming, and precision technologies. Academic institutions, government research organizations, and implementation in the private business form collaborative networks that help to share knowledge and develop innovations and implement AI applications on a large scale.

2.5 Technological Drivers Enabling AI Adoption in Agriculture

The development of digital sensing technologies, computational platforms, and data analytics structures has contributed to the adoption of artificial intelligence in agriculture. The high-resolution and multi-temporal information about crop health, soil conditions, and water availability created by remote sensing systems, such as satellite pictures and unmanned aerial vehicles. Field sensor networks can be used to measure environmental and physiological parameters, which form a continuous stream of real-time data that can be used to support predictive and prescriptive agricultural interventions. With the integration of such data sources with the cloud-based storage and processing systems, it possible to analyze such large volumes of data, whereas high-performance computing systems will allow training complex machine learning and deep learning models that are specific to agro-climatic conditions. The relationship between the acquisition, storage, and processing technologies of data constitutes the basis of intelligent decision-making in contemporary farming.

The further evolution of communication infrastructure, edge computing, and Internet of Things (IoT) ecosystems also supports the implementation of AI-enabled solutions. IoT-enabled devices enable the automated check-up of soil moisture, soil temperature, nutrient availability, and pest activity, thus eliminating the need to manually observe and control them. Edge computing systems minimize latency through processing the data close to the source to help in real-time reactions to the ever-changing agricultural conditions. The high-speed development and adjustment of intelligent models of smallholder and large-scale farming systems can be done by using open-source software libraries, common data formats, and scalable AI algorithms.

2.6 Challenges and Barriers to AI Adoption in Agriculture

Although artificial intelligence technologies offer significant opportunities for improving agricultural productivity, several barriers continue to limit their large-scale adoption. One major challenge is the limited digital infrastructure in rural regions, where access to reliable internet connectivity and advanced technological tools remains constrained.

In addition, high implementation costs associated with sensors, drones, and AI-based decision-support systems may prevent smallholder farmers from adopting smart farming technologies. Another challenge involves the availability and quality of agricultural data, as machine learning models require large datasets that are often fragmented or inconsistent across agricultural regions.

Farmers may face technical skill gaps, making it difficult to interpret AI-based recommendations without appropriate training and support. Addressing these challenges requires coordinated efforts from governments, research institutions, and technology providers to develop inclusive digital agriculture ecosystems.

3. DATA SOURCES AND ANALYTICAL FRAMEWORK FOR SCIENTOMETRIC EVALUATION

3.1 Scopus Database as a Knowledge Source for Agricultural AI Research

The Scopus database an all-inclusive and reliable source of bibliographic data; it the place where one can get peer-reviewed journal articles, conference proceedings, and review papers in various fields. It has a broad scope of science, engineering, computer science, and agricultural research, which makes it possible to find high-quality publications applicable in the creation of artificial intelligence in the agricultural sector. Detailed metadata, including authorship, institutional affiliations, the year of publication, keywords, abstracts, and citation counts, are available in indexed records and allow for conducting robust quantitative analysis as well as correctly mapping knowledge structures. The broad area of disciplinary focus and selective content of the database enable comparative assessment of subdisciplines, suggesting a clear insight into research trends, focus areas of thematic focus, and intellectual contributions in the field of Indian agricultural AI research.

The precision of the Scopus indexing system, enabling the formulation of a query and extraction of the data in specific ways, enables the retrieval of the publications that would outline the innovations in precision agriculture, prediction of the crop yield, detection of different diseases, livestock management, and related topics. The exported datasets can undergo scientometric processing such as author name normalization, institution standardization, and keyword harmonization. The combination of bibliometric measures, including the frequency of publication, the impact of citation, and co-authorship networks, gives information about the level of productivity and impact of researchers and organizations.

3.2 Bibliographic Data Collection and Preprocessing Strategies

The acquisition of bibliographic data to be used in scientometric evaluation based on well-designed search strategies that will aid in covering the entire scope of literature. Scopus the main source of data because it provides ample coverage of peer-reviewed articles, conference proceedings, and review articles on a wide variety of disciplines, such as agricultural sciences, computer science, and engineering. The search queries were formulated to include publications on the topic of artificial intelligence, machine learning, deep learning, precision agriculture, smart farming, and other related fields in the Indian context by the use of keywords. The search narrowed down to English-language documents, and irrelevant publications were eliminated with the help of Boolean operators, truncation, and field-specific filters, resulting in a high-quality dataset that provides a representative picture of the research situation.

The bibliographic dataset preprocessed by cleanup, normalization, and standardization to improve the accuracy of the further analysis. Duplicates were detected and deleted, author names and institutional affiliations were normalized to eliminate variations and inconsistencies, and keywords were standardized by combining synonyms and similar words. The records that did not contain the necessary metadata were omitted so as to retain analytical rigor. Such systematic methodology guarantees dependability in scientometric measures like the number of publications, citation analysis, co-authorship mapping, and keyword co-occurrence networks, which constitute a solid platform for discovering research trends and thematic organization and developing fronts in the AI-driven agricultural research.

The bibliometric dataset used in this study obtained from the Scopus database, which provides comprehensive coverage of peer-reviewed scientific publications across multiple disciplines. A structured search query combining keywords such as "Artificial Intelligence," "Machine Learning," "Deep Learning," and "Agriculture" used to retrieve relevant publications related to agricultural AI research in India.

The search limited to English-language journal articles and conference papers published between 2010 and 2024. After data cleaning and removal of duplicates, the final dataset consisted of 1,245 documents.. The dataset analyzed using scientometric tools such as VOSviewer and Bibliometrix, which enabled visualization of publication trends, citation networks, co-authorship patterns, and keyword co-occurrence clusters.

3.3 Scientometric Indicators and Performance Metrics

Scientometric indicators give a numerical basis for assessing the development, influence, and organization of research in a given field. The number of publications, the number of citations, and the h-index are used as the main metrics for evaluating the productivity of research and the scholarly influence of authors, institutions, and countries. Citation-based indicators capture the credit and significance of the contributions of individual researchers among the overall academic community, whereas normalized indicators consider the difference in the years of publication and the area of study and thus allow comparison of the performance of research. Time series analysis of these indicators shows how the field has evolved through time, including patterns of emergence of research, maturity, and changes in research areas of focus.

Network metrics are an addition to the traditional metrics of performance, as they can be used to chart the links and partnerships within the research ecosystem. Co-authorship networks are used to show the trends of scientific collaboration between researchers and institutions, whereas co-citation analyses are used to show clusters of related studies and seminal works that can be recognized as the intellectual backbone of a discipline. Bibliographic coupling enables one to track thematic relationships between the publications and identify a sphere of concentrated research activity and areas of emergence.

Figure 4. Scientometric Indicators and Performance Metrics

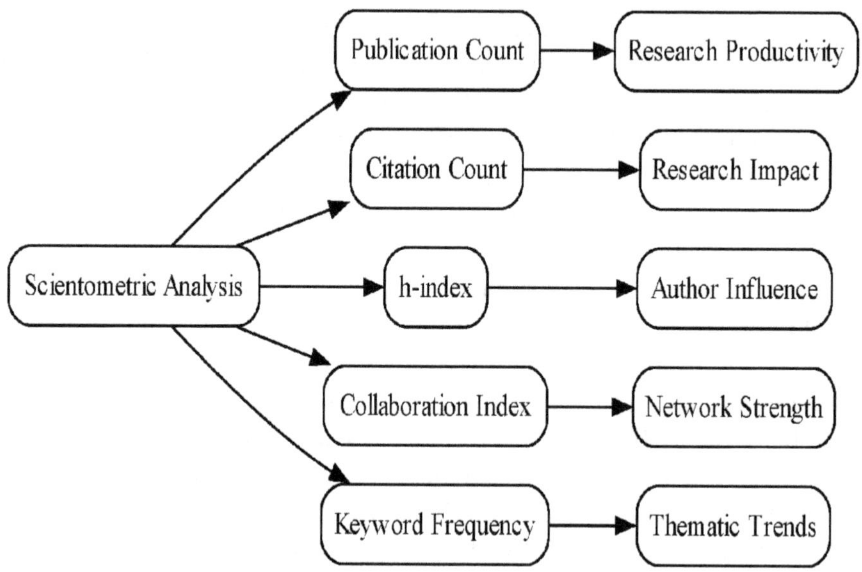

Precise scientometric measures of performance are used to determine research fronts, future trends, and scholarly contributions in a field. The citation burst, author centrality, and betweenness of co-authorship or co-citation networks are measures that reflect influential publications and key researchers who influence the process of knowledge building. This density- and centrality-based thematic mapping can be used to visually distinguish between well-established clusters of research and nascent and developing topics in terms of their strategic outlook on knowledge gaps and as a source of innovation.

3.4 Science Mapping Techniques for Knowledge Structure Analysis

The science mapping methods give a systematic method of visualizing and analyzing the intellectual architecture of a research field. Co-authorship networks, co-citation networks, and keyword co-occurrence networks are all forms of bibliometric networks that demonstrate collaboration patterns, diffusion of knowledge, and the concentration of topics within a field. Co-authorship networks emphasize how the scholarly input shared between researchers and institutions, which draws on the levels of collaboration and power centers. Co-citation analysis reveals the presence of foundational and emerging trends that are determined by the frequency of co-citation of the publications, whereas keyword co-occurrence networks discover thematic groups, defining the prevailing and interdisciplinary research agendas. The visualization of these networks makes it possible to find central nodes, clusters, and areas of research under development, as well as learn about the structural and conceptual development of the domain.

Figure 5. Science Mapping Techniques for Knowledge Structure Analysis

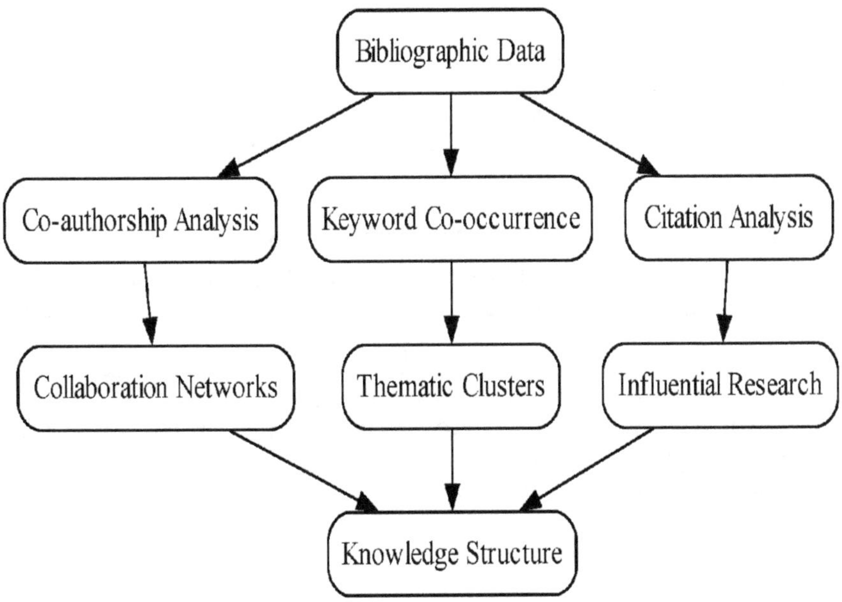

More complex techniques of network analysis go beyond the traditional bibliometrics and are used to measure ties as well as community formations in the scientific literature. The methods of modularity clustering, multidimensional scaling, and density-based visualization lead to the identification of research clusters that are closely related and the identification of new fronts of knowledge. Temporal mapping of networks depicts changes in research agenda, advancements in methodologies, and adoption of new technologies in the discipline.

3.5 Tools and Software for Bibliometric Visualization

The bibliometric visualization tools present a crucial tool for analyzing the massive amounts of scholarly data and identifying patterns of structural research areas. They can be used to generate knowledge maps, co-authorship networks, citation networks, and keyword co-occurrence structures, giving raw bibliographic data understandable visualizations. The ability to single out key authors, major institutions, journals, and new findings clusters enables one to formulate an overall picture of the thought arena of a scientific area. The visualization frameworks allow quantitative and qualitative interpretations and enable identifying the research patterns, networks of collaboration, and theme development over time.

Figure 6. Tools and Software for Bibliometric Visualization

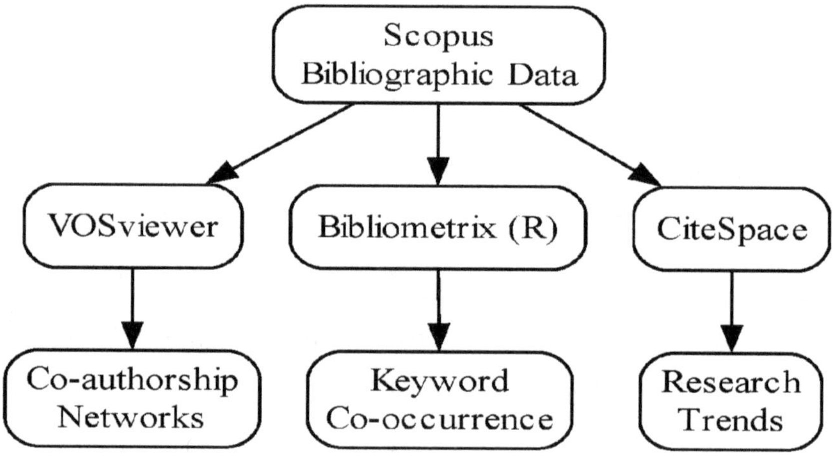

Two of the most popular bibliometric visualization tools include VOSviewer and Bibliometrix, both of which are very strong with the consideration of their capabilities. VOSviewer allows network-based analyses and provides interactive maps of co-authorship, co-citation, and keyword co-occurrence studies to allow the exploration of relationships between research in high resolution. Bibliometrix written in the R programming language and enables bibliographic data to be preprocessed in large amounts, statistically analyzed, and represented graphically, giving flexibility in analysis of the trend of publications, the impact of citation, and thematic clusters.

The use of visualization tools will also assist in decision-making when planning the research project and policy formulation by showing the untapped spaces of research, the possible partnering opportunities, and the future movement of the upcoming research directions. Network diagrams, density maps, and cluster representations facilitate better understanding of more complicated bibliometric relations and emphasize the patterns that could be hidden in text or tabular data. The power to deal with visual parameters, to filter nodes, and to browse changing relationships enables one to make a sensitive evaluation of scientific output, research impact, and thematic development. Using these tools, therefore, constitutes a part of the scientometric evaluation and giving a systematic and enlightening presentation of the AI-enabled innovations in Indian agriculture and other related industries.

3.6 Limitations of the Scientometric Approach

While scientometric analysis provides valuable insights into research trends and knowledge structures, several limitations should be acknowledged. First, the analysis relies primarily on publications indexed in the Scopus database and may therefore exclude relevant studies published in regional or non-indexed journals. Second, citation counts may not fully capture the practical impact of research, as recently published studies may receive fewer citations due to limited exposure. Finally, bibliometric indicators emphasize quantitative publication patterns and may not fully reflect the qualitative significance of individual research contributions.

3.7 Analytical Workflow

The scientometric analysis follows a structured workflow consisting of data collection, preprocessing, analysis, and visualization. Initially, bibliographic data were extracted from Scopus using predefined search queries. The dataset then cleaned and normalized to remove inconsistencies in author names, affiliations, and keywords. Subsequently, bibliometric indicators such as publication count, citation analysis, and h-index were computed. Network analysis techniques including co-authorship, co-citation, and keyword co-occurrence were applied using VOSviewer. Finally, thematic mapping and trend analysis were performed to identify emerging research fronts and knowledge structures.

4. PUBLICATION GROWTH AND CITATION TREND ANALYSIS

4.1 Temporal Evolution of AI-Driven Agricultural Research in India

The history of AI-based studies in the field of Indian agriculture shows that the development of the field has been growing steadily in the first part of the 2010s, with an apparent sharp increase in the second half of the decade. The early research mostly on exploratory uses of machine learning in estimating crop yield, detecting diseases, or analyzing soils, with small datasets and local studies. The growing access to satellite images of high resolution, field sensors, and historic agricultural information helped to target more advanced modeling methods and implement them in a wide range of agro-climate areas. Initial research stages preconditioned the interdisciplinary integration, the combination of agricultural sciences, computer science, and engineering principles to solve complex problems in the resource management and optimization of production.

Figure 7. Temporal Evolution of AI-Driven Agricultural Research in India

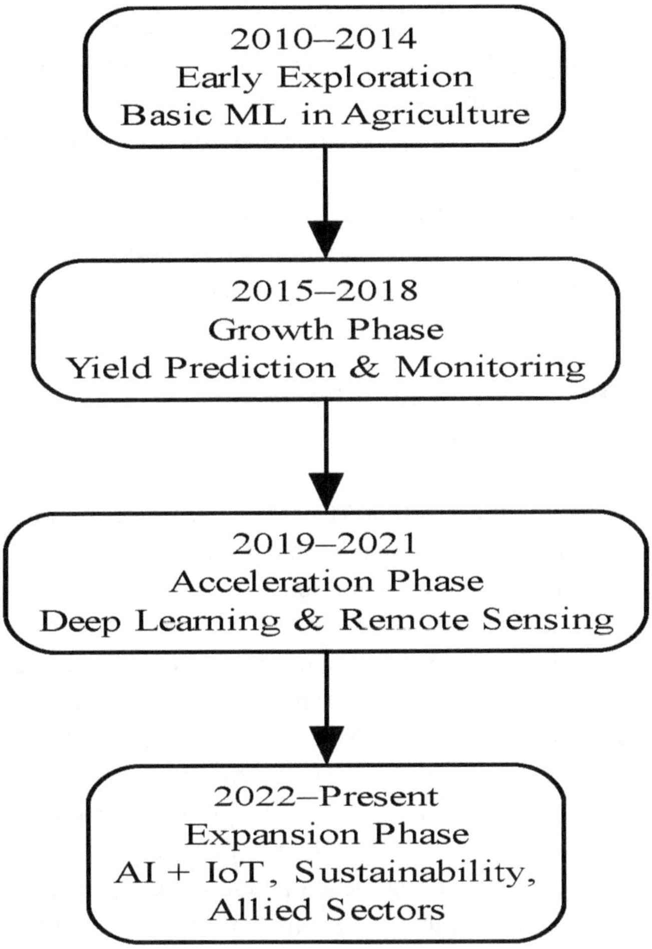

The year after 2016 has seen an accelerated growth in articles that represent the implementation of deep learning, hybrid artificial intelligence models, and predictive analytics in Indian agriculture. The national programs of digital agriculture, precision farming, and technology-based rural development helped to increase the volume of research and the level of global cooperation. There also an increasing trend of acknowledgement of the Indian contributions as high-impact publications have appeared in the journals concerned with the fields of agricultural informatics, artificial intelligence, and environmental sustainability.

The observed growth in publications after the mid-2010s can be attributed to several technological and policy developments. The increasing availability of satellite imagery, agricultural sensor networks, and open agricultural datasets has enabled researchers to apply advanced machine learning and deep learning models to agricultural problems. In addition, government initiatives promoting digital agriculture and precision farming have encouraged interdisciplinary research collaborations between computer scientists, agronomists, and environmental scientists. These factors have significantly accelerated the adoption of artificial intelligence techniques in agricultural research and contributed to the rapid growth of scientific publications in this domain.

The observed exponential growth after 2016 indicates a technological inflection point driven by increased data availability and computational advancements. This trend suggests a transition from experimental research to large-scale implementation, highlighting the maturity of AI applications in agriculture.

4.2 Annual Publication Growth Patterns and Phases of Development

Comparison of publication trends in AI-based agricultural research shows a slow but steady rise in annual publication in the last ten years, indicating the development of increased interest in the use of computational solutions to agricultural problems. The initial work on basic machine learning methods done, and there little publication output because of the lack of data and computing power. The era characterized by investigatory works on the estimation of crop yields, pests, and optimization of irrigation with simple predictive models. Work done at this stage mainly by academic establishments and specialized agricultural research centers that formed the first base of knowledge on the application of AI in Indian agriculture and the related fields.

Another significant acceleration in the rate of publication has been observed in the mid-to-late 2010s, along with the development of sensor technologies, access to large-scale agricultural data, and the expansion of access to high-performance computing tools. This period characterized by the increase in the interdisciplinary research that combines machine learning, remote sensing, and IoT platforms, leading to more sophisticated and practical works. Networks of research increased, with the involvement of several institutions, both domestic and foreign, resulting in superior citation impact as well as worldwide distribution of study outcomes.

The last several years have been the period of the blistering development associated with the acceptance of more sophisticated deep learning architectures, hybrid intelligence, and explainable AI frameworks. The trends of publication output are exponential, indicating a stronger interest of scholars and policy-based emphasis on digital agriculture programs. According to the citation analysis, the works involving high citation have frequently dealt with predictive modeling, precision agriculture, and climate-smart farming intervention, which highlights the role of this particular research in shaping the further investigations.

The three phases of research development identified in this analysis exploratory, expansion, and technological maturity indicate the gradual integration of artificial intelligence technologies into agricultural systems. Early research primarily focused on experimental applications of machine learning algorithms, while later studies emphasize scalable solutions and real-time agricultural monitoring systems. This transition reflects the broader digital transformation occurring in agriculture, where data-driven technologies are increasingly used to improve productivity, optimize resource use, and support sustainable farming practices.

The rapid increase in publications after the mid-2010s reflects the growing integration of digital technologies within agricultural research. Improvements in remote sensing systems, cloud computing platforms, and agricultural data collection have enabled researchers to apply advanced machine learning and deep learning techniques to large agricultural datasets. In addition, government initiatives promoting digital agriculture and smart farming have encouraged interdisciplinary collaborations between agricultural scientists, computer engineers, and environmental researchers.

4.3 Citation Impact and Research Visibility Analysis

Citation impact used as a quantitative concept to determine the value and recognition of research publications in a research community. A semantic study of the number of citations will reveal a list of seminal literature, leading authors, and the best journals that have significantly influenced the evolution of AI-driven agricultural research in India. Patterns of citation contribute to the understanding of the relative significance of particular topics, methodologies, and the area of application over the years, showing the evolution of knowledge dissemination over crop management, precision agriculture, related industries, and the new research fields. The citation analysis also indicates the level of international interest and involvement with the research output in the country and the international exposure and academic coverage of the research done in the nation.

Research visibility goes further than the metrics of numerical citations because it encompasses the existence and the availability of publications in digital databases, open-access journals, and interdisciplinary journals. High visibility increases the spread of methodological progress, experimental findings, and field-specific innovations, which allow sharing of knowledge among the researchers, policymakers, and industry practitioners. The citation analysis in combination with the visibility assessment provides an in-depth view of scholarly influence to make strategic decisions in the priorities of research priorities, collaboration, and resource distribution.

The increasing citation impact of AI-based agricultural studies also reflects the growing recognition of digital technologies as critical tools for addressing agricultural challenges such as climate variability, food security, and resource management.

Highly cited publications often introduce novel algorithms, large-scale datasets, or integrated decision-support systems that provide practical solutions for farmers and policymakers. As a result, citation patterns not only indicate academic influence but also highlight the practical relevance of AI applications in modern agricultural systems.

4.4 Highly Cited Publications and Influential Contributions

The review of highly cited articles gives an understanding of the foundation and transformational roles that have influenced AI-based agricultural studies in India. These magazines usually present new methodology, new applications, or large datasets that make substantial contributions to knowledge in crop management, precision farming, and related fields. Citation metrics demonstrate how this work affected the way further research done, pointing to the works that have become standards of how the algorithms were developed and data incorporated and the implementation of AI techniques at the field level. The significance of the research in the local and international scientific circles also demonstrated by the number of citations, which serve as the measure of knowledge diffusion outside of local and disciplinary localities.

The contributions made by influential people are usually methodologically sound, interdisciplinary, and practical. Research on crop yield prediction with machine learning, image-based deep learning models to detect diseases and pests, and optimization of irrigation and nutrient management has developed a common set of practices that develop subsequent studies.

Mapping the research frontiers and new thematic areas also made possible through the identification of highly cited works. Citation networks reveal groups of studies that are especially influential and expose the ties of collaboration of authors and institutions that lead to innovation. Citation pattern analysis demonstrates knowledge gaps, trends in methodologies, and the current trend in the development of AI applications that are not based on experiments but on a real-world application. The insights are useful in the strategic formulation of research agendas in the future to direct future research to high-impact areas, facilitate cross-disciplinary research, and enhance the overall contribution of AI-driven innovations to the field of Indian agriculture and related industries.

4.5 Growth Comparison Across Agriculture and Allied Sectors

Development of AI-based studies in the agricultural field shows different trends by different subdomains, due to variations in data, technology, and priorities of the sector. The highest number of publications has been drawn to crop production and precision farming due to the need to predict yield, control diseases and pests, and optimize various irrigation and nutrient applications. Other related industries, such as livestock, fisheries, and forestry, have realized relatively relative growth in research, which tends to center on monitoring, resource management, and resource efficiency. The trends in publications have shown that disciplines that have a longer history, have established surveillance, and whose output measures indicate a quantifiable output have faster growth in research output, whilst disciplines with a difficult ecological dynamic and sparse digitized information exhibit slower growth.

Figure 8. Growth Comparison Across Agriculture and Allied Sectors

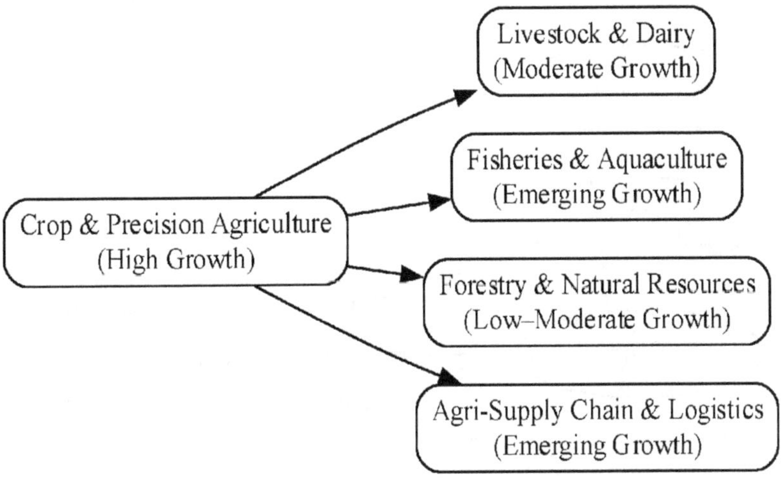

Analysis of citations has shown that the research on high-impact studies focused on crop-specific and precision agriculture, which has shown a strong focus on both methodology and practical applicability. Areas of livestock analytics and fisheries management have a lower citation density, although they are steadily growing, indicating that the scholarly interest and possibilities remain unexplored.

5. AUTHOR, INSTITUTION, AND COUNTRY-LEVEL PRODUCTIVITY ANALYSIS

5.1 Prolific Authors and Research Leadership in Agricultural AI

The author productivity analysis shows that there a focused group of researchers that put a lot of effort into developing and implementing AI in Indian agriculture.

Top researchers have a steady stream of scholarly publications in journals and conferences and have published in the areas of crop modeling, disease detection, precision irrigation, and similar sector applications, including livestock monitoring and fisheries management. These authors are also marked by citation patterns, which suggest that their work does not only present some new methodological ideas but also offers some practical frameworks that further research has implemented. The interdisciplinary integration that can occur with the help of collaborative practices amongst these researchers related to connecting computer science, agricultural engineering, and environmental sciences to solve complex agricultural problems.

Figure 9. Prolific Authors and Research Leadership in Agricultural AI

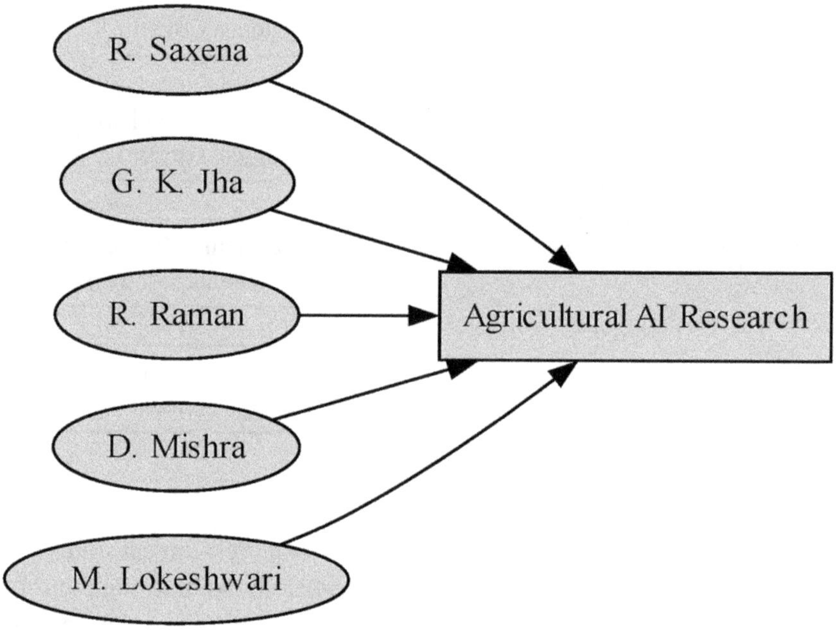

The central role of research leadership institutional affiliation, as significant knowledge production centers are represented by major universities, research institutions, and technology centers. The proportion of high-impact publications in Indian Institutes of Technology, state-level agricultural institutions, and national research centers high, which indicates the availability of computational facilities, detailed knowledge of a field, and cross-disciplinary networking of contacts.

The concentration of highly productive authors within a relatively small group of institutions indicates the presence of specialized research clusters focusing on artificial intelligence in agriculture. These research leaders often act as knowledge hubs, promoting interdisciplinary collaboration and guiding new researchers entering the field. Their work contributes to the development of standardized methodologies and encourages the wider adoption of AI technologies across different agricultural domains.

5.2 Institutional Knowledge Hubs and Research Output Distribution

Knowledge centers in the institutions are at the core of AI-based agricultural research in India and act as a locus of methodological innovation, cross-disciplinary teamwork, and technology. Most major academic centers, agricultural colleges, and national research centers play a big role in the number of publications, with a large percentage of high-impact studies being published in precision agriculture, crop yield prediction, and resource optimization. These institutions usually have dedicated research institutions and laboratories that combine computer science, agricultural engineering, and environmental sciences, allowing them to make broad studies of AI uses in crop and allied industries.

Figure 10. Institutional Knowledge Hubs and Research Output Distribution

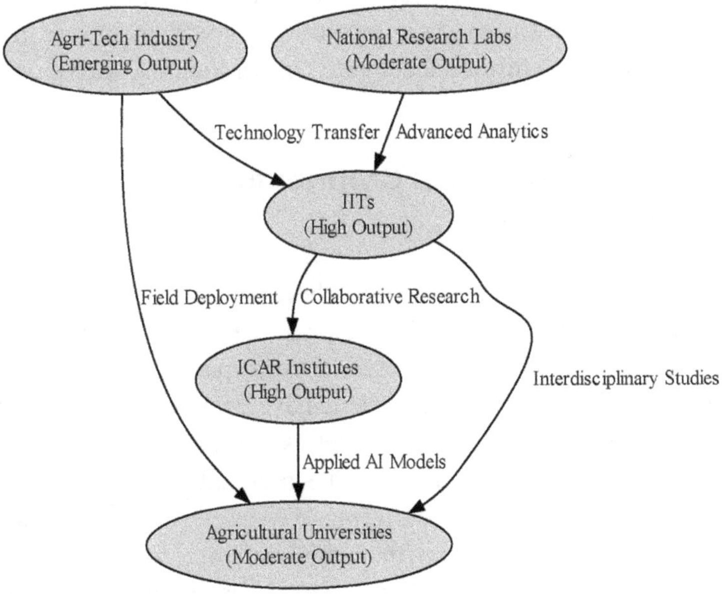

The dissemination of research output indicates the concentration of academic activity in a small number of major institutions, which are also the major sources of knowledge and also partners in multi-institutional projects. State agricultural universities and institutes of technology are characterized by high regional interactions, as they tend to work more on local crop systems, climatic conditions, and farmer-centric solutions.

Cross-country comparison suggests that Indian institutions are becoming more and more involved in the global research networks, where they are publishing together with international partners in high-impact journals and conferences as well. The institutional linkages will help in creating access to datasets as well as collaborating in experiments and publishing co-authored articles, reinforcing the presence and impact of Indian-driven AI-powered agricultural research. The geographic and organizational concentration and underrepresentation of research output by regions point to the opportunities to invest in the development of new centers, capacity building, and development of institutional partners that will bring AI innovation to less-researched agricultural and other allied areas.

The concentration of research output within a limited number of institutions indicates the emergence of specialized research hubs focusing on artificial intelligence applications in agriculture. These institutions play an important role in promoting interdisciplinary collaboration and technological innovation by integrating expertise from agricultural sciences, computer science, and environmental studies. Strengthening collaborations between academic institutions, government research centers, and industry partners could further accelerate innovation in AI-driven agricultural technologies.

5.3 National and International Contributions to Indian Agricultural AI

Through Indian agricultural AI research, there a high level of national input to the scientific outputs, with some of the major academic institutions, agricultural universities, and government research organizations making up a large percentage of the scientific outputs. Major organizations have grown to be proficient in machine learning, remote sensing, and data analytics, utilizing them to track crops and determine yields and optimize resources. The regional cooperation between the institutions in India has resulted in the sharing of datasets, computational facilities, and methodological innovation through the establishment of focal centers of research activity that support both foundational research and practical solutions throughout a wide range of agro-climatic regions.

International partnerships can supplement the domestic research activities in the form of introducing methodological soundness, access to sophisticated computational technologies, and cross-cultural insights into the AI challenges in agriculture. The collaborations with the institutions in technologically developed nations help to publish jointly, co-author the studies, and take on the knowledge exchange program, increasing the quality of the Indian research and its presence on international platforms. The trends in publication patterns point to the fact that the most frequently cited works tend to be the results of the projects of the combination of national knowledge with international cooperation, which the outcome of the synergy of intellectual and technological resources.

5.4 Research Productivity and Impact Assessment

The productivity of research in AI-driven agricultural studies evaluated, and it discovered that there a high variance among the authors, institutions, and countries adding to the body of work. Top authors are those who maintain the publication rate and have a high level of co-authors, which indicates their expertise in the domain as well as their interaction within the interdisciplinary teams. The access to highly advanced computational resources, institutional support, and involvement in cross-national research are often related to high productivity. There also an uneven distribution of institutional contributions, with the larger share of high-impact publications produced by the premier agricultural universities and technology-oriented research centers and the smaller institutions showing emerging but less frequent contributions. This distribution emphasizes the presence of scholarly power in the knowledge centers, which predetermine the general trend of AI implementation in Indian agriculture.

Analysis at the country level reveals that India has one of the highest contributions towards the AI-based agricultural research, and the international collaboration contributes to the visibility and the citation impact. Cooperation with nations that have developed computational facilities, remote sensing, and well-founded AI skills promotes the exchange of procedures, training strategies, and experimental design. The results of collaboration are likely to have a greater number of citations and a wider range of interdisciplinary applicability due to the synergy of the merging of localized knowledge of agriculture with the global technological expertise.

Impact assessment through citation gives the visibility of the impact and presence of published work within the scientific community. The most frequently cited ones tend to present new AI algorithms, new data integration systems, or even proven enhancements in crop management, livestock surveillance, and related industry uses. Citation metrics, in combination with volume of publication, provide a moderately accurate view of productivity and scholarly impact, including who are core contributors as well as influential institutions. The research impact assessment using such measures allows identifying the most active knowledge generators, directing the resource distribution, and aligning policy interventions to enhance AI-empowered agricultural research networks and to ensure technology adoption to be sustainable in various agricultural settings.

5.5 Interdisciplinary Participation in AI-Driven Agricultural Studies

The research in artificial intelligence applied to agriculture, and it shows considerable interdisciplinary interaction, involving the application of the knowledge of agricultural sciences, computer science, engineering, environmental studies, and data analytics. The interdisciplinary interactions between researchers working in the field of crop physiology, soil science, and agronomy and those working in the field of machine learning, image processing, and optimization algorithms have enabled the realization of smart systems that can solve complex agronomical issues. This intersection has facilitated the transfer of theoretical AI models to practical uses, such as detecting a disease with the help of a deep-learning model, automating irrigation with the help of predictive modeling, and making nutrient management as precise as possible with the help of sensor networks. These studies are interdisciplinary, which increases the rigor of the methodology and applicability of the research results to real-world agricultural settings.

The institutional and country-level analysis displays the patterns of collaboration on the basis of interdisciplinary participation. Most major research institutions, such as agricultural universities, engineering and technology institutes, and national research laboratories, are frequently forming collaborative projects where different expertise used. The collaborations on an international level broaden the borders of research by involving globalized methodology, comparison of data, and new methods of computation.

6. JOURNAL AND SOURCE ANALYSIS

6.1 Core Journals Publishing AI-Based Agricultural Research

The scholarly productivity of AI-based agricultural research being clustered in a small number of central journals, which function as important sources of dissemination of methodological progress as well as practical innovations. Precision agriculture, agricultural informatics, and computational intelligence journals offer a well-organized system where predictive modeling, machine learning models, and artificial intelligence-supported decision support systems are presented. Reports in such sources are common on crop monitoring, yield prediction, disease and pest detection, and smart resource management and have absorbed theoretical contributions as well as practical applications. The popularity of these journals helps to gain visibility and impact in the scientific community, which will have future research and shape the creation of new technologies in the field of agriculture.

Figure 11. Core Journals Publishing AI-Based Agricultural Research

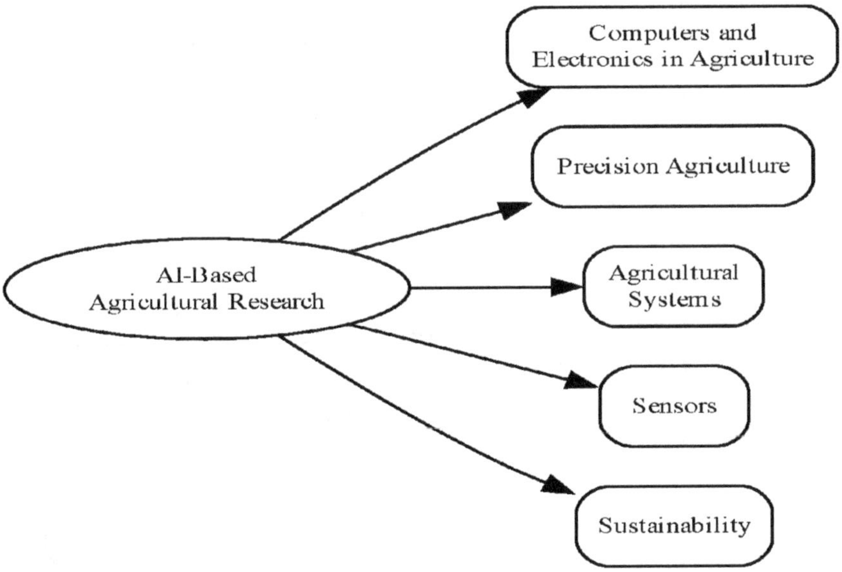

Interdisciplinary journals that combine computer science, engineering, and agricultural sciences include citation patterns and publication volumes indicating that the field of great use to domestic and international researchers. Environmental monitoring, remote sensing, and geospatial analytics journals are the extensions of the conventional sources on the subject of agriculture, as they offer the sophisticated analytical models that can be applied in AI applications.

6.2 Source Impact and Thematic Orientation

The review of the sources of publications shows that there a strong difference in the impact and thematic focus of publications about the distribution of AI-based agricultural research. Products of advanced computational techniques combined with applied agricultural solutions that are more focused on precision farming, predictive modeling, and decision-support systems are most often published in high-impact journals. These journals are also open to local and international researchers, and this a sign of acknowledging methodological rigor and practical relevance. The patterns of references have shown that the journals have more scholarly impact when they have interdisciplinary methods and rigorous AI methods, as they are the major reference sources to the new trends of research.

Figure 12. Source Impact and Thematic Orientation

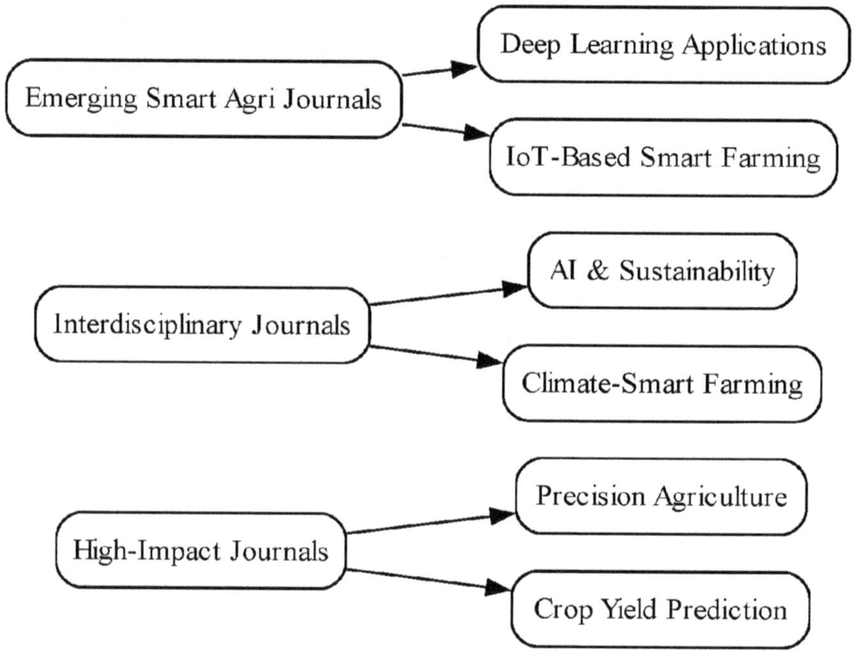

Thematic orientation of the journals a systematic mapping of the research orientation to the priorities of the sector. Articles in the field of agricultural informatics, remote sensing, and environmental monitoring are also more likely to focus on applied AI to crop management, irrigation optimization, and resource efficiency. Computer science, data analytics, and engineering journals emphasize the creation of algorithms, the validation of models, and hybrid AI, which facilitate the methodology development of the field.

A longitudinal study of journal impact reveals gradual congruency of themes of publication with the priorities of national agriculture and the technological improvements. Publications that were published earlier accentuated exploratory AI models and pilot studies that were carried out on a small scale; recent sources have been concerned with scalability, explainable AI, and integration across various subdomains of agriculture. The impact of the source and thematic orientation gives a structure for comprehending the impact of the scholarly communication channels on research visibility, knowledge diffusion, and development of AI applications in agriculture. High-impact sources are also not only determining research agendas but are also a place where collaboration, spreading of innovations, and recognition of new fronts of research in the sector are achieved.

6.3 Interdisciplinary Journals Bridging AI and Agriculture

Interdisciplinary journals are important in publishing research incorporating artificial intelligence into agricultural sciences and availing platforms of studies that involve computational innovation with practical farming implementations. The journals are inclusive of research work in the field of precision agriculture, remote sensing, IoT-based farm surveillance, and predictive modelling, and they cross-pollinate with computer science, agronomy, environmental engineering, and data analytics. Articles in these respective venues tend to focus on not only the methodological aspect but also the implementation results, providing information about the design, validation, and implementation of AI-based systems in various agricultural and related settings. These journals are interdisciplinary, which promotes the integration of theoretical and practical views to obtain a holistic view of AI solutions in agriculture.

Interdisciplinary journals make these fields more visible and collaborate efficiently, soliciting academic institutions, research, and industry practitioners to submit their contributions. Studies published in these journals have been shown to have a great impact on citation considering that they are applicable in a wide range of fields such as computational intelligence, sustainable agriculture, and environmental management. The sharing of knowledge based on such journals helps in identifying new areas of research, trends in methods of approach, and the application of AI integration as a solution to urgent agricultural issues.

6.4 Publication Distribution Across Journal Quartiles

The quality and visibility of scholarly contributions are represented by the distribution of AI-driven agricultural research by quartiles of the distribution. A significant percentage of the publications are in the high-impact journals in the top quartiles, which shows that they are rigorously peer reviewed, have high standards of methods used, and are relevant to the global research community. The journals are usually focused on interdisciplinary research combining AI practices with agriculture, focusing on predictive modeling, precision farming, and sustainability-related innovations. The fact that high-quality studies are concentrated on high-quality journals emphasizes the increased chances of unity of AI applications in increasing productivity, resource saving, and decision support in Indian agriculture.

The products of research in mid-tier journals are on new fields and experimental research that used to study new algorithms, hybrid models, and localized implementations. Such journals tend to concentrate on applications to specific regions, pilot project field experiments, and comparative results of machine learning or deep learning applications to crop management, soil evaluation, and related industries, including livestock and fisheries. The mid-quartile journals distribution implies that people actively experiment and develop context-sensitive AI solutions, which can be further used to create scalable applications and then published in higher-impact media.

Although lower quartile journals also contribute a smaller proportion of publications, they are significant contributors to the publication of preliminary findings and methodological revising of the study and applied case studies tackling niche or underrepresented issues. Such studies can offer good information on the practical issues, data constraints, and operational constraints affecting the adoption of AI on the field level. The distribution of totals across the journal quartile shows a well-balanced research ecosystem with the presence of high-impact studies that provide the methodological standards and the presence of mid- to lower-quartile articles that fill an applied knowledge gap, innovation diversity, and further development of AI-based agricultural research.

The concentration of highly productive authors within a relatively small group of institutions indicates the presence of specialized research clusters focusing on artificial intelligence in agriculture. These research leaders often act as knowledge hubs, promoting interdisciplinary collaboration and guiding new researchers entering the field. Their work contributes to the development of standardized methodologies and encourages the wider adoption of AI technologies across different agricultural domains.

6.5 Emerging Publication Venues in Smart Agriculture Research

The recent years have seen the emergence of dedicated publication outlets concerned with smart agriculture, digital farming, and AI-based agricultural technologies. Conferences and journals focused on agricultural informatics, precision agriculture, and computational intelligence are becoming a viable means of publication of innovative studies in the fields of crop monitoring, soil management, optimization of irrigation, and related applications in the sector. These new platforms open up prospects for interdisciplinary research and can contribute to such integration of data science, sensor networks, remote sensing, and decision support systems. The fact that the best writings are presented in these sites evidence of the increasing awareness of agricultural research based on AI and the need to have high-quality, specialized research.

The creation of new journals and special issues on digital agriculture assists in the quick spreading and dissemination of knowledge as well as innovating the methods. The new sources of publications focus more on feasible application, sustainability, and scalability of AI applications, which why research focusing on the local-specific problems in Indian agriculture draws attention. The inclusion in these sites will help promote the interdisciplinary work, provide the international collaborations, and stimulate the comparative investigations among agro-climatic regions. The move towards dedicated outlets highlights the fact that smart agriculture indeed a mature area of research with formalized knowledge-sharing avenues and the creation of evidence-based practices that promote resource efficiency and productivity in agriculture and associated fields.

The emergence of specialized journals focusing on digital agriculture and agricultural informatics reflects the increasing academic recognition of AI-driven agricultural research. These publication venues support interdisciplinary collaboration and encourage the integration of computational technologies with agricultural sciences. As digital agriculture continues to expand, such journals will play an important role in disseminating innovative research and supporting knowledge exchange among researchers, policymakers, and technology developers.

7. DISCUSSION AND FUTURE RESEARCH DIRECTIONS

The findings of this study highlight a significant transformation in agricultural research driven by artificial intelligence technologies. The transition from traditional statistical models to advanced deep learning and hybrid AI approaches reflects the increasing complexity of agricultural challenges and the need for data-driven decision-making systems.

The dominance of crop-focused research indicates a strong emphasis on food security and yield optimization, while the relatively lower representation of allied sectors such as fisheries and livestock suggests potential research gaps. The emergence of explainable AI and climate-smart agriculture as key thematic areas demonstrates a shift toward sustainable and transparent AI solutions.

The lack of standardized agricultural datasets, limited digital infrastructure in rural regions, and scalability issues hinder widespread adoption. Furthermore, the fragmented nature of research across disciplines limits the development of integrated agricultural intelligence systems.

Comparatively, global research trends show increasing emphasis on AI-driven automation, robotics, and climate resilience, indicating that Indian research must expand toward these areas to remain competitive.

Overall, the study underscores the need for interdisciplinary collaboration, policy support, and technological innovation to fully realize the potential of AI in agriculture.

Managerial and Practical Implications

The findings of this study have significant implications for policymakers, researchers, and agricultural practitioners. Policymakers can utilize scientometric insights to identify priority research areas and allocate funding toward high-impact domains such as precision agriculture and climate-smart farming.

For researchers, the identification of emerging research fronts provides guidance for future investigations and interdisciplinary collaborations. Practitioners and agribusiness stakeholders can leverage AI-based solutions for improving productivity, optimizing resource use, and enhancing decision-making processes.

The study highlights the importance of developing scalable and cost-effective AI solutions tailored to smallholder farmers, ensuring inclusive and sustainable agricultural development.

8. CONCLUSION

The implementation of artificial intelligence in Indian agriculture and other related industries has transformed the ancient farming models and made them precise, efficient, and sustainable. Scientometric analysis shows a fast-growing research environment, with a significant increase in the number of publications, a high citation impact, and interdisciplinary collaboration. The predominance of crop-based studies in the scholarly literature most evident, but closely related fields like livestock, fisheries, and forestry are experiencing increased involvement, and this indicative of technological capability, as well as a rise in scholarship interest. Thematic analyses also show that the traditional machine learning models were replaced by more sophisticated deep learning models, hybrid intelligence models, and explainable AI models, as the needs of more complex agricultural problems and interpretable as well as data-driven decision-making are increasing. Scientific networks provide an insight into how academic institutions, research organizations, and international collaborations have become the key to innovation and the translation of computational methods into practice. Both keyword co-occurrence and clustering research papers highlight the high-priority areas of research as precision agriculture, climate-smart farming, resource optimization, and related sector analytics as contributing to sustainable intensification and improved productivity.

In spite of impressive advances, there are significant gaps in the scaling of AI applications to a variety of agro-climatic environments and the quality and interoperability of data, as well as the establishment of the holistic solution that combines the management of crops and other related sectors. There a lack of discussion on multi-dimensional and interdisciplinary approaches, and future research can enable the enhancement of the efficiency and sustainability of the system. The scientific insights are strategic to the policymaker, the researchers, and the funding agencies, as they provide the information on the strategic prioritization of the areas with high impact and a possibility of technology adoption. The further emphasis on interdisciplinary engagement, scalable AI-based solutions, and productive knowledge sharing will also be necessary to translate research progress into the working agricultural systems. Thorough mapping of the tendency of publications, evolution of the topic, and research networks creates an organized idea of the intellectual environment, which points out both accomplishments and opportunities.

Future research should explore the integration of artificial intelligence with emerging technologies such as Internet of Things sensors, big data analytics, and climate-smart agricultural systems. The development of explainable AI models for agricultural decision-making and scalable digital platforms accessible to smallholder farmers represents an important research direction. Interdisciplinary collaboration between computer scientists, agronomists, policymakers, and industry partners will be essential for developing sustainable and inclusive AI-enabled agricultural ecosystems.

9. CASE STUDY: AI APPLICATIONS IN INDIAN AGRICULTURE

One notable example of artificial intelligence implementation in agriculture is the use of deep learning models for crop disease detection. Mobile applications based on convolutional neural networks can analyze images of plant leaves captured by farmers and automatically identify diseases such as leaf blight or rust. These applications provide farmers with early warnings and recommendations for treatment, reducing crop losses and improving productivity.

AI technologies are also being used in drone-based crop monitoring, smart irrigation systems, and predictive yield forecasting, demonstrating the growing role of digital technologies in transforming agricultural management practices.

10. ETHICAL AND DATA GOVERNANCE CONSIDERATIONS

The increasing use of artificial intelligence in agriculture also raises important concerns regarding data governance and ethical AI practices. Agricultural data collected through sensors, drones, and digital platforms often contain sensitive information related to farm productivity and environmental conditions. Issues such as data privacy, ownership, and security must be addressed to ensure responsible use of AI technologies. Additionally, algorithmic bias and lack of transparency in AI models may lead to unequal access to technological benefits among farmers. Therefore, the development of ethical AI frameworks, transparent decision-support systems, and inclusive data governance policies is essential for sustainable and equitable adoption of AI in agriculture..

Responsible AI frameworks that emphasize transparency, explainability, and fairness will be essential to ensure that AI technologies benefit farmers equitably and contribute to sustainable agricultural development.

REFERENCES

Akhavan, M., Alivirdi, M., Jamalpour, A., Kheradranjbar, M., Mafi, A., Jamalpour, R., & Ravanshadnia, M. (2022). Impact of Industry 5.0 on the Construction Industry (Construction 5.0): Systematic Literature Review and Bibliometric Analysis. Buildings 2025, 15, 1491. Artificial Intelligence.

Arora, J., Joshi, A., & Ray, R. C. (2024). *Transforming agriculture residues for sustainable development*. Springer., DOI: 10.1007/978-3-031-61133-9

Boutracheh, H., Mejjad, N., & El Bouhadioui, M. (2024a, August). for Trends and Opportunities. In GIS, Applied Computing and Data Science for Water Management

Boutracheh, H., Mejjad, N., El Bouhadioui, M., & Moumen, A. (2024b, August). Water Research in the Age of AI: A Bibliometric Heuristic Analysis for Trends and Opportunities. In International Conference GIRE3D Participatory and Integrated Management of Water Resources in Arid Zones (pp. 3-45). Cham: Springer International Publishing. DOI: 10.1007/978-3-031-63038-5_1

Fernberg, P. J. (2024). *Artificial Intelligence in Landscape Architecture: A Survey of Theory*. Culture, and Practice.

Goel, A., Raut, G., Sharma, A., & Taneja, U. (2024). Artificial Intelligence and Sustainable Business: A Review. *South Asian Journal of Business and Management Cases, 13*(3), 340–365. DOI: 10.1177/22779779241302146

Jayadatta, S. (2024). A Study on AI-Driven Agricultural Innovations for Rural and Industrial Development in the Indian Context. Journal of Rural and Industrial Development, 12(1), 01-16.

Jena, J. R., Biswal, S. K., Panigrahi, R. R., & Shrivastava, A. K. (2024). Investigating the potential areas in artificial intelligence and financial innovation: A bibliometric analysis. *Journal of Scientometric Research, 13*(1), 71–80. DOI: 10.5530/jscires.13.1.6

Lokeshwari, M., Jha, G. K., Praveen, K. V., & Bharadwaj, A. (2024). Artificial intelligence for crop yield prediction: a bibliometric analysis. Current Science (00113891), 126(10).

Mishra, D., Muduli, K., Raut, R., Narkhede, B. E., Shee, H., & Jana, S. K. (2023). Challenges facing artificial intelligence adoption during COVID-19 pandemic: An investigation into the agriculture and agri-food supply chain in India. *Sustainability (Basel), 15*(8), 6377. DOI: 10.3390/su15086377

Naeem, S., Nabeel, N., Beg, W., Ali, S., Kanojiya, R. N., Mandawade, S. S., & Patil, A. V. (2024). A Review on Exploring Artificial Intelligence Applications, Advancements, Issues, and Future Challenges. *POLYTECHNIC JOURNAL, 14*(2), 96–113.

Raman, R., Pattnaik, D., Lathabai, H. H., Kumar, C., Govindan, K., & Nedungadi, P. (2024). Green and sustainable AI research: An integrated thematic and topic modeling analysis. *Journal of Big Data, 11*(1), 1–28. DOI: 10.1186/s40537-024-00920-x

Saxena, R., Pant, D. K., Pant, S. C., Joshi, L., Paul, R. K., & Singh, R. (2023). Artificial intelligence-led innovations for agricultural transformation: A scoping study. *Agricultural Economics Research Review, 36*, 1–18.

Skubis, I., Wolniak, R., & Grebski, W. W. (2024). AI and Human-Centric Approach in Smart Cities Management: Case Studies from Silesian and Lesser Poland Voivodships. *Sustainability (Basel), 2024*(16), 8279. DOI: 10.3390/su16188279

Subasi, A., Qaisar, S. M., & Nisar, H. (Eds.). (2024). *Artificial Intelligence and Multimodal Signal Processing in Human-Machine Interaction*. Elsevier.

Wu, M. (2023). *An Intelligent Bibliometric System for Knowledge Association and Hierarchy Discovery*. University of Technology Sydney.

Chapter 6
AIoT-Based Flood Alert System Using Rainfall + River Sensors?
Targeting Flood-Prone Regions in Assam, Kerala

Sandu Venkata Varun
http://orcid.org/0009-0003-8649-6034
Vel Tech Rangarajan Dr. Sagunthala R&D Institute of Science and Technology, India

Yaswanth Kumar Reddy Bollavaram
http://orcid.org/0009-0003-2462-0490
Vel Tech Rangarajan Dr. Sagunthala R&D Institute of Science and Technology, India

Dhanunjay Reddy Sagili
Vel Tech Rangarajan Dr. Sagunthala R&D Institute of Science and Technology, India

D. Prabhu
Vel Tech Rangarajan Dr. Sagunthala R&D Institute of Science and Technology, India

ABSTRACT

Floods in India cause significant socio-economic, infrastructure, and environmental damage. Traditional monitoring systems are inadequate for real-time flood prediction and risk mitigation. This chapter introduces an AIoT-based flood warning system that utilizes IoT sensors for continuous data acquisition on rainfall and river stream

DOI: 10.4018/979-8-2600-2383-9.ch006

© 2027, IGI Global Scientific Publishing. Copying or distributing in print or electronic forms without written permission of IGI Global Scientific Publishing is prohibited. Use of this chapter to ■erative artificial intelligence (AI) technologies is expressly prohibited. The publisher reserves all rights to license its use for generative AI training and machine learning model development.

levels, combined with machine learning models for accurate flood prediction. The system uses edge and cloud computing for low-latency processing, ensuring timely alerts. Custom regional models cater to Assam's riverine floods and Kerala's flash floods, improving lead times and emergency responses. Integrated with disaster management platforms, the system enhances flood resilience and offers a scalable, intelligent solution for climate-sensitive areas.

1. INTRODUCTION

One of the most destructive natural risks to human settlements, agriculture, infrastructure, and ecosystems, floods continue to be a major problem around the globe (Boulouard et al., 2022; Wu, 2024). In India, the frequency and intensity of floods have increased in the last few decades because of climate variability, cutting of forests, and unplanned urbanization (ElMougy et al., 2023; Dhaya and Kanthavel, 2022). The states of Assam and Kerala are especially susceptible, with a unique flood pattern due to the specifics of the hydrology and geology of the area. Assam was mainly riverine, with the Brahmaputra river system having massive riverine floods during the monsoon seasons. The intensity of rainfall and the presence of sediments in the channels of rivers lead to extensive inundation in extensive floodplains (Kong et al., 2024). Conversely, the steep slope and limited river basin topography of Kerala cause flash floods after a short time with high-intensity precipitation (Nerella & Ahmed, 2023; Tham et al., 2023). Advanced flood monitoring and predictive systems are necessary to issue early warnings in time to reduce socio-economic and environmental losses in both regions, which include repeatedly damaged property, loss of life, and disruption of livelihood (Adli et al., 2023; Mitra et al., 2022).

In India, the traditional methods of flood monitoring are mainly based on the manual measurements of river gauges, rainfall records, and centralized hydrological models (Wu, 2024; Pourroostaei Ardakani and Cheshmehzangi, 2023). The weaknesses of these approaches are that have poor spatial coverage, slow data acquisition, and insufficient real-time analytics. Remote and inaccessible regions are not monitored enough, and, therefore, communities are very vulnerable to floods (Chen et al., 2023a). Localized terrain variations, drainage patterns, and fine-scale river morphology are often not represented in centralized hydrological models and decrease the accuracy of their predictions (Bhatia and Jaglan, 2024). Traditional early warning systems rely on fixed threshold levels that are not able to respond to sudden bursts of rainfall or fast river reactions, especially in flash flood-prone regions (Halubanza, 2024; Zhang, 2024). Lack of ability to present precise, localized, and punctual flood predictions adds to the emergency reaction issues, slows the evacuation, and adds human and economic damages, which clarifies the pressing need for innovative approaches to monitoring.

Artificial Intelligence (AI) and the Internet of Things (IoT) can be combined, creating the so-called AIoT, which will serve as an innovative solution to the constraint of traditional flood monitoring systems (Boulouard et al., 2022; ElMougy et al., 2023). Sensors on rainfall and river water distributed networks allow a continuous and high-resolution hydrometeorological data collection (Wu, 2024; Kong et al., 2024). The use of edge and cloud computing infrastructure enables real-time processing of data, which enables quick evaluation of the dynamics of rivers and the intensity of rainfall (Tham et al., 2023). The datasets used in machine learning algorithms are able to identify abnormal patterns, predict increases in river level, and predict the onset and intensity of flooding (Nerella and Ahmed, 2023; Dhaya and Kanthavel, 2022). In Assam, gradual riverine floods and sudden flash floods of Kerala are distinguished by region-specific modeling to provide different predictive analytics to local hydrological circumstances. AIoT-based systems improve the situational awareness, help to implement proactive measures aimed at flood control, and support timely evacuation and resource distribution (Adli et al., 2023; Mitra et al., 2022).

Localized predictive flood monitoring provides an opportunity to intelligently combine with the disaster management platform and enhance the efficiency of the work of the emergency response authorities (Bhatia and Jaglan, 2024; Halubanza, 2024). Alerts that are produced by AIoT systems can be shared at various levels, such as mobile applications, cloud dashboards, and outdoor sirens, and offer practical information to the authorities and vulnerable groups (Chen et al., 2023b; Wu, 2024). High-risk areas can be prioritized with multi-level risk assessment, and this information was useful in effective allocation of relief resources and targeted evacuations. Connections with the current resilience infrastructure of the disaster management systems contribute to cross-agency cooperation, increasing community and institutional resilience (ElMougy et al., 2023; Nerella and Ahmed, 2023). The integration of real-time sensing with predictive intelligence will enhance the effectiveness of local governments and stakeholders to predict the occurrence of floods and take actions to mitigate the response of the event on human, economic, and environmental damages by AIoT-based systems.

Flood risk was also increased by climate change and changes in land use, making it essential to have adaptive and scalable systems of long-term resilience to floods (Pourroostaei Ardakani and Cheshmehzangi, 2023; Zhang, 2024). The flood alert systems built on AIoT offer the basis to combine past flood data with satellite data, hydrological modeling, and real-time data gathered by sensors, allowing the continuous learning and improvements in models (Maraveas, 2022; Chen et al., 2023b). Scalability can be used to deploy in non-homogenous areas that are prone to floods, which can handle morphological differences in rivers, intensity of rainfall, and patterns of urbanization (Boulouard et al., 2022; Tham et al., 2023). AIoT analytics can be explained and transparent, which will build trust with local and state stakeholders. Using AIoT technologies, flood-prone areas like Assam and Kerala will be able to move beyond reactive response to disaster management, which was based on the collected data, which will minimize the exposure of risk and enhance the resilience of vulnerable communities, as well as reduce the risk of floods (Kong et al., 2024; Nerella and Ahmed, 2023).

1.1 Research Gap and Contribution

Despite significant advancements in flood monitoring using IoT and AI techniques, existing systems exhibit several limitations. Most traditional approaches rely on centralized architectures, lack real-time responsiveness, and fail to capture localized hydrological variations. Furthermore, many machine learning-based models do not incorporate region-specific flood characteristics, reducing their predictive accuracy in diverse environments such as riverine and flash flood zones.

To address these gaps, this chapter proposes a novel AIoT-based flood monitoring framework that integrates distributed sensing, edge-cloud computing, and adaptive machine learning models. The key contributions of this work are as follows:

- A region-specific predictive modeling approach tailored for Assam (riverine floods) and Kerala (flash floods)
- A multi-sensor data fusion mechanism combining rainfall and river-level data
- An adaptive thresholding model for dynamic flood risk detection
- A hybrid edge-cloud architecture for real-time flood prediction and alert dissemination

This contribution moves beyond descriptive analysis and provides a structured, intelligent framework for next-generation flood management systems.

Literature Review

Recent studies have explored the integration of Artificial Intelligence (AI) and Internet of Things (IoT) technologies for flood monitoring and prediction. AIoT-based systems enable real-time data acquisition, predictive analytics, and automated alert generation.

Boulouard et al. (2022) proposed an AIoT-based framework for river flood prevention; however, their approach lacks localized predictive modeling. Wu

(2024) developed ANN-based hydrological models, but these models are limited in handling real-time data streams. Tham et al. (2023) introduced wireless mesh networks for disaster monitoring, yet their system does not incorporate adaptive learning mechanisms.

Overall, existing approaches suffer from limitations such as lack of region-specific modeling, insufficient real-time processing, and absence of adaptive thresholds. These gaps motivate the development of an integrated AIoT-based predictive flood monitoring system.

Table 1. Comparative Analysis of Existing Systems

Study	Method	Strength	Limitation
Boulouard et al. (2022)	AIoT monitoring	Real-time sensing	No regional adaptation
Wu (2024)	ANN model	Accurate prediction	Limited real-time capability
Tham et al. (2023)	Wireless IoT	Scalable network	No ML-based prediction
Nerella & Ahmed (2023)	Alert system	Early warnings	Limited integration

This comparison highlights the lack of integrated, adaptive, and region-specific flood prediction systems, which motivates the proposed approach.

2. REGIONAL FLOOD CHARACTERISTICS AND HYDROLOGICAL CONTEXT

2.1 Flood Typology and Seasonal Patterns in Assam and Kerala

The hydrology of Assam and Kerala was characterized by hydrology type and season characteristics that are based on geographical, climatic influences, and river system dynamics. Assam was mostly characterized by riverine floods, which are related to the Brahmaputra basin, where sustained flood increases are caused by long monsoon rainfall between June and September. The eastern Himalayan snowmelt coupled with severe upstream rainfall leads to prolonged flood periods in large floodplains. Peak monsoon periods cause a change in the morphology of the river as a result of the high sediment transport, which causes channel shifting and overbank flow of the river. Seasonal flooding has a predictable yearly pattern, with pre-monsoon precipitation leading to a slow rise in water level, with peak water inundation taking place in the mid- to late-monsoon periods. The consequence of this type of flood was extensive flooding over a big region with a slower development, leaving the floods in the agricultural and residential areas lasting for very long periods.

Kerala has a different flood typology, which was controlled by flash floods that are associated with short periods and high-intensity rainfall during the southwest monsoon season. The steep soil, small river basins, and high surface runoff all lead to sudden increases in water levels within hours of heavy precipitation. Rainfall concentration during the seasons of June to August initiates sudden responses to rivers, whereas localized cloudbursts enhance the severity of floods in upstream catchments. Flood patterns during extreme rainfall episodes are also caused by interactions at the coastal backwaters and releases of the reservoir spillways. The frequency of occurrence of floods was largely episodic as opposed to protracted inundation, with the outcomes of floods being very sudden and intense as of a limited spatial extent.

2.2 Brahmaputra River Basin Dynamics and Floodplain Vulnerability in Assam

The Brahmaputra River basin in Assam was one of the most intricate and dynamic systems of fluvial systems in South Asia due to the high levels of monsoonal precipitation, the large high inputs of catchment in the upstream, and the high computations of sediment transport in the region of the Himalayas. The large discharge variability produced by seasonal rainfall patterns results in frequent overbank flow occurring on large floodplains. The channel braiding and the erosion of the banks and subsequent lateral movement are the most prevailing geomorphological activities in the basin, as are continually shifting the morphology of rivers together with their floodplains. Massive heaps of sediment deposition raise the riverbeds over a period of time, decreasing channel carrying capacity and increasing the frequency of floods in lower reaches between the high bank and the main stream. This hydrological behavior exposes rural colonies and agricultural states, as well as important infrastructure spread along the riverbed, in a sustained manner.

Another factor that predisposes the Brahmaputra basin to floods was the close relationship between the natural hydrological processes and the human settlement patterns. A high concentration of population resides in flood-prone areas because of the presence of fertile alluvial soils, which harbor agriculture-based lives. The yearly occurrence of the events of recurrent inundation affects the cropping cycle and destroys transportation networks as well as displaces communities. Structural stress on embankment systems built to control floods was evident, as the systems cannot withstand sustained high flows, resulting in breaches that further increase the effects of floods in downstream areas. Quick shifts in flow direction and deposition patterns also make the prediction and risk assessment of floods more complicated and create more uncertainty in disaster preparedness and mitigation planning.

2.3 Flash Flood Mechanisms in the River Basins of Kerala

The mechanisms of flash floods in the Kerala river basins are a product of a special scheme of convergence of geomorphology, climatic controls, and land-surface interactions. The state receives heavy rainfall that was short-lived and was related to the southwest monsoon and extreme rainfall that was caused by convective cloud structures. The steep Western Ghats terrain speeds up surface runoff, and the length and width of rivers are short and have limited attenuation of natural flows. The filtration speed of the lateritic and clay soils leads to a slower infiltration rate, and thus the surplus rainfall transforms into overland flow rapidly. Reservoir releases, riverbank intrusions, and changed drainage patterns further increase the speed of flow in river channels, which cause abrupt and large-scale increases in the water level across settlements down the river.

Figure 1. Flash Flood Mechanisms in the River Basins of Kerala

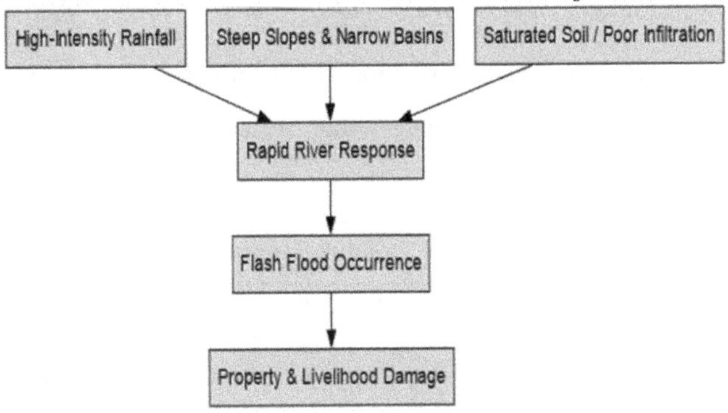

The hydrological response of Kerala river basins indicates a small lag time between rainfall and flood occurrence in the river basins, hence little time for manual intervention or delayed warning measures. The thick settlement of people in the floodplains and the midland areas exposes them to the impact of sudden increases of floods, and the urban development erodes the natural buffers of floods like wetlands and paddy fields. Peak flows cause significant sediment transmission that changes the geometry of the channels, thereby enhancing turbulence and focal overflow situations. The properties of such characteristics require constant high-resolution rainfall and river-level measurements that are able to record rapid hydrodynamic changes. The role of intelligent sensing and predictive analytics thus takes critical significance in the river systems of flash floods in Kerala, whose real-time sensing and instantaneous alert sharing was the only indicator of the efficacy of flood risk control measures.

2.4 Impact of Climate Variability and Extreme Rainfall Events

The effect of climate variability and extreme rainfall events has a prevailing role in flood behavior over the river basins of Assam and Kerala, affecting hydrological response at both seasonal and event scales. Changes in the time, duration, and location of the onset of the monsoons are also variations in the irregular precipitation patterns, which have increased the generation of runoffs and changed the regime of river discharges. The high monsoonal rainfall in Assam was augmented by the upstream precipitation of transboundary catchments in the Brahmaputra system and increases the river stage. The frequent severe rain events enhance the transfer of sediments and the channel aggradation, as well as decrease the carrying capacity and augment the frequency of overbank flows. The episodes of short-duration, high-intensity rainfall occur in Kerala, which causes fast surface runoff over steep landscapes and creates sudden increases in river levels within short response duration frames.

Figure 2. Impact of Climate Variability and Extreme Rainfall Events

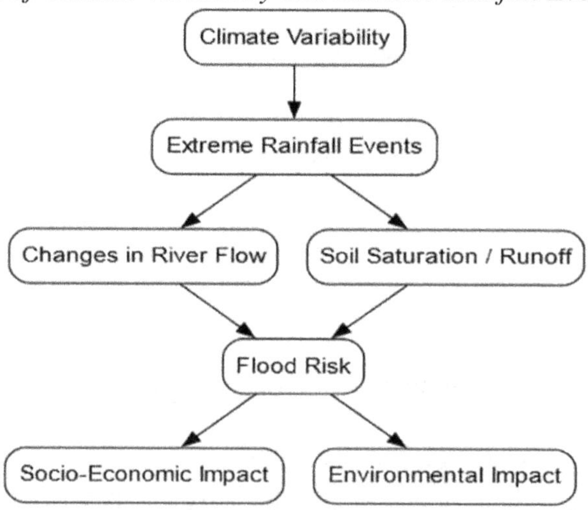

Climate variability causes extreme rainfall events, which increase the magnitude of floods by precipitation cumulative effects and saturation-based runoff processes. Repeated rainfall on days leads to the increase in the level of soil moisture, infiltration capacity inhibition, and direct surface flow to the river channels. In Assam, which was mostly affected by floodplains, the process maintains high levels in rivers over long durations and therefore causes prolonged flooding in farm and residential areas. In Kerala, heavy rains that are centered on limited catchments create flash flood conditions with sharp hydrographs and steep recession. These dynamics induced by rainfall contravene traditional assumptions used in hydrology about stationary climate behavior and require dynamic monitoring approaches that are able to record real-time variation.

The increasing number of extreme rainfall events because of shifting climatic conditions increases uncertainty in flood forecasting and early warning systems. The nonlinear relationships between the intensity of precipitation, the catchment response, and the river hydraulics make the flood prediction accuracy difficult when the alert was generated by the use of static thresholds. Diversity of spatial distribution of rainfall also accentuates localized flood risk, especially in urbanized and low-lying areas. Climate-driven volatility in hydrological systems in Assam and Kerala was becoming a common phenomenon, which supports the necessity to develop intelligent and data-driven flood monitoring solutions. The combination of rainfall variability analysis into high-tech sensing and predicting systems enhances the situational awareness and aids in the effective management of flood hazards in climate-prone areas.

2.5 Socio-Economic and Environmental Consequences of Recurrent Flooding

Repeat floods in flood-prone areas like Assam and Kerala generate far-reaching socio-economic impacts that go beyond the direct physical destruction. Farming livelihoods are also disrupted repeatedly due to submerging standing crops, changing soil fertility, and delaying planting activities, thereby causing low productivity and income instability to farming communities. Rural and semi-urban communities, which are situated by the riverbanks, experience frequent displacement, destruction of residential buildings, and the destruction of other valuable resources, contributing to the aggravation of economic risks and social inequality. The transportation systems, such as roads, railways, and bridges, are regularly damaged by long-term waterlogging and erosion, and this limits market, health, and educational access.

Figure 3. Socio-Economic and Environmental Consequences of Recurrent Flooding

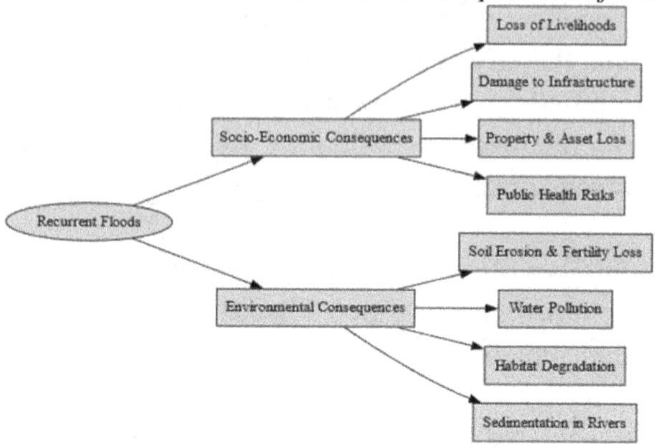

The environmental impacts of frequent flooding also pose equally complicated problems in the river and coastal ecosystems. Extensive flooding has changed the morphology of rivers by increasing erosion, deposition, and channel migration to redefine floodplains and impacting aquatic life. The water quality in freshwater ecosystems was affected by the influx of pollutants, sewage, and agricultural runoff, thereby degrading biodiversity and endangering fisheries, which form the basis of local economies. Floods increase soil erosion and loss of nutrients in highlands and overload the lowlands, with the end result of land being less useful eventually. Although wetlands are natural flood buffers, experience ecological stress when flood frequency surpasses the natural recovery cycles, which reduce the ecosystem services, including water purification and habitat support.

However, despite these advancements, existing systems fail to provide localized and adaptive predictions. To address this limitation, the proposed framework is introduced in the following section.

3. LIMITATIONS OF CONVENTIONAL FLOOD MONITORING AND EARLY WARNING SYSTEMS

3.1 Traditional Hydrological Monitoring Approaches and Their Constraints

The conventional hydrological monitoring systems are mostly based on manual river gauge stations, periodic rainfall stations, and central hydrological models, which are managed by government bodies. These systems rely heavily on human intervention and predetermined monitoring points to collect data, which means that a large amount of spatial coverage of large and complex river basins was not covered. In highly hindered areas like Assam and Kerala, where the river behavior changes extremely fast along tributaries and catchments, the dense distribution of sensors inhibits the possibility of capturing localized hydrological changes. Data acquisition and reporting delays in time decrease the situational awareness capacity in case of a heavy rain event, and the quality of early warning generation was limited. The traditional flood indicators based on a fixed threshold are used in the majority of traditional systems, which are not adjusted to the morphology of rivers, land-use dynamics, and shifting climate, which decreases predictive accuracy.

The presence of operational limitations to conventional monitoring systems also undermines the preparation and response to floods. The centralized data processing architectures also bring about latency in the cases of extreme weather conditions when the communication networks are overloaded or paralyzed. Repeated repair and maintenance plans of old infrastructure systems and using manual methods to calibrate instruments contribute to the possibility of inaccuracies in measurements and lack of information. The hydrological models employed in the traditional forecasting are usually based on simplistic assumptions and averages, which restrict their sensitivity to abrupt rainfall events and fast changes in river levels. The lack of real-time combination of data and dynamic analytical system limits the traditional flood monitoring systems from providing prompt, localized, and forecastive early warning services in dynamic flood settings.

3.2 Limitations of Manual River Gauge and Rainfall Measurement Systems

River gauge and rainfall measurement systems are manual systems, which have inherent operational and structural limitations as far as their abilities to be effectively utilized in flood monitoring and early warning efforts. The use of human intervention in recording, verifying, and reporting of data was a key factor in observation processes, and thus data would not be available at the time it was needed the most, at the time of critical floods. Measurements of gauges that are made at regular intervals do not record fast changes in hydrology, especially when there was heavy precipitation and sudden rising of rivers. Spatial coverage was also limited by the limited density of gauges, and significant areas of river basins and upstream catchment are not measured. Such limitations decrease the situational awareness and undermine the capacity to predict the onset of floods with adequate lead time.

It was exposed to environmental and operational variables to make the measurement accuracy in the manual monitoring system susceptible. The reliability of data in high-flow conditions of gauges would be compromised by physical damage due to debris, sedimentation, and extended submergence. Local precipitation variability over diverse terrains (e.g., mountainous catchments, floodplains, etc.) was regularly underrepresented in rainfall measurements made at ground stations. Inconsistencies in data can be attributed to the wearing of instruments, calibration drift, and recording errors and bring about uncertainty in the process of flood assessment. In flood-affected areas characterized by frequent extreme weather, gaps in data due to inaccessible monitoring sites cause even more problems to the continuity of hydrological records.

Figure 4. Limitations of Manual River Gauge and Rainfall Measurement Systems

There are serious issues of communication and coordination challenges in early warning dissemination based on manual measurement systems. The exchange of data was based on hierarchical reporting systems that postpone the interpretation and regional/local decision-making. The warning criteria that are based on the static threshold are not flexible to the changing hydrological conditions and regional variation, which makes the alerts slow or erroneous. The situation was a failure of emergency response planning, as flood warnings come when water levels are close to or at the critical points. The compounding effect of latent data collection, low spatial resolution, and rigid warning systems emphasizes the inefficiency of the manual river gauge and rainfall estimation system in present-day flood management.

3.3 Challenges in Centralized Flood Forecasting Models

Aggregated hydrological and meteorological information that was processed in national or regional control centers was the basis of centralized flood forecasting models, which bring structural limitations of responsiveness and spatial accuracy. The low density of observed stations used in data acquisition leads to a low representation of localized variability in rainfalls and quick response of rivers, especially in complex terrains and highly networked river basins. Sequential transporting and validating data transmission and model execution pipeline delays cause processing delays, thus decreasing the usefulness of flood warnings, which are time-sensitive.

Projections of forecasts are usually based on fixed thresholds and generic basin parameters and do not provide flexibility to immediate hydrological variations and local flood drivers. Bottlenecks in communication between centralized systems and local authorities add to the ineffectiveness of the timeliness and usefulness of the alerts issued during extreme flood events.

Figure 5. Challenges in Centralized Flood Forecasting Models

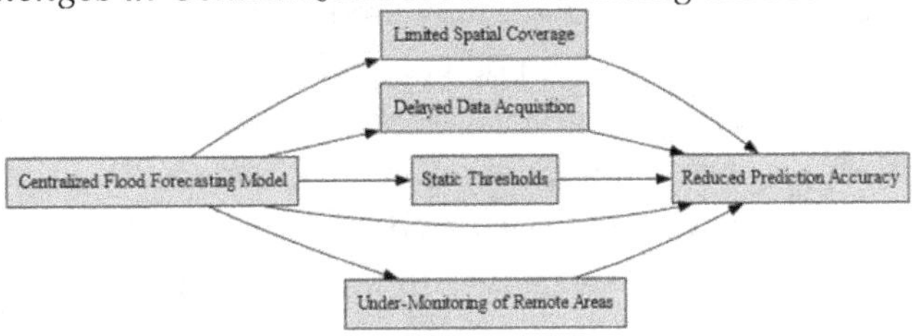

Good performance using models in centralized forecasting systems has further problems because little real-time and high-resolution sensor information was integrated and little region-specific hydrological behavior was considered. Homogenous modeling assumptions used in a wide range of flood regimes lower predictive power in regions with flash flood occurrence, river processes that are driven by sediments, or when the land use was changing rapidly. System scalability was limited because when multiple basins are experiencing flood events at the same time, then computations become overloaded and the outputs take longer to arrive. The reliance on continuous communication systems adds to the vulnerability in times of harsh weather when network failures are the norm. All of these constraints limit the capacity of centralized flood forecasting models to provide localized, precise, and timely early warning that would be useful in mitigating the risks of a flood.

3.4 Delays and Inaccuracies in Existing Alert Dissemination Mechanisms

There was a lot of delay in the flood alert dissemination systems that exist in the flood-prone areas in Assam and Kerala, which restricts the success of the early warning systems. Data obtained in river gauge stations and meteorological stations are also prone to reporting lines before reach the local governments and the vulnerable groups. The resultant time lags between collecting of flood data and informing people about it accumulate during this chain of data flow, decreasing the possible response time to evacuation and emergency preparedness. The infrastructures of communication in remote and rural communities also delay the process of transmitting alerts, especially during a severe rainfall when power failures and lack of network are common. Consequently, early warning systems have failed to reach the intended purpose because warnings often come when floodwaters have already started to affect settlements.

The flaws in current alert systems are based on the use of fixed thresholds and the roughness of spatial coverage of surveillance systems. Alarm on floods was often delivered on the basis of predetermined water levels, which do not keep up with the case of faster hydrological fluctuations through localized precipitation or inflows along the upper stream. This strategy produces false alarms in one place and overestimates the severity of floods in others, which undermines people's belief in warning tools. A low level of integration between rainfall data and river response measurements leads to poor situational awareness, especially in areas with high flash floods where the water level rises swiftly. Lack of adaptive modeling was one of the reasons that leads to variability in the reliability of the alert in various flood situations.

Figure 6. Delays and Inaccuracies in Existing Alert Dissemination Mechanisms

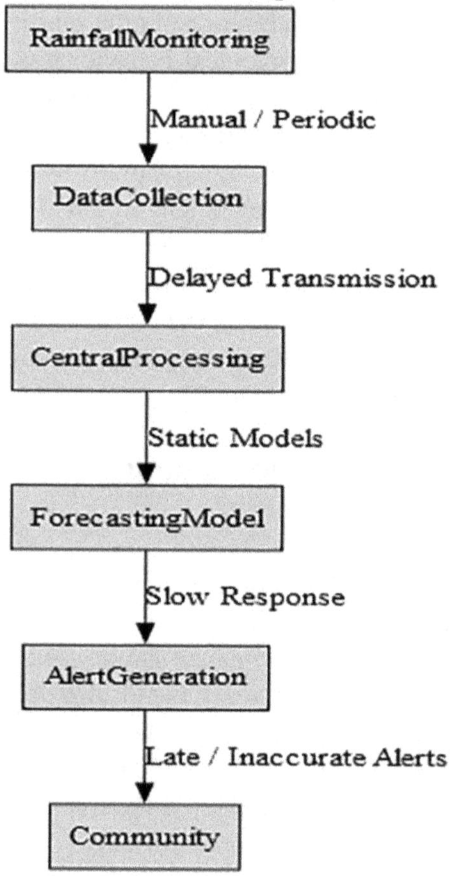

There are also issues associated with message clarity, message targeting, and coordination among the stakeholders in alert dissemination practices. Generalized warnings do not give specific guidance on the location, and the residents get confused as to the extent and urgency of the threat. The fragmentation between the meteorological agencies and water resource departments and disaster management authorities was an institutional issue resulting in inconsistency in the messages and negatively affecting the time to make decisions. Restrictions on the outreach of alerts to vulnerable populations are language barriers and the lack of access to digital communication tools. Such limitations of the systems in totality undermine the ability of traditional flood alert systems to help in timely and effective responses to the disaster.

3.5 Need for Real-Time, Localized, and Predictive Flood Monitoring

The usual flood monitoring methods are usually manualized river gauges, regular rainfall measures, and centralized hydrological models, which makes them slow in data collection and have a restricted geographical scope. The characteristics of flood development, particularly in places that are prone to flash flooding, develop quickly, and conventional processes fail to capture the sudden variations in the amount of water in the river or the intensity of rain or local rainfall. The remoteness or inaccessibility of the community exposes them to greater dangers, as the monitoring networks do not offer information on time. Lack of prediction features within these systems impairs the ability to take proactive decisions, as the authorities have to rely on reactive actions that are usually inadequate to reduce economic losses and provide safety to the population.

Localized, predictive, and real-time flood monitoring provides these essential gaps, allowing collection of data and intelligent analysis in fine space and time. A combination of distributed rain and river water level sensors creates real-time information about the hydrology dynamics, and both abnormal dynamics and areas vulnerable to floods can be detected at the earliest. The process of predictive models receives these streams of data and predicts the onset, duration, and severity of the floods and then enables the delivery of timely warnings to those at risk. The small-scale changes in topography, river morphology, and urban drainage are considered through localized monitoring, which enhances more accurate and relevant warnings.

However, despite these advancements, existing systems fail to provide localized and adaptive predictions. To address this limitation, the proposed framework is introduced in the following section.

4. IoT-ENABLED HYDROMETEOROLOGICAL SENSING INFRASTRUCTURE

4.1 Rainfall Sensor Technologies and Measurement Principles

The monitoring of rainfall was an essential element of hydrometeorology, which was used to obtain vital information on flood prediction and water resources. The types of principles that modern rainfall sensors use encompass tipping-bucket, optical, capacitive, or piezoelectric-based sensing. Tipping-bucket precipitation sensors are instruments that measure precipitation by gathering water in a small tank, which tips when full, creating discrete signals depending on the intensity of rainfall. Raindrops are detected by optical and laser-based sensors that use the fact that light interruption or scattering can be used to measure the raindrop presence and velocity to enable the measurement of the temporal variations of the raindrop with high precision. Capacitive and piezoelectric sensors measure the weight or hit of rain droplets and convert the mechanical variations to electrical signals, which give the accurate information on the rate and the precipitation amounts. The technologies have their own peculiar benefits in the form of sensitivity, maintenance, and adaptability to various environmental conditions.

The location and configuration of rainfall sensors was critical in sound hydrological evaluation of flood control areas. Sensor networks should be able to measure spatial variation in rainfall around catchments and take into consideration the topography of the area, vegetation, and urbanization, which determine the distribution of precipitation and rate of runoff. Connection with the IoT platform allows automatic data transfer in remote areas, which facilitates the constant real-time monitoring process and immediate access to rainfall records for predictive modeling. Detection of sudden rainspikes, which lead to flash floods, was also possible due to high temporal resolution obtained with such sensors, and long-term rainfall data was used to forecast the flows of rivers, as well as assess the risk of floods on the basin scale.

4.2 River Water-Level and Flow Sensors for Flood Detection

River level and flow sensors represent an important part of hydro-meteorological monitoring systems in flood-managed areas. These are sensors that give real-time data on river stage, discharge, and flow velocity, making it possible to identify abnormal hydrological conditions that precede the occurrence of flooding. Quality water-level records record the temporal behavior of river channels to the occurrence of precipitation, such as sudden increases and slow increases, which are indispensable in determining the risk of floods. The approaches usually include ultrasonic, radar, or pressure-based measurement devices that are chosen based on the situation of site-related needs, the geometry of the river, and the conditions of the environment. Temporal data at high resolution enables tracking of rapid changes in the behavior of rivers, which was of importance especially in areas that experience flash floods, like Kerala, or in the vast floodplains of Assam.

Figure 7. River Water-Level and Flow Sensors for Flood Detection

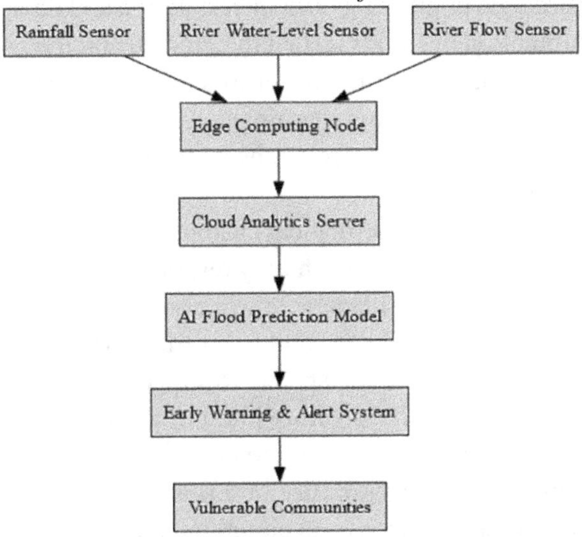

Sensor networks installed on rivers are spatially distributed to improve the situational awareness of the catchments. There are several sensors installed on the upstream, midstream, and downstream, which provide a multi-sampler dataset through which changes in flow patterns and water volume accumulation across the basin are recorded. A combination of these measurements with rainfall information makes it possible to establish cause-and-effect relationships between precipitation and river behavior. Information gathered on remote and inaccessible locations was relayed using the IoT-driven communication infrastructure to centralized or edge computing locations, which guarantee timely tracking and analysis. Such a stream of information helps in predictive models and automatic early warning, minimizing the use of manual observations that are always delayed.

4.3 Sensor Placement Strategies for Riverine and Flash Flood Regions

Sensor positions are the key to good flood monitoring systems, especially where different areas have varied hydrological patterns. In a riverine floodplain, e.g., the Assam floodplain, the progressive increase in water levels and changes in the flow are monitored by the sensors placed along the upstream and downstream parts of the larger rivers. The strategic points are confluences, bends, and those points that have the historical accumulation of floods so that the gradual and the peak flow patterns are captured. The combination of rainfall sensors in the upstream catchments with river-level measurements gives early warnings on possible inundation, which the predictive models can use to draw correlations on the intensity of precipitation and the level of river stage alteration.

The high responsiveness to hydrological changes and steep topography of a location such as Kerala demand dense and highly responsive sensor networks because of the flash-flood-prone regions. Placement was concerned with small tributaries, slopes of hills, urban drainage systems, and low valleys where sudden water accumulation was taking place. Harmonic spatial coverage means localized events are identified before getting beyond the limits into downstream floods. Redundancy within the critical areas enhances high fault tolerance and continuity of data during extreme weather conditions. Accessibility in terms of maintenance, resilience to environmental risks, and integration with communication networks was also taken into account as an optimal positioning of a robust infrastructure that can be used to deliver real-time flood prediction and disseminated early warnings.

4.4 Communication Protocols for Reliable Data Transmission

The high-quality data transfer was a significant part of the IoT-based hydrometeorological sensing system, as in this way, real-time data obtained by rainfall and river level measurements are sent to platforms of processing and analysis without failures. The problems associated with remote sensors in flood-prone areas usually include power limitation, poor weather conditions, and geographical remoteness. Low-powered, long-range, and resilient communication protocols allow continuously connected distributed nodes to centralized or edge-computing systems. Procedures like LoRaWAN and NB-IoT have a long-range transmission capability with minimum energy consumption and thus can be used in locations with low infrastructure.

Communication protocols of hydrometeorological IoT networks are designed in terms of reliability, latency, and fault tolerance. The sensor nodes produce constant floods of data, which need to be sent on the fly to facilitate predictive flood modeling and generation of alerts. The mechanisms used in protocols to ensure data integrity in changing network conditions are automatic repeat requests, error correction, and packet prioritization. The topology of the network, such as star, mesh, and hybrid structure, affects the reliability of the transmission by offering alternatives, i.e., redundancy. Flexible and adaptive communication policies dynamically modify the rate of transmission and the size of the information packet to address network congestion, energy limitations, or emergency situations and deliver high-priority hydrological data in time.

Figure 8. Communication Protocols for Reliable Data Transmission

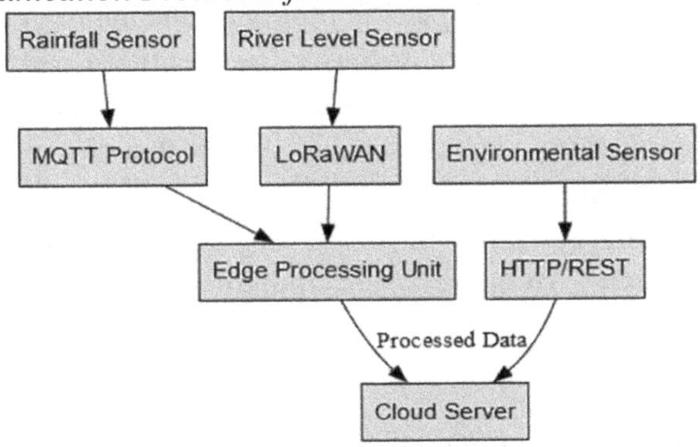

Embarkation of communication protocols and edge and cloud computing improves the general performance and resilience of the system. Edge nodes are used to process sensor data, which decreases the amount of data sent over the network and also minimizes the load on the network, at the same time providing low-latency responses in flood prediction models. The cloud-based solutions gather information across various locations and can analyze this data on a large scale and compare trends over time. Sensitive data related to the environment and infrastructure was secured through the use of encrypted and authenticated communication channels that ensure sensitive data was not tampered with or accessed by unauthorized individuals.

4.5 Energy-Efficient and Fault-Tolerant Sensor Network Design

In designing an energy-efficient and fault-tolerant sensor network to monitor floods, power management, network topology, and hardware reliability must be carefully looked upon. Remote area sensors or floodplain sensors need to be able to work long hours with little maintenance and require low-power communication protocols and energy collection schemes like solar sources or kinetic sources. Duty-cycling techniques are used to optimize the sensor and enable it to be in low-energy states during inactive periods and to be broadcast with a high-priority event. The network arrangement should be struck between coverage and redundancy so that any specific points of riverbanks, tributaries, and low-lying communities are always covered without the needless energy wastage.

Fault tolerance was required to ensure that the flood monitoring system was not compromised when there are severe environmental conditions. Sensor nodes are also vulnerable to water submersion and impact of debris and corrosion, and communication links fail due to poor weather conditions or topographical impediments. Redundancy of the pathways and self-healing routing algorithms make sure that data was still delivered to the central processing devices or edge devices in case of failure of individual nodes. Consistent diagnostic programs and anomaly identification on the network can make malfunctioning sensors be identified early and can trigger automatic recalibration or alert maintenance personnel. This energy-efficient and fault-tolerant system creates a robust sensing system that can withstand real-time observation in case of extensive floods.

However, despite these advancements, existing systems fail to provide localized and adaptive predictions. To address this limitation, the proposed framework is introduced in the following section.

5. PROPOSED AIoT-BASED FLOOD MONITORING FRAMEWORK

The proposed system follows a multi-layered architecture consisting of sensing, communication, edge computing, cloud analytics, and application layers.

- The sensing layer includes rainfall and river water-level sensors deployed across flood-prone regions
- The communication layer ensures reliable data transmission using IoT protocols such as LoRaWAN and NB-IoT
- The edge layer performs real-time preprocessing and anomaly detection to reduce latency
- The cloud layer enables large-scale data storage and predictive modeling using machine learning algorithms
- The application layer provides visualization dashboards and automated alert systems for decision-makers

This integrated architecture enables continuous monitoring, real-time prediction, and efficient dissemination of flood alerts.

Figure 9. Edge Computing for Low-Latency Flood Risk Assessment

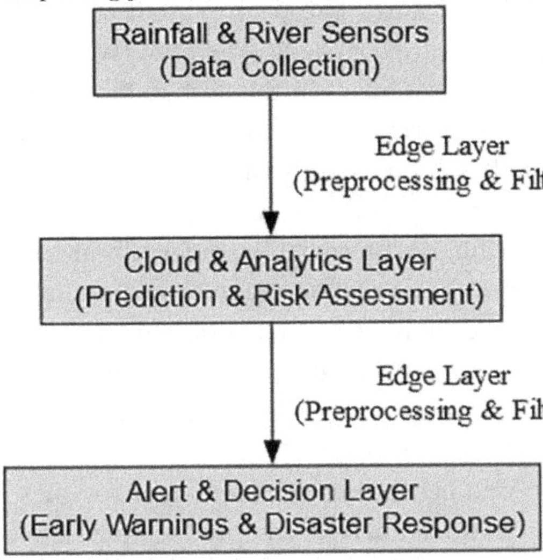

Figure 10. Data Integration and Sensor Fusion Techniques

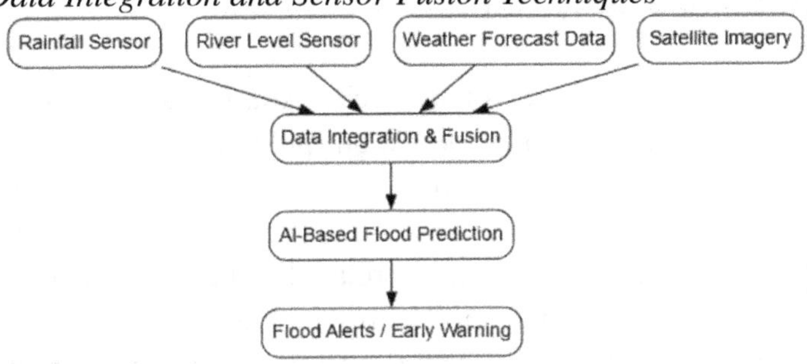

6. AI-DRIVEN FLOOD PREDICTION AND RISK ASSESSMENT MODELS

6.1 Time-Series Analysis of Rainfall and River Level Data

Predictive flood modeling was based on time-series analysis of rainfall and river level data, as it allows detection of temporal patterns, trends, and correlations of hydrological behavior. Sequential flows of data provided in real time by distributed sensors capture changes in the intensity and stage of a river and precipitation measurements. These sequences are subjected to statistical and machine learning methods to obtain important features like peak rainfall and rate of river rise and lag between the time that rainfall occurs and when the river responds. This methodology would enable the measurement of the probability of flood onset plus the estimation of critical values of various river basins and catchment areas upon which early warning systems are based on a data basis.

The identification of the temporal dependencies in the data can assist in predicting the occurrence of floods in the short-term and in the medium-term, which helps the alert systems to respond more efficiently. Autoregressive and chronological-moving-average models, as well as recurrent neural network structures, are used to learn the intricate nonlinear interactions between the past and the present hydrological state to make correct predictions of impending river level increases. The combination of rainfall and river stage sequences allows the evaluation of all the cumulative effects on the flood risk, taking into account antecedent water regimes and the intensity of the present precipitation. Such analysis can produce predictive flood maps and risk indices that can be used to make proactive decisions and allocate resources in the risky regions.

6.2 Machine Learning Techniques for Flood Probability Estimation

Machine learning methods offer robust mechanisms for estimating the probability of a flood based on complex dependencies among precipitation, river height, and past instances of floods. Decision trees, random forests, and gradient boosting methods are some of the supervised learning algorithms that could be trained on previous hydrometeorological data to detect patterns related to the occurrence of floods. The observed floods are overlaid with input features, including cumulative rainfall, river stage trends, soil moisture, and upstream flow conditions that allow the system to categorize the conditions into risk levels that have varying degrees. This was an inference-based methodology that nonlinear dependencies that are not well captured by traditional statistical models are better reflected in this method, enhancing the precision of estimation of flood probabilities in a myriad of hydrological settings.

Unsupervised learning (e.g., clustering and anomaly detection algorithms) was used to supplement supervised models (in identifying abnormal hydrological behavior, which can occur before the occurrence of a flood). These techniques identify anomalies in the normal distribution of rainfall and river level without specified flood events to give early warnings of possible dangers. Continuous high-resolution data was also provided by sensor networks and was critical in identifying sudden changes in rainfall or rapid rises in river stage. Both supervised and unsupervised methods can be combined to make probabilistic measurements of the risk of flooding by determining the probability of inundation in real-time and improving disaster preparedness and resource allocation decision-making.

Figure 11. Machine Learning Techniques for Flood Probability Estimation

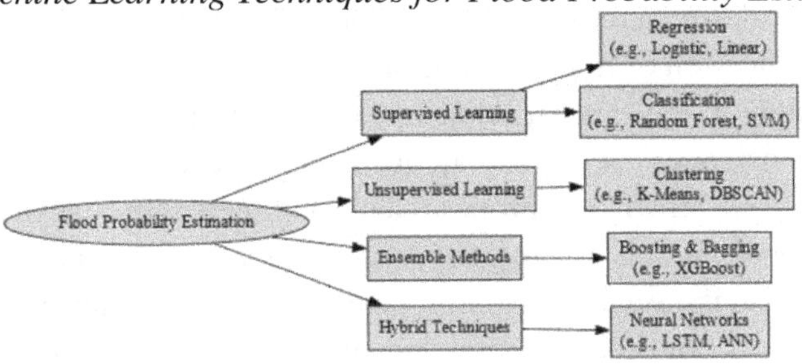

Ensemble learning techniques also increase the robustness of the model, as represent the joint prediction of multiple algorithms that decrease uncertainty and the effect of a single model bias. These models produce probabilistic outputs, which are used in threshold-based alert systems, which allow the risk levels of a particular geographic area to be classified as low-risk, moderate-risk, and high-risk in floods. Repeated retraining based on real-time sensor measurements makes the model parameters adaptively refined in order to enhance the predictive performance as the environment changes. The integration of machine learning-based flood probability estimation into AIoT systems can help to provide localized and timely warnings and support proactive actions to save communities and reduce economic and environmental impacts.

6.3 Deep Learning Models for Flood Onset and Severity Prediction

Deep learning models have substantial potential to forecast the onset of floods and the intensity of floods due to their ability to identify complex and nonlinear relationships between rainfall patterns, changes in the water level, and hydrological processes in rivers. Sequence-based architectures, i.e., Long Short-Term Memory (LSTM) networks and Gated Recurrent Units (GRUs), consume temporal data of rain and river level gauges to detect early signs of increase in water level. CNNs on spatial data such as river basin topography and satellite images can be used to increase the capacity of the system to identify vulnerable areas and determine the level of inundation. These models are constantly being trained on historical and real-time data, and are therefore better at making predictions and offer a scalable framework to be used with varying hydrological situations in Assam and Kerala.

The estimation of the severity of floods was based on the combination of different streams of data by predictive models that are able to distinguish between minor, moderate, and extreme events. The probabilistic outputs provided by deep learning frameworks are a measure of the probabilities and severity of floods; thus, risk-based classification of impacted regions was possible. The models can be used to allow variations in the rainfall intensity, soil saturation, river morphology, and catchment responsiveness, and generate localized forecasts and predictive powers and are specific to specific flood-prone areas. Combined with edge or cloud computing platforms, these models are able to be used to make near real-time predictions, which can be used to issue early warnings, prioritize resources, and plan evacuations. The synergistic effect of both temporal and spatial learning also assures timely and actionable predictions of flood occurrences and severity with a stronger resilience of floods as a whole.

6.4 Adaptive Thresholding and Anomaly Detection Mechanisms

The fraudulent mechanisms of adaptive thresholding and anomaly detection constitute an essential part of the AI-driven flood prediction systems because can dynamically detect the presence of unusual hydrological processes. The fixed thresholds of river heights or rainfall severity usually do not reflect the vagaries of various flood-prone areas and hence cause delays in alerts or false alerts. Adaptive techniques keep on processing the sensor's incoming data and changing the threshold levels in accordance with the trends in the time frame, seasonal changes, and hydrological conditions of the upstream. These mechanisms can be used to improve the precision of the early warning systems by having the alert thresholds adjusted based on the river basin-specific behavior to prevent the timely detection of the potential floods.

The method of anomaly detection was used with the adaptive thresholding to recognize the trends within the data that do not conform to the typical hydrologic patterns. The algorithms described as machine learning, such as clustering, support vector machines, and neural network-based models, are trained to identify the normal relationship between rainfall and runoff and changes in river stage. The deviations of the learned patterns, like the sudden increase of water levels or unexpected collection of rainfall, cause alerts to continue the analysis. Such methods can be observed especially when it comes to flash floods, which have very small durations of time and cannot be detected through the traditional monitoring methods.

6.5 Region-Specific Model Customization for Assam and Kerala

The models of flood prediction are more accurate when are developed according to the peculiarities of hydrology and climate of certain regions. Riverine floods in Assam are gradually formed, as there was a long monsoon rainfall and great variation in discharge in the Brahmaputra basin. The predictive models of this area revolve around accumulation of rainfall over a long period of time, river stage dynamics, and the dynamics of sediment transport. Addition of historical records of past floods, topography of the basin, and the patterns of past precipitation upstream enables the models to simulate the slow onset nature of riverine floods, which gives adequate lead time to the community to carry out warning and mobilization of resources.

Figure 12. Region-Specific Model Customization for Assam and Kerala

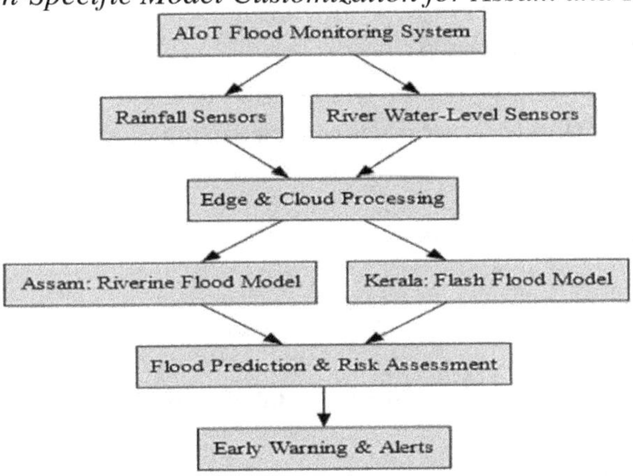

It was the opposite of the situation in Kerala, as flash floods are caused by rainfall of high intensity but short length over steep roads. Their models tailored to this area give more emphasis on high-resolution time data and quick river response modelling. The prediction of short-term rainfall, runoff coefficient, and local drainage properties are combined to model the development of flash floods and the maximum water velocity realistically. The predictive framework focuses on dynamic thresholding and real-time signal integration of dense sensor networks in order to provide an explanation of abrupt hydrological variations. Regionalization of model architectures to the local flood patterns also makes alerts accurate, timely, and actionable to aid emergency responses and reduce the socio-economic and environmental effects in both states.

7. MATHEMATICAL FORMULATION OF FLOOD PREDICTION MODEL

Flood prediction can be expressed as a multivariate function of hydrometeorological parameters:

$$F(t) = f(R(t), WL(t), SM(t), \Delta t)$$

where:

$R(t)$ denotes rainfall intensity at time t,
$WL(t)$ represents river water level,
$SM(t)$ indicates soil moisture content, and
Δt is the temporal lag between rainfall and hydrological response.

Flood Probability Estimation

The probability of flood occurrence is modeled using a logistic regression function:

$$P(\text{Flood}) = \frac{1}{1 + e^{-(w_1 R + w_2 WL + w_3 SM + b)}}$$

where:

w_1, w_2, w_3 are weight coefficients associated with each variable, and

b is the bias term.

Adaptive Thresholding Mechanism

Dynamic flood thresholds are computed as:

$$T_{\text{adaptive}} = \mu + k\sigma$$

where:

μ is the mean water level,
σ is the standard deviation, and
k is a scaling factor controlling sensitivity.

Temporal Prediction Using Deep Learning

Temporal dependencies in hydrological data are captured using Long Short-Term Memory (LSTM) networks:

$$h_t = \mathrm{LSTM}(x_t, h_{t-1})$$

where:

x_t represents the input feature vector at time t, and
h_t denotes the hidden state capturing temporal patterns.

8. PROPOSED ALGORITHM FOR FLOOD PREDICTION

Step 1: Collect real-time rainfall and river-level data from IoT sensors
Step 2: Preprocess data using edge computing (noise filtering, normalization)
Step 3: Perform feature extraction (rainfall intensity, rate of water rise)
Step 4: Apply machine learning model to estimate flood probability
Step 5: Compare results with adaptive threshold values
Step 6: Generate flood risk level (Low, Medium, High)
Step 7: Disseminate alerts via cloud-based systems and mobile applications

This workflow ensures continuous monitoring and real-time decision-making.

9. DISCUSSION

The proposed AIoT-based flood monitoring system enhances predictive accuracy and response time compared to traditional methods. The integration of real-time sensing and machine learning enables early detection of flood events, particularly in flash flood-prone regions such as Kerala.

Challenges remain in large-scale deployment, including sensor reliability, communication infrastructure, and data quality issues. The dependency on continuous data streams and the need for model retraining under changing climate conditions are additional considerations.

Future work may include integration with satellite data, digital twin models, and hybrid AI approaches to further improve prediction capabilities.

CONCLUSION

Unpredictable weather patterns, unplanned urbanization, and complicated hydrology have led to floods in areas like Assam and Kerala, which have continued to be a great threat to human life, economic and environmental sustainability. Conventional monitoring and early warning systems are proved to be rather ineffective based on their reliance on manual measurements, centralized approaches, and fixed threshold-based forecasts, which lead to delayed response and poor preparedness. The Artificial Intelligence (AI) and the Internet of Things (IoT) system that was known as an AIoT-based flood alert allows addressing these shortcomings by allowing continuous and real-time monitoring, predictive analytics, and a machine-based risk measurement. High-resolution data can be collected and processed in low latency through distributed rainfall and river water-level sensors and edge and cloud computing infrastructures to support the accurate identification of deviant hydrological changes and possible floods.

The tailoring of the model to specific regions will improve the predictive accuracy because it will consider unique hydrological factors of Assam and Kerala. Assam Riverine floods are trend-driven, cumulative rainfall processes that require trend-based cumulative rainfall models, whilst Kerala flash floods are fast-responsive prediction processes that can handle high tonnage and short-period precipitation information. The AIoT system can be used to identify localized flood behavior by adapting to it through machine learning and deep learning models and analyzing anomalies and risk measures. The capabilities can deliver accurate and actionable early warnings, emergency planning, evacuation plans, and effective distribution of disaster response resources.

Linkage with disaster management systems also reinforces the effectiveness of the systems through the ability to disseminate alerts on multiple levels, coordination across agencies, and situational awareness on community levels. It has high temporal and spatial resolution information to help make informed decisions under fast-changing flood situations to minimize socioeconomic losses and minimal impact on the environment. The fault-tolerant and energy-saving design of the system will provide resilience to its operation, even in remote areas with limited infrastructure, and will ensure that it continues to operate during extreme weather conditions.

Flood monitoring using AIoT was a resilient, scalable, and intelligent system in climate-prone areas, which will offer a platform through which the potential of calamities can be reduced proactively. Real-time sensing, predictive analytics, and automated alerting come together and create a paradigm shift in flood management, which was based on reaction rather than anticipation. The future developments that combine satellite images, the use of digital twins, and hybrid physics-based AI will be able to improve the accuracy of prediction and system adaptability even more.

Such a strategy does not only ensure the protection of local communities but also promotes sustainable resource management and planning infrastructure and long-term climate resilience, which was why AIoT-based flood warnings are an important part of intelligent disaster response measures in India and other settings around the world.

Performance Evaluation Metrics

The performance of the proposed AIoT-based flood prediction model is evaluated using standard metrics:

- Accuracy: Measures correct flood predictions
- Precision: Indicates reliability of positive flood alerts
- Recall: Measures ability to detect actual flood events
- RMSE (Root Mean Square Error): Evaluates prediction error
- MAE (Mean Absolute Error): Measures deviation from actual values

These metrics ensure that the model minimizes false alarms while maintaining high sensitivity to flood events. The evaluation framework enables comparison with existing flood prediction systems.

Case Study: Application in Assam and Kerala

To validate the proposed framework, a conceptual case study is considered for two flood-prone regions.

In Assam, the system analyzes cumulative rainfall and gradual river level rise to predict riverine flooding. The model provides early warnings with sufficient lead time, enabling evacuation and resource allocation.

In Kerala, the system detects high-intensity rainfall spikes and rapid increases in river levels. Real-time predictions allow immediate alerts for flash flood-prone areas, significantly reducing response time.

This case study demonstrates the adaptability and effectiveness of the proposed AIoT system in handling diverse flood dynamics.

REFERENCES

Adli, H. K., Remli, M. A., Wan Salihin Wong, K. N. S., Ismail, N. A., González-Briones, A., Corchado, J. M., & Mohamad, M. S. (2023). Recent advancements and challenges of AIoT application in smart agriculture: A review. *Sensors (Basel), 23*(7), 3752. DOI: 10.3390/s23073752 PMID: 37050812

Bhatia, V., & Jaglan, V. (2024). Integrating Artificial Intelligence and IoT for Smart Cities: Applications and Challenges. Reshaping Intelligent Business and Industry: Convergence of AI and IoT at the Cutting Edge, 267-278.

Boulouard, Z., Ouaissa, M., Ouaissa, M., Siddiqui, F., Almutiq, M., & Krichen, M. (2022). An integrated artificial intelligence of things environment for river flood prevention. *Sensors (Basel), 22*(23), 9485. DOI: 10.3390/s22239485 PMID: 36502187

Chen, H. S., Tam, K. I., Zhao, Y. L., Yuan, L., Wang, W., Lin, M., & Chiang, P. C. (2023a). Development of environmental action plans for adaptation to climate change: A perspective of air quality management. *Aerosol and Air Quality Research, 23*(10), 220377. DOI: 10.4209/aaqr.220377

Chen, L. B., Huang, X. R., Chen, W. H., Pai, W. Y., Huang, G. Z., & Wang, W. C. (2023b). Design and implementation of an artificial intelligence of things-based autonomous mobile robot system for cleaning garbage. *IEEE Sensors Journal, 23*(8), 8909–8922. DOI: 10.1109/JSEN.2023.3254902

Dhaya, R., & Kanthavel, R. (2022). IoT based urban flooding high definition surveillance using concurrent multipath wireless system. *Earth Science Informatics, 15*(3), 1407–1416. DOI: 10.1007/s12145-022-00817-4

ElMougy, S., Alrahmawy, M., El-dosuky, A. D. M., & Hamed, A. E. (2023) COMPREHENSIVE SURVEY OF AIOT APPROACHES FOR WATERLOGGING CRISIS MONITORING 19.

Halubanza, B. (2024). A framework for an early warning system for the management of the spread of locust invasion based on artificial intelligence technologies (Doctoral dissertation, The University of Zambia).

Kong, S. M., Yoo, C., Park, J., Park, J. H., & Lee, S. W. (2024). AIOT monitoring technology for optimal fill dam installation and operation. *Applied Sciences (Basel, Switzerland), 14*(3), 1024. DOI: 10.3390/app14031024

Maraveas, C. (2022). Incorporating artificial intelligence technology in smart greenhouses: Current State of the Art. *Applied Sciences (Basel, Switzerland), 13*(1), 14. DOI: 10.3390/app13010014

Mitra, A., Vangipuram, S. L., Bapatla, A. K., Bathalapalli, V. K., Mohanty, S. P., Kougianos, E., & Ray, C. (2022). Everything you wanted to know about smart agriculture. arXiv preprint arXiv:2201.04754.

Nerella, O., & Ahmed, S. M. (2023, March). Advance warning and alert system for detecting lightning risk to reduce human disaster using AIoT Platform—A proposed model to support rural India. In International Conference on Communications and Cyber Physical Engineering 2018 (pp. 925-935). Singapore: Springer Nature Singapore.

Pourroostaei Ardakani, S., & Cheshmehzangi, A. (2023). Big data-enabled time series analysis for climate change analysis in brazil: an artificial neural network machine learning model. In *Big Data Analytics for Smart Transport and Healthcare Systems* (pp. 141–156). Springer Nature Singapore. DOI: 10.1007/978-981-99-6620-2_8

Somashekhar Gaddanakeri, V. S., Singh, S., Samal, I., Dubey, V. K., & Bhoi, T. K. (2024). An In-Depth Analysis of Artificial Intelligence-Based Crop Pest Management and Water Supply. Artificial Intelligence Techniques in Smart Agriculture, 151.

Tham, M. L., Wong, Y. J., Kwan, B. H., Ng, X. H., & Owada, Y. (2023, January). Artificial intelligence of things (AIoT) for disaster monitoring using wireless mesh network. In *Proceedings of the 2023 6th International Conference on Software Engineering and Information Management* (pp. 234-239). DOI: 10.1145/3584871.3584905

Wu, S. J. (2024). Modeling an evaluation framework for adding IoT water-level sensors based on ANN-derived 2D inundation simulations. *Journal of Hydroinformatics*, *26*(9), 2261–2288. DOI: 10.2166/hydro.2024.097

Zhang, F. (2024). Design and Application of Microcontroller-Based Tunnel Construction Environment Monitor. *Tehnicki Vjesnik (Strojarski Fakultet)*, *31*(4), 1060–1068.

Chapter 7
Dynamic Multi-Hazard Forecasting Addressing Cascading Disasters Through Machine Learning-Based Approaches

K. Sandhiya Rajeshwari
http://orcid.org/0009-0004-4086-0264
Department of Computing Technologies, SRM Institute of Science & Technology, India

ABSTRACT

This chapter presents a multi-hazard prediction model that addresses the complexity of contemporary disaster scenarios, where multiple hazards interact and cascade in real time. Traditional disaster risk assessments, focused on single hazards and non-dynamic modeling, fail to capture the dynamic interdependencies in such situations. Leveraging advancements in data availability and computational intelligence, this model integrates machine learning, deep learning, and ensemble-based predictions to forecast cascading disaster mechanisms. By combining past data, real-time sensor measurements, satellite images, and socio-economic data into an adaptive, scalable pipeline, the model enhances forecasting accuracy, reduces latency, and increases sensitivity to risk trends. The approach enables proactive disaster risk mitigation and supports emergency management with actionable insights, advancing both theoretical and practical disaster management frameworks.

DOI: 10.4018/979-8-2600-2383-9.ch007

© 2027, IGI Global Scientific Publishing. Copying or distributing in print or electronic forms without written permission of IGI Global Scientific Publishing is prohibited. Use of this chapter to [ge]nerative artificial intelligence (AI) technologies is expressly prohibited. The publisher reserves all rights to license its use for generative AI training and machine learning model development.

INTRODUCTION

Natural disasters pose significant threats to human life, infrastructure, and economic stability, particularly in the context of rapid urbanization and the increasing interdependence of natural, technological, and socio-economic systems. In recent years, hazard events have evolved into complex cascading processes, where an initial disturbance propagates through interconnected systems, triggering multiple secondary impacts across spatial and temporal scales. For instance, floods can induce landslides, earthquakes may disrupt energy and communication infrastructures, and extreme weather conditions can exacerbate technological failures. These interconnected dynamics substantially amplify risk and challenge the effectiveness of conventional disaster preparedness and mitigation strategies (He et al., 2025; Akiyama, 2025).

Traditional disaster forecasting and risk assessment approaches predominantly rely on statistical and deterministic models that assume independence among hazards and temporal stability. Although such methods provide interpretability and foundational insights, they are inherently limited in capturing nonlinear interactions, feedback mechanisms, and evolving hazard behaviors. Cascading disasters involve complex dependencies, delayed responses, and compound risk pathways that are difficult to represent using linear formulations. As a result, these limitations reduce predictive accuracy, particularly under dynamic and uncertain environmental conditions (Afandi et al., 2025; Zhao et al., 2024; Jeddi et al., 2022). The increasing complexity of hazard systems necessitates a transition toward adaptive and data-driven approaches capable of supporting real-time decision-making.

Recent advancements in data acquisition technologies have significantly transformed disaster monitoring and analysis. High-resolution satellite imagery, real-time sensor networks, and geospatial information systems generate continuous streams of environmental and infrastructural data, while socio-economic datasets provide critical insights into exposure and vulnerability patterns. The integration of these heterogeneous data sources enables a more comprehensive understanding of hazard dynamics and risk evolution. However, challenges related to temporal synchronization, data uncertainty, and computational scalability must be effectively addressed to fully leverage their potential (Razavi-Termeh et al., 2025; Geiß, 2025; Wang, 2023).

Machine learning and artificial intelligence techniques have emerged as powerful tools for modeling complex disaster systems characterized by high dimensionality and nonlinear interactions. Ensemble learning methods such as Random Forests provide robustness against noise and feature variability, while deep learning models effectively capture temporal dependencies essential for understanding hazard evolution and cascading effects (Wang et al., 2023; Yasmin et al., 2025; Valarmathi et al., 2025). Furthermore, hybrid and ensemble frameworks enhance predictive performance by combining the strengths of multiple algorithms, thereby improving accuracy and computational efficiency in multi-hazard forecasting scenarios (Ferrario et al., 2025; Palanikkumar et al., 2025; Pugliese Viloria et al., 2024).

In addition, recent studies emphasize the importance of adaptive forecasting models that incorporate real-time data assimilation and online learning capabilities. Such models can dynamically update predictions as new data becomes available, enabling more responsive and accurate disaster risk assessments. The integration of machine learning with Internet of Things (IoT) systems and intelligent agents further enhances real-time predictive intelligence and supports effective disaster management strategies (Masud et al., 2025; Hamid & Abedlmajid, 2025; Fearnley, 2024). Emerging technologies such as digital twins and multi-agent systems are also gaining attention for enabling real-time simulation, resource allocation, and decision support in disaster scenarios (Shen & Zhou, 2025; Wu & Zou, 2025; Yang et al., 2025).

Recent research also highlights the importance of integrating social vulnerability, infrastructure resilience, and multi-modal intelligence into disaster risk frameworks. These approaches emphasize the need to consider human, environmental, and technological factors simultaneously to improve resilience and response strategies (Zhang et al., 2023; Wang, 2025; Sharma et al., 2025). Multi-hazard and multi-risk assessment frameworks further demonstrate the potential of combining diverse data sources and analytical techniques to enhance situational awareness and predictive accuracy (De Maio et al., 2024; Ferrario et al., 2025).

Despite these advancements, existing approaches predominantly focus on single-hazard scenarios and often fail to capture the interdependencies and cascading nature of modern disaster systems. Additionally, challenges related to data heterogeneity, scalability, and model interpretability remain insufficiently addressed. These limitations highlight a critical research gap and underscore the need for comprehensive multi-hazard forecasting frameworks capable of integrating diverse data sources and adapting to dynamic environmental conditions.

This chapter presents a machine learning-based framework for multi-hazard disaster forecasting, with a focus on improving predictive accuracy and enhancing decision-support capabilities. The proposed approach aims to integrate heterogeneous datasets, model complex hazard interactions, and provide a scalable and adaptive solution for real-world disaster management applications..

OBJECTIVES

- To investigate weaknesses of the conventional one-hazard forecasting models and develop the necessity of the dynamic and multi-hazard models, which may be used to consider cascading disaster processes.

- To create a conceptual and methodological basis of integration of heterogeneous sources of data, such as historical data, real-time sensor data, satellite data, and socio-economic data, into scalable forecasting pipelines.
- To examine the purpose of machine learning, deep learning, and ensemble-based methods in modeling non-linear interactions and temporal dependencies existing in multi-hazard disaster environments.
- To compare hybrid and adaptation forecasting models by considering predictive accuracy, computational efficiency, and real-time responsiveness in cascading disaster occurrences.
- To illustrate the use of predictive outputs in converting them into actionable intelligence to mitigate disaster risks, early warning systems, and decision-support applications.

POLICY, GOVERNANCE, AND DECISION-MAKING IN MULTI-HAZARD FORECASTING

Beyond technological advancements, effective disaster forecasting requires strong policy support and governance frameworks. Multi-hazard forecasting systems must be integrated into institutional decision-making processes to ensure timely and coordinated disaster response. Governments and disaster management agencies play a critical role in translating predictive insights into actionable strategies, such as evacuation planning, resource allocation, and early warning dissemination.

The success of data-driven forecasting systems depends on inter-agency collaboration, data-sharing policies, and standardized operational protocols. The integration of forecasting models into decision support systems can enhance situational awareness and enable proactive risk management. However, challenges such as data privacy, institutional silos, and limited technical capacity in developing regions may hinder implementation.

In effective communication of uncertainty in model predictions is essential for informed decision-making. Policymakers must be equipped with interpretable and transparent forecasting outputs to balance risk, cost, and response strategies. Therefore, aligning technological innovations with governance frameworks is crucial for maximizing the societal impact of multi-hazard disaster forecasting systems.

CONCEPTUAL FOUNDATIONS OF CASCADING AND MULTI-HAZARD DISASTERS

Understanding Cascading Disaster Mechanisms in Interconnected Hazard Systems

Cascading disasters are a result of complex interactions in hazard systems in which natural, technological, and socio-economic processes interact to produce chains of compounding effects. The difference in such systems is that the triggering hazard causes a secondary process to occur due to physical proximity, functional dependency, or systemic weakness, and the effect propagates in the various sectors. An example of this interdependent behavior is earthquakes that cause landslides, floods that damage energy networks, or extreme weather events that cause instability of transport and communication systems. These interactions take place in space and time and have their outcomes, which are greater than the sum of the effects of the separate hazards, which are individually taken into account.

The interdependence of hazard systems is that they have feedback mechanisms that increase the spread of a risk and the degree of disruption on a system-wide level. Ecological systems, human settlements, and infrastructure networks exist as closely interlocked systems in which stress in one domain propagates swiftly into others in an infrastructure network. The cascading pathways are dynamic processes that vary with changes in the environmental conditions and with variation in the pattern of exposure and vulnerability in the human response. The disaster procession that has taken place thereof portrays nonlinear behavior, uncertainty, and changing dependence and makes a fixed or chronological hazard model incomprehensible to overall risk understanding.

Figure 1. Understanding Cascading Disaster Mechanisms in Interconnected Hazard Systems

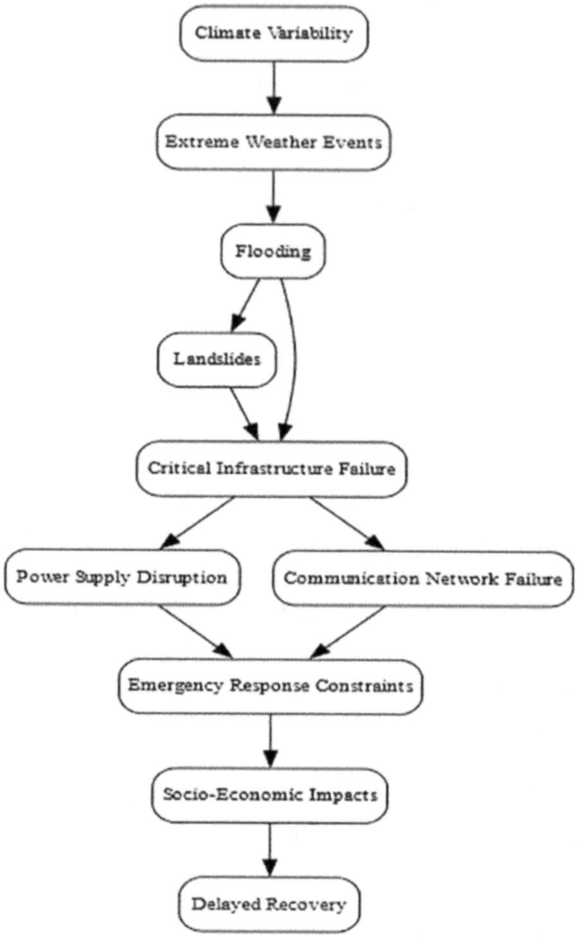

Conceptualization of cascading disaster mechanisms involves the realization that hazards are interdependent processes that exist as parts of adaptive systems. Risk is an outcome of interactions between hazards, vulnerabilities, and capabilities to respond as opposed to individual incidents. A system-based approach aids in identifying strategic slip-over points where small-scale crises start growing into crises of large scale. This kind of understanding provides a foundation for dynamic multi-hazard forecasting methods, which attempt to capture changing disaster behavior in an interconnected hazard environment.

Multi-Hazard Risk Theory: From Independent Events to Interacting Systems

The multi-hazard risk theory has developed from the early models of analytic theory, where hazardous events were viewed as isolated occurrences, to integrated views that acknowledge the interdependent and sequential behavior of hazards. Older risk models focused on the single estimation of the probability of hazards and their effect and frequently made the assumption of independence of hazards and stability over time. These assumptions were also consistent with a lack of data and computer power, but that offered limited explanatory power in complex disaster scenarios, where cascading effects dominate the results.

Modern disaster systems are characterized by close relationships between the natural processes, constructed environments, and socio-economic systems. Earthquakes may trigger landslides, infrastructure collapse, and secondary flooding, and severe weather conditions may lead to compound events caused by floods, heatwaves, and disruption of the power system. Such environments give rise to risk in terms of dynamic feedback loops and amplification of cross-hazards as opposed to cause-effect chains. Multi-hazard risk theory hence views disasters as dynamic systems that are governed by time dependency, spatial interrelations, and system vulnerability.

This conceptual change justifies the need for forecasting models that will reflect the aspects of interaction, propagation, and escalation within the hazard domains. The methods of its static assessment do not reflect the dynamic process of the primary-to-secondary hazards transition or the systemic stress gradualness between the interrelated elements. The view of the cascading disasters as part of an interacting system positions the risks as an ever-changing analytical space where the environment acts as a forcing factor, and structural interdependence is acknowledged as a driving factor. These theoretical underpinnings give reason to the dynamic machine learning-driven forecasting methods that have the potential to deal with cascading behavior of disasters in the operational decision-support scenarios.

Systemic Vulnerability and Feedback Loops in Disaster Propagation

Systemic vulnerability in the context of disaster environments is a result of structural interdependence among natural processes, man-made infrastructures, and socio-economic systems. Contemporary settlements are highly interconnected systems with the provision of energy, transport, communication, and water supply and emergency services being highly interconnected. The impact of disruption in a single network often spills over to the functioning of several industries, exposing them further and causing a speedy rise in risks. These interconnections allow spreading the consequences of disasters beyond the original sources of hazards and converting local disruptions into extensive crisis situations. Such interactions are complex enough to put a strain on linear risk assessment methods and require perspectives based on systems that can capture the interrelated failure pathways.

Figure 2. Systemic Vulnerability and Feedback Loops in Disaster Propagation

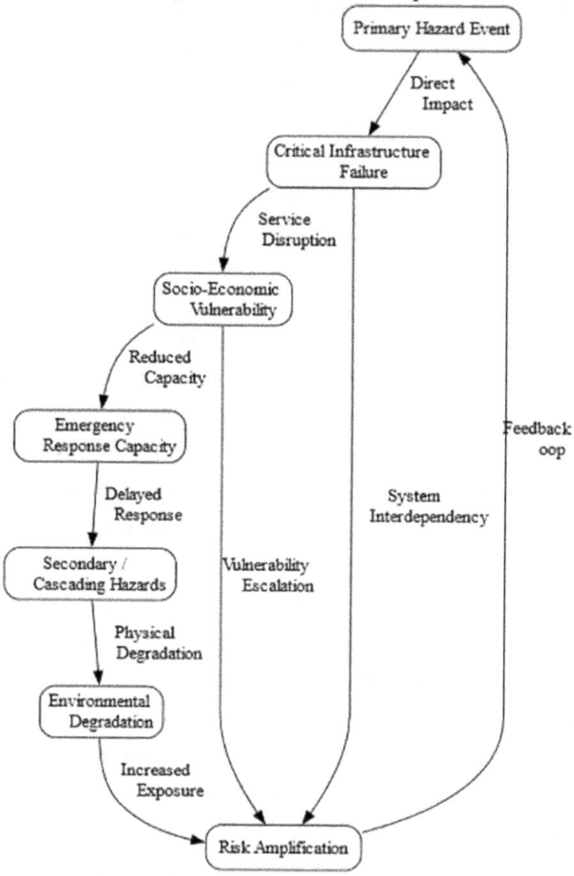

The importance of feedback loops in the control of multi-hazard environments is central to the control of disaster propagation. Primary hazard effects tend to create secondary conditions that make the area more vulnerable, e.g., a damaged infrastructure, displacement of population, or environmental destabilization. These secondary effects bolster exposure to other risks, posing reinforcing loops that increase the overall destruction and increase the recovery times. As an illustration, when power outages are caused by flooding, the emergency response is limited, and this increases vulnerability to health crises and other secondary risks. These recursive interactions have a role in nonlinear disaster dynamics, in which small perturbations have disproportionate results both spatially and temporally. Identification of such feedback processes is considered the key to further development of forecasting models aiming at capturing the cascading disasters behavior in dynamic risk systems.

The proposed approach follows a structured methodology involving data collection, preprocessing, feature extraction, and model development. Initially, heterogeneous datasets are gathered from relevant sources and preprocessed to remove inconsistencies and noise. Subsequently, feature selection techniques are applied to identify significant parameters influencing disaster occurrence. The processed data is then used to train machine learning models capable of capturing spatial and temporal patterns.

Challenges in Modeling Compound and Sequential Hazard Interactions

Simulating compound and sequential hazard interactions constitutes an important scientific and methodological issue in the research of multi-hazard disasters. The nature of cascading disasters is that an action by one hazard changes environmental, infrastructural, or social conditions in a manner that results in subsequent hazards in space and time. These interactions provide nonlinear behavior, feedback, and delayed effects, which cannot be well-modeled using traditional analytical constructs. The changing quality of the relationships between hazards means that prediction undertakings are difficult, with interaction pathways being based on dynamic exposure patterns, infrastructure susceptibility, and time overlap between hazard occurrences.

The sequential hazard interactions also complicate the models by having time-dependent propagation mechanisms. Triggering events commonly alter the states of a baseline system and produce dynamic risk landscapes, which evolve over the duration of the disasters. Static modeling models do not represent such dynamic dependencies, as they cause the partial model of cascading risk. The ability to make data available, run hazard indicators at different times, and the variability of the duration of occurrence pose high uncertainty to the creation of common forecasting frameworks. The sound representation of the hazards of compounds would thus require modeling solutions with the ability to learn dynamic patterns and the ability to capture the uncertainty, scale variance, and transformation of the system in real time.

Limitations of Single-Hazard Paradigms in Contemporary Disaster Risk Assessment

One faction of scientific practice and research in disaster risk assessment, single-hazard paradigms, represents the tradition of analysis that assumes the independence and isolation of risks as hazards. These paradigms construct disasters as defined by close, physical processes, habitually focused on the estimation of probability and impact of sets of natural, technological, and socio-economic systems. In the changing and tricky environmental and city environments, this small-scale framing does not reflect the chronological activation processes and cross-sectoral disturbances that define cascading calamity events. Consequently, risk appraisals based on single-hazard standpoints often under-scale the impacts of compounds and give a partial account of the disaster dynamics in real life.

Figure 3. Limitations of Single-Hazard Paradigms in Contemporary Disaster Risk Assessment

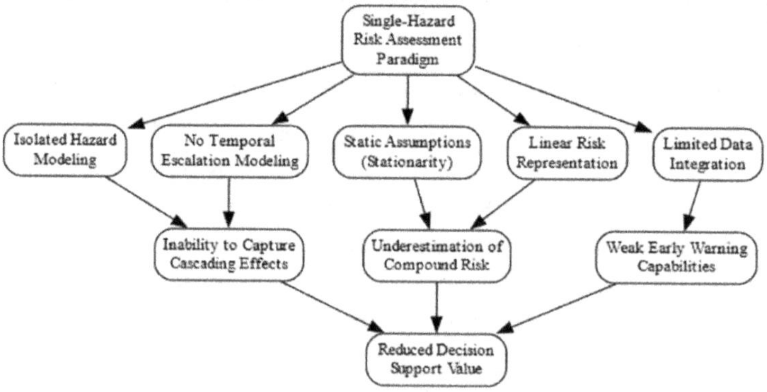

The increasing interdependence of critical infrastructure, extremes caused by climate, and human systems reveal some underlying vulnerabilities in the single-hazard risk assessment paradigms. Such frameworks have limited ability to address feedback, time escalation, and space spillover effects due to the interaction of hazards because of linear modeling conditions and the use of a static exposure representation. In disaster situations where the risk rapidly changes, the use of isolated hazard assessments limits proactive planning and undermines the ability to use early warning. The conceptual change in developing disaster risk science thus needs to be multi-hazard and integrated to capture dynamic interactions and cascading effects among complex socio-environmental systems.

This discussion of conceptual foundations leads naturally into the need for advanced data integration techniques, which are explored in the following section.

HETEROGENEOUS DATA INTEGRATION FOR DYNAMIC DISASTER FORECASTING

Fusion of Historical, Real-Time, and Remote Sensing Data for Multi-Hazard Analysis

The combination of historical data, on-site monitoring, and distance sensing data is one of the foundational prerequisites of successful multi-hazard analysis in dynamic disaster forecasting systems. Given the long-term contextual information on frequency, magnitude, and patterns of disasters, the historical data on disasters enables the characterization of baseline risks and identification of trends. In comparison, real-time sensor streams provide changing environmental conditions at a high rate, which can be used to support situational awareness throughout the unfolding events. This information space is supplemented with remote sensing data, which provides important data on land use change, atmospheric conditions, surface deformation, and environmental stressors associated with cascading hazard development, and that is provided in large volumes and continuously over space.

These heterogeneous data sources are important to be integrated because they assist in a more detailed representation of the disaster processes in time and space. Spatial harmonization and temporal alignment enable correlations amid antecedent circumstances and emergent hazards to be analyzed through a single framework of analysis. The data fusion increases the ability to identify the initial indicators of the hazard interaction, track the escape routes, and assess the risk scenarios of the compounds. In the dynamic forecasting environment, these combined datasets can provide adaptive model reaction and enhance predictive power, facilitating informed decision-making in complicated multi-hazard situations.

Managing Data Uncertainty and Noise in Disaster Prediction Systems

Dynamic systems of disaster prediction are based on streams of data that have different points of origin, which differ in their spatial resolution, their temporal frequency, and their reliability. Measurements of sensors in real-time, satellite data, past data, and socio-economic data are usually inconsistent due to measurement errors, missing data, delays in the transmission of data, and interference of the environment. In multi-hazard forecasting situations, this type of uncertainty and noise infects interconnected data streams and affects the process of feature extraction and model learning. Data flaws that have not been resolved cause bias in predictive results and make forecasting systems that run under the unstable conditions of a disaster less stable.

Figure 4. Managing Data Uncertainty and Noise in Disaster Prediction Systems

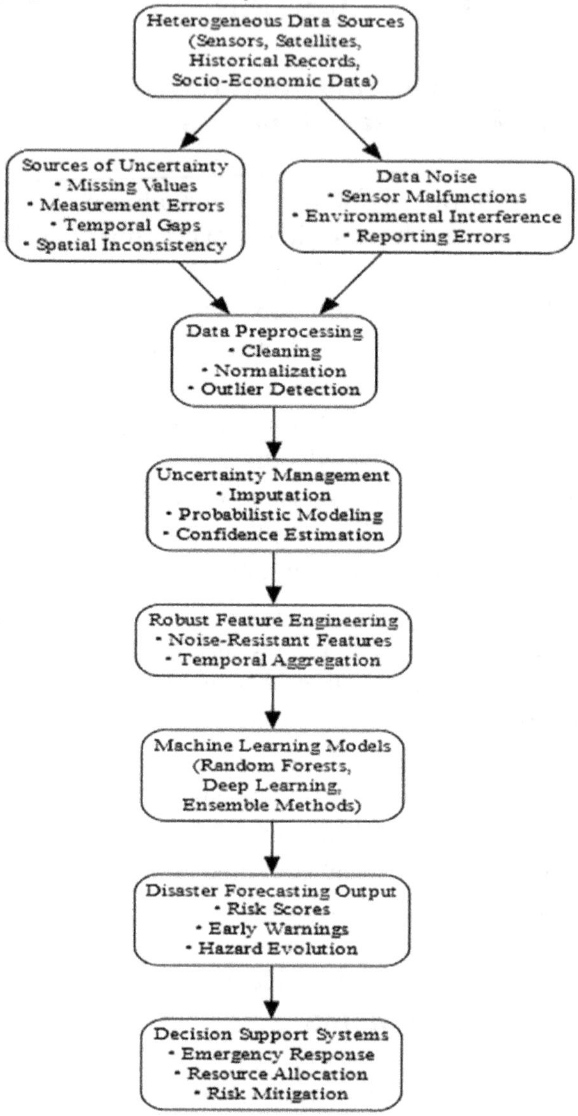

The ability to handle uncertainty and noise in data is one of the key problems of creating trustworthy disaster forecasting measures. Strong preprocessing schemes, data fusion schemes that are sensitive to uncertainty, and adaptive learning systems help in reducing the effects of erroneous or missing observations. Temporal matching and spatial coherence are highly important in maintaining meaningful correlations between different datasets, whereas noise-tolerant modelling strategies increase predictive stability when dealing with ambiguous situations. Enhancing the management of uncertainty in dynamic forecasting systems can help in creating more reliable predictions of hazards and enhance the reliability of the decision-support products in complex, evolving disasters.

Temporal and Spatial Synchronization of Multi-Source Disaster Data

Time and space synchronization is one of the key problems of dynamic disaster prediction systems based on different information sources. The sources of information related to disasters are highly varied, as they can be provided through satellite measurements, ground-based measurements, historical data, and socio-economic data, each having a particular timing resolution and a specific spatial resolution. Sporadic sampling rates, lateness of reporting, and spatial mismatch bring fragmentation to integrated datasets, which restrict predictive model coherence. In the absence of proper synchronization mechanisms, forecasting systems will be prone to generate distorted images of the changing hazard conditions and low situational awareness upon time-sensitive disasters.

Proper time and space synchrony allows specification of multi-source data to a single analytical framework capable of representation of the dynamic hazard interactions. Temporal techniques, which aid in aligning real-time streams with the past, and the spatial techniques, which aid in consistency across the georeferenced datasets, are supported by temporal and spatial registration, respectively. This harmonization increases the ability of the forecasting models to identify occurring patterns, trace the cascading progress of hazards, and provide useful predictions under the fast-changing conditions. Strong synchronization plans are thus a supportable condition to the adaptive and scalable multi-hazard forecasting platforms.

Feature Engineering Strategies for High-Dimensional Hazard Datasets

The complexity and dimensionality of multi-source hazard datasets make feature engineering an important part of the dynamic disaster forecasting process. Huge volumes of heterogeneous variables are created by high-resolution sensor feeds, satellite observations, historical data, and socio-economic indicators, which must be structured and transformed prior to predictive modeling. Successful feature engineering facilitates the extraction of meaningful patterns by minimizing noise, issues with scale, and the latent relationships between spatial and temporal dimensions. With temporal aggregation, statistical transformation, and domain-inspired encoding, raw hazard data are transformed to representations that tend to depress the changing disaster processes and interdependent risk behavior.

Figure 5. Feature Engineering Strategies for High-Dimensional Hazard Datasets

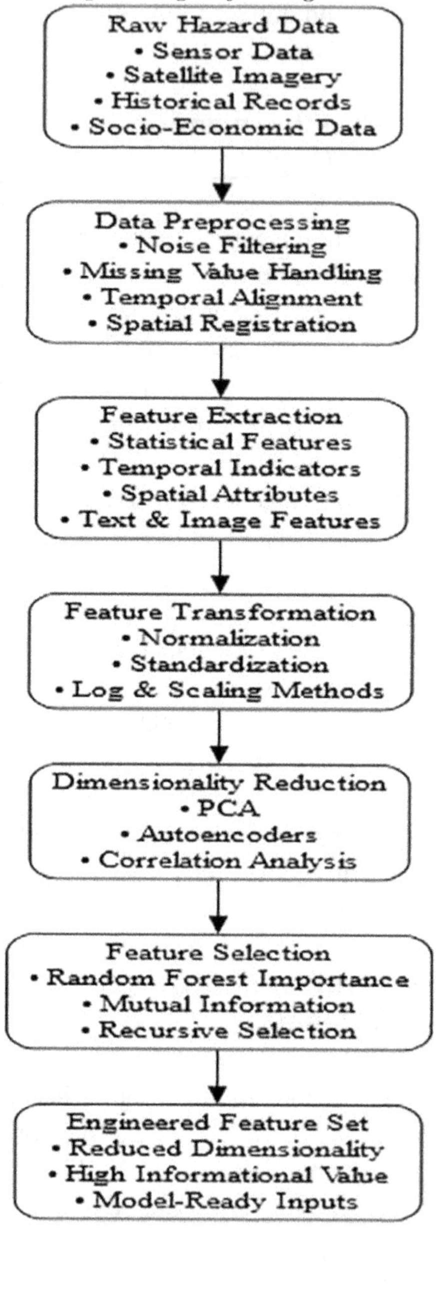

In hazardous high-dimensional feature spaces, feature strategies should be carefully designed to assist in predictive accuracy and computational efficiency. Dimensionality reduction methods and interaction-based features, as well as time-sensitive representations, increase the sensitivity of the model to cascading effects without putting undue strain on the learning architectures. The interpretability is also reinforced through context-aware feature selection that focuses on variables that are connected to physical processes and the vulnerability of a system. In dynamic forecasting systems, feature engineering thus plays an extremely important role as an interface between the integration of heterogeneous data and high-confidence real-time hazard prediction.

Scalable Data Pipelines for Real-Time Multi-Hazard Forecasting Applications

Real-time multi-hazard forecasting has a foundation in constant ingestion and processing of data produced by a wide variety of sources, such as environmental sensors, satellite systems, and socio-economic surveillance systems. Scalable data pipelines are a core part of the operation of such forecasting in supporting high-volume flows of data, fast preprocessing, and coordinated integration between different spatial and temporal resolutions. In a dynamic disaster setting, it is necessary that data pipelines should be able to handle the changing velocity of data and remain consistent and reliable so that incoming observations can be translated into actionable data for predictive models without latency-related degradation.

PIA environments Energy forecasting operational sites require the architecture of pipelines that can adapt to the growing data streams and changing hazard scenarios. It is possible to use the distributed processing frameworks and cloud-based infrastructures to permit the concept of elasticity, where the forecasting systems can scale the computational power based on the demand in real time. Scalable pipelines make possible the continuity needed for multi-hazard analysis by automatically validating data, normalizing, and extracting features, as well as allowing real-time features to be updated to models. Such architectures provide the basis of dynamic forecasting systems to transform heterogeneous data streams in real time into intelligence to respond to emergencies and mitigate risks.

This discussion of conceptual foundations leads naturally into the need for advanced data integration techniques, which are explored in the following section.

MACHINE LEARNING TECHNIQUES FOR MULTI-HAZARD FORECASTING

Ensemble-Based Machine Learning Models for Non-Linear Hazard Prediction

The ensemble-based machine learning models are suitable to deal with the non-linear relations, which are complex to predict the processes of multi-hazard disasters. Enhancing the robustness and generalization of the ensemble approaches in diverse hazard conditions, through their combination of several learning algorithms into a single predictive structure, is achieved. These models facilitate the combination of different feature representations based on environmental, infrastructural, and socio-economic information to permit complex patterns of interaction to be formed in the manner of collective learning behavior. Ensemble learning also helps to maintain consistent prediction performance in the presence of uncertainty and minimize sensitivity to noise and data imbalance, especially in the multi-hazard forecasting setting.

Figure 6. Ensemble-Based Machine Learning Models for Non-Linear Hazard Prediction

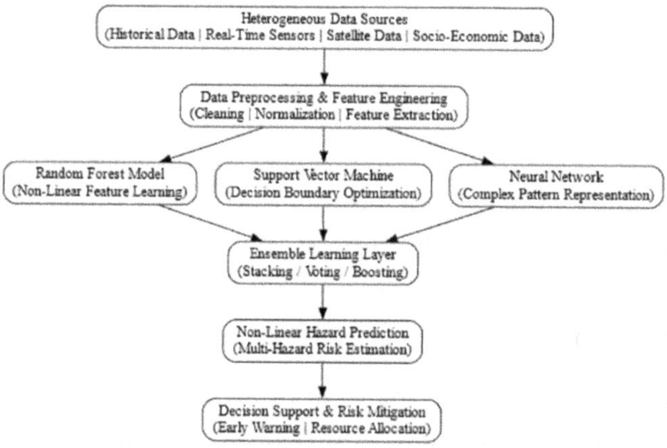

The non-linear hazard dynamics require modeling models that are able to capture complex space and time dependence. Ensemble-based methods obtain this ability by means of aggregation schemes that trade off bias and variance between individual learners. The group conclusion created by the group modeling of the ensemble model enhances predictive validity in the conditions of cascading disasters, when the interactions of hazards change quickly and irregularly. Ensemble-based machine learning models that can be trained and updated on demand, allowing real-time forecasting applications with a limited need for computational resources, make ensemble-based forecasting methodologies an essential building block of operational multi-hazard prediction systems.

Feature Selection and Interpretability Using Random Forest Algorithms

RF algorithms provide an effective method of feature selection in a multi-hazard forecasting framework that will operate on high-dimensional and heterogeneous data. Random Forest models find influential variables by building an ensemble of decision trees using randomized sampling, where the variables with a high contribution to the performance of the prediction are selected. This internal mechanism of feature selection helps to distinguish between the primary drivers of a hazard and the redundant or weak predictors, which then enhances the stability of the models in noisy and uncertain disaster conditions. This is necessary in the context of multi-hazard forecasting, where the interplay of environmental, infrastructural, and socio-economic variables needs a systematic prioritization.

Interpretability is a core prerequisite to working systems of disaster prediction, especially in the decision-support mode of operation. The interpretability of Random Forest models can be provided by the measures of variable importance and systematic decision processes, which allow the analysts to reconstruct predictive results in terms of hazard indicators. Such transparency improves the confidence in data-driven predictions and assists in the domain knowledge of cascading disaster processes. Random Forest algorithms have predictive performance and interpretability features that can be reconciled, and as a result, their application in practitioners of real-time multi-hazard risk assessment is justified.

Data-Driven Modeling of Hazard Interactions Using Supervised Learning

The learning of structured relationships between measured inputs and recorded disaster outcomes over time and space is aided by supervised learning to model the relationships between hazard interactions in a data-driven way. In multi-hazard environments, such models represent non-linear relationships between environmental variables, environmental stressors (infrastructure), and indicators of socio-economic variables. Labeled disaster data can be used to allow supervised algorithms to identify patterns of interaction that indicate escalation routes where the presence of a particular hazard condition increases the probability or severity of other events to follow. This learning enhances prospective knowledge of cascading behavior under the independent signals of hazardous events.

Figure 7. Data-Driven Modeling of Hazard Interactions Using Supervised Learning

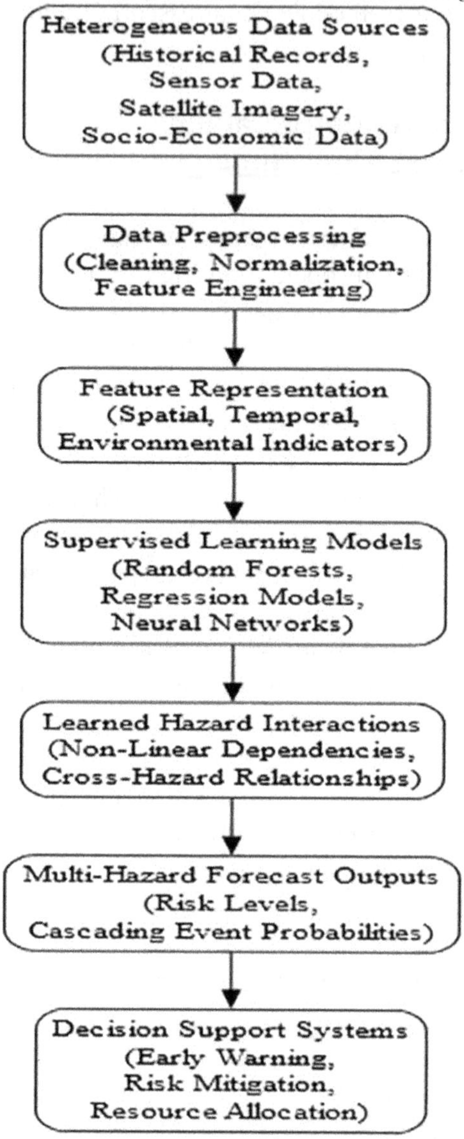

Supervised models increase the interaction dynamics in different combinations of hazards and context environments through exposure to a variety of training examples. Prediction of hazard transitions, severity of impacts, and the degree of risk of compounds in the changing conditions are aided by classification and regression systems. Multi-source features can be integrated to enable supervised learning to capture interdependent system behavior of learning, as opposed to independent effects of variables. These types of data modeled can improve the prediction power by transforming past evidence of interactions into predictive risk representations that can be applied in the dynamic disaster management environment.

Addressing Overfitting and Bias in Multi-Hazard Machine Learning Models

The machine learning models used in multi-hazard forecasting are used in data environments that are imbalanced, sparse, and highly correlated amongst hazard variables. These conditions make the models more vulnerable to overfitting, whereby the model is simply learning noise or event-specific relationships instead of generalizable relationships between hazards. Architectures that are over-parameterized and that have been trained using small amounts of disaster observations tend to give strong signals of rare events, leading to poor predictive performance when exposed to new combinations of hazards. This practice defeats the accuracy of forecasting and discredits model-based early warning systems, especially in cascading disaster situations.

The bias in multi-hazard learning systems is manifested by the asymmetric representation of data, the predominance of certain hazard types, and the spatial or temporal concentration of sampling. The effects of the training procedures on the historically common hazards are inclined to the marginalization of the low-frequency but high-impact events and cause systematic distortion in estimating risks. Strong overfitting and bias mitigation thus necessitate disciplined model regularization and balanced training methods as well as validation using a variety of hazard scenarios. These safeguards, in the context of dynamic forecasting systems, are useful in enabling stable generalization under changing conditions of disaster scenarios and in lending greater credibility to outputs of machine learning-based risk assessment.

Comparative Evaluation of Traditional and Machine Learning-Based Forecasting Approaches

Historical-based statistical and deterministic formulations are the main predictive tools used as a traditional method in disaster risk assessment based on historical observations and pre-established physical relationships. These methods are based on interpretability and mathematical simplicity and provide formatted descriptions of hazard behavior in unchanging conditions. In multi-hazard complex settings, the inability to embrace changing relationships between hazards, exposure, and vulnerability is limited by the dependence on linear assumptions and parameterization. Under such circumstances, the accuracy of the forecasts is frequently adversely affected due to facing non-stationary patterns, excessive variability, and interdependent hazard dynamics.

Figure 8. Comparative Evaluation of Traditional and Machine Learning-Based Forecasting Approaches

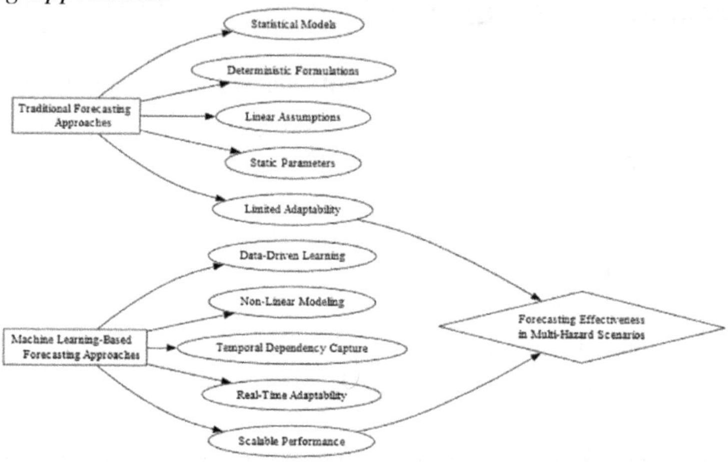

Forecasting methods that use machine learning provide a data-oriented paradigm of learning non-linear relationships and adaptive trends directly on large and heterogeneous datasets. The machine learning models of cascading hazard behavior and dynamic environmental change help to better represent it by utilizing ensemble algorithms and temporal learning architectures. The comparative assessment has shown that the data-driven approaches have better predictive sensitivity and robustness in multi-hazard scenarios, especially when real-time updates to the models are supported by the streams of data. This change in fixed formulations to the adaptive learning models is part of a fundamental change in the approach of forecasting such complex disaster systems.

This discussion of conceptual foundations leads naturally into the need for advanced data integration techniques, which are explored in the following section.

DEEP LEARNING AND TEMPORAL MODELING OF CASCADING DISASTERS

Temporal Dependency Modeling in Cascading Hazards Using LSTM Networks

The temporal dependency modeling is very important in the study of the cascading processes of disasters, because hazard development is realized by a series of exchanges over time. Cascading hazards have delayed response and cumulative and feedback-driven escalation that cannot be forecasted using a static or memoryless forecasting model. Long Short-Memory The long short-term memory networks handle these problems by storing information over long periods of time, allowing the identification of changing hazard trends and transition states. LSTM architectures learn the temporal correlations occurring within sequential disaster data through gated memory structures and thus forecast hazard evolution, as opposed to the occurrence of a single event.

In a multi-hazard environment, the temporal modeling of LSTM networks can be used to represent both the short-term variability and long-term interactions of an interacting set of hazards. Sequential learning allows identifying the precursor signals and changing risk pathways, which lead to the secondary or tertiary disasters. Real-time data streams that are integrated into the LSTM-based frameworks enhance the responsiveness to the dynamism of the environmental condition but maintain the learned time context. These abilities place deep temporal learning as a baseline of dynamic forecasting systems intended to forecast cascading disaster behavior in complex and rapidly evolving conditions.

Sequential Learning Approaches for Dynamic Disaster Evolution Prediction

Sequential learning methods are important in the time-dependent modeling of disasters by learning sequential dependencies within the data of hazard progression. The cascading disasters develop in time-connecting phases, in which initial indications affect later hazard conditions and the impact patterns. Sequence modeling architecture Deep learning models can be used to extract long-term time-varying features of continuous data streams, including sensor measurements, meteorological history, and satellite-based indicators. Sequential models can be used through the temporal memory processes to anticipate the escalation of hazards as opposed to detecting solitary incidents.

Figure 9. Sequential Learning Approaches for Dynamic Disaster Evolution Prediction

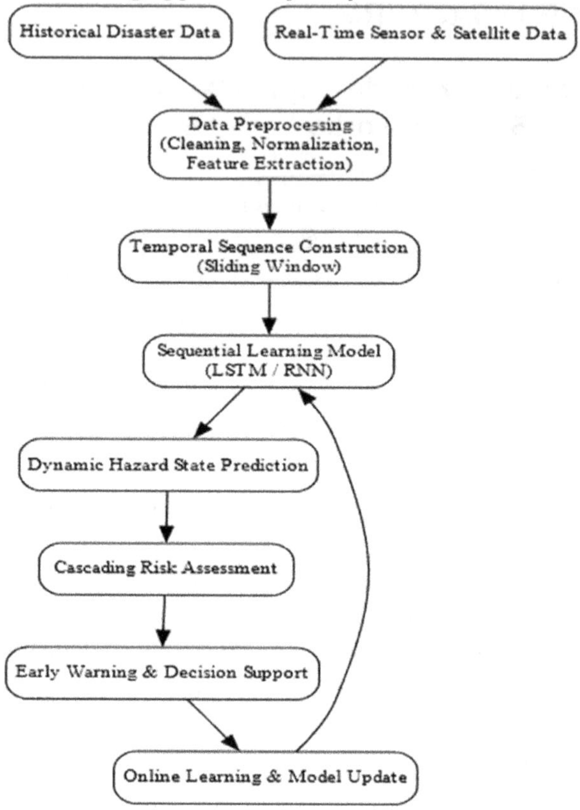

The advantageous features of the dynamic disaster evolution prediction are the learning structures that evolve with new observations as the unfolding events occur. Sequential learning promotes ever-refining the patterns, which enables changing hazard paths to affect the predictive results over consecutive time. In multi-hazard scenarios, this type of temporal intelligence can provide better early warning abilities by identifying precursor states and transition phases relating to cascading effects. Such an ability to capture time-varying relations makes sequential deep learning methods critical elements in high-level forecasting systems that will operate in complex disaster scenarios.

Capturing Long-Term Hazard Interactions with Recurrent Neural Networks

Cascading disasters: These develop in chains of connected hazard processes that emerge over long periods of time. Representation of these dynamics is only possible by analytical models that are able to maintain the historical context and be constantly updated with new information. The recurrent neural networks offer an organized way of capturing temporal relationships in sequential disaster data that allows the learning of patterns that relate to delayed reactions, gradual growth, and feedback-driven interaction among hazards. Recurrent networks, with the help of memory-based architectures, allow representing long-term dependencies, which describe the evolution of complex disasters.

Neural networks Temporal learning models constructed on recurrent neural networks improve the ability of the forecasting system to predict the occurrence of secondary and tertiary hazards. These models represent transitions and cumulative effects not available to short-window or static predictive models because of maintaining internal state representations. The use of recurrent architectures in multi-hazard forecasting helps in earlier detection of cascading pathways of risks and enhances predictive coherence over long periods of time, which reinforces the predictive basis of managing disaster risks dynamically.

Computational Trade-Offs in Deep Learning Models for Real-Time Forecasting

Temporal modeling deep learning architectures offer effective mechanisms of modeling successive dependencies and changing patterns in cascading disaster processes. Recurrent and sequence-based networks allow forecasting systems to discover long-range temporal dependencies between hazard indicators and enhance susceptibility to early indications and progression directions. In real-time forecasting situations, the requirements are very high in terms of computational efficiency in that the latency, memory, and model size have a direct impact on the ability to run. The depth of the network combined with the size of the time horizon increases representational capacity and at the same time heightens the computational load, which leads to a balance of crucial proportions between predictive performance and execution speed.

Figure 10. Computational Trade-Offs in Deep Learning Models for Real-Time Forecasting

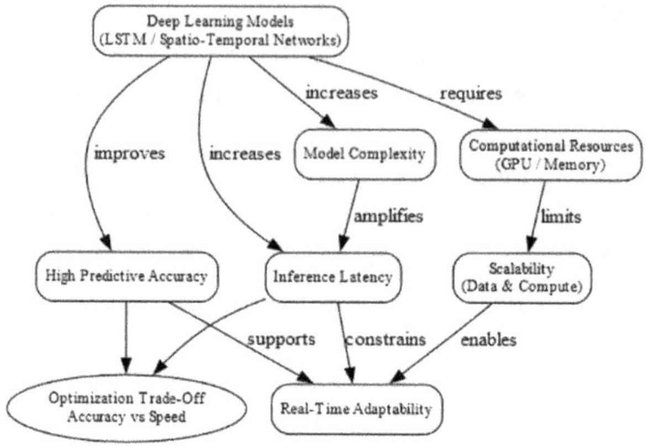

Use of deep learning models in operations needs to take into consideration the computational trade-off to maintain real-time responsiveness. Data streams at very high frequencies and continuous updates to the model put processing constraints on large-scale architectures. Pruning of models, parallel computation, and adaptive update policies are some of the optimization strategies that minimize the inference time without compromising the ability to learn over time. Good trade-off management is necessary so that deep learning-based forecasting systems can be responsive to time-sensitive disaster conditions, which is likely to lead to timely decision-making in disaster cascading hazards.

Advances in Spatio-Temporal Deep Learning for Multi-Hazard Risk Analysis

Spatio-temporal deep learning methods have transformed the analysis of multi-hazard risks by facilitating the joint capturing of spatial relationships and temporal dynamics of highly complex disaster systems. Sophisticated neural structures combine space feature-reading and sequential learning parts, enabling an attempt to model patterns of hazard spreading throughout geographic areas and through time. These techniques facilitate the discovery of the latent relationships between interacting hazards, networks of infrastructure, and environmental conditions and provide better sensitivity to cascading effects that manifest themselves in dynamic disaster events.

The current methodological innovations focus on the combination of the spatial context and long-term temporal memory, which enhances the ability of deep learning models to predict the escalation of hazards and the effects of compounds. Spatio-temporal models can improve the forecasting resolution and reaction to rapidly changing conditions by learning dynamic risk signatures over incoming streams of data. This development of integrated spatial-temporal learning models marks a major improvement on multi-hazard risk analysis, which should help predict and respond to disasters better with more effective management plans.

This discussion of conceptual foundations leads naturally into the need for advanced data integration techniques, which are explored in the following section.

HYBRID, ENSEMBLE, AND REAL-TIME ADAPTIVE FORECASTING FRAMEWORKS

Hybrid Machine Learning Frameworks for Integrated Multi-Hazard Forecasting

Integrated multi-hazard forecasting Hybrid machine learning structures are important in the whole-hearted combination of paradigmatic modeling models in a common forecasting framework. The feature-induced models that reflect more intricate correlation between environmental and socio-economic variables, whereas sequence-based learning frameworks reflect temporal dynamics and hazard developments. A combination of the above modeling strategies helps provide a more holistic representation of the cascading disaster processes so that the systems of forecasting can capture the instantaneous state of affairs as well as the long-term patterns of risk in the interconnected areas of hazards.

Figure 11. Hybrid Machine Learning Frameworks for Integrated Multi-Hazard Forecasting

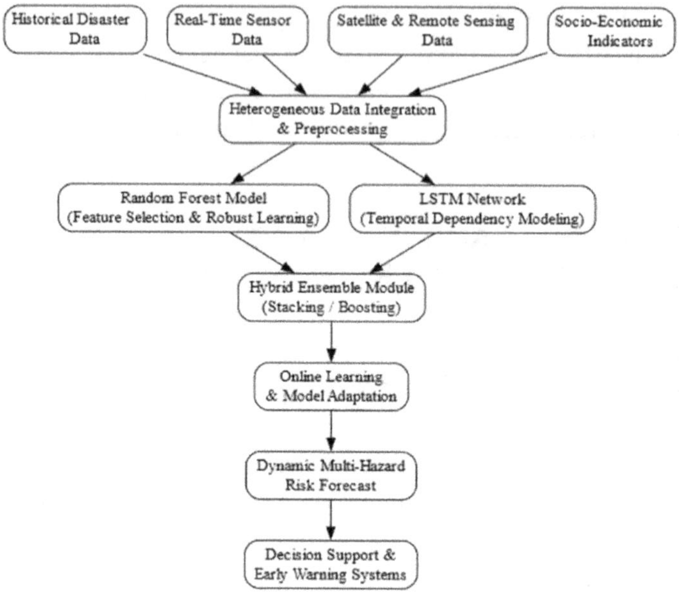

In dynamic forecasting conditions, the hybrid structures are effective in improving robustness and predictive stability through balancing model specialization and generalization. Ensemble integration processes combine the results of various learning components and minimize uncertainty and increase resilience to variability and noise in data. These frameworks make it possible to have adaptive forecasting pipelines, which can react to incoming real-time data streams and be computationally efficient, which adds to the operational relevance of multi-hazard forecasting systems in disaster risk management scenarios.

Ensemble Learning Strategies for Balancing Accuracy and Computational Efficiency

The ensemble learning strategies are of great importance to multi-hazard forecasting that involves the combination of outputs of several predictive models to enhance reliability in a complex setting of disasters. Ensemble structures are able to decrease noise, uncertainty, and local data anomalies through aggregation of complementary learning mechanisms in the real-time hazard context. These strategies permit more generalization in a variety of hazard conditions and maintain stability in a very dynamic series of disasters, which facilitates more predictive performance in both space and time.

The control of predictive accuracy and computational efficiency is a very crucial goal in real-time forecasting systems. Ensemble setups that are based on the parallel implementation, weighted averaging, and gradual model activation enable quick inference with a low computational cost. Ensemble learning frameworks enable maintaining timely forecasting cycles by spreading the predictive responsibility among the lightweight models and high-capacity models and can be robust in performance. This is a trade-off of efficiency and accuracy that enhances applicability in operations infrastructures in emergency management and decision support.

Online Learning and Model Adaptation in Real-Time Disaster Scenarios

Real-time disaster situations require forecasting structures that can continuously learn despite the high dynamics of the environmental conditions. Through online learning methods, the predictive model can be updated in-flight as new data streams are obtained, which allows the forecasting system to be aware of the current state of hazards instead of relying on past trends that are not current. In a multi-hazard setting, these adaptive processes allow risk signals and changing interactions between hazards, exposure, and vulnerability to be recognized timely to enhance the early warning system during cascading disaster events.

Real-time forecasting models can be made more resilient to non-stationary data distributions and changing hazard dynamics through model adaptation. Adaptive learning mechanisms re-tune predictive behavior to the evolving properties of the inputs to ensure the relevance of the forecasts even in long-lasting or increasing disaster events. By facilitating stability of the model and responsiveness, online learning integrates into hybrid and ensemble architecture, which allows balanced performance of the multi-hazard forecasting systems in decision-critical situations to enhance their operational efficiency.

Decision-Support-Oriented Forecasting Systems for Emergency Management

The decision-support-oriented forecasting systems are taking center stage in modern-day emergency management since they convert predictive results into operationally applicable intelligence. Hybrid and ensemble forecasting systems augment this role by incorporating complementary strategies in modeling that emphasize the spatial variability and temporal development as well as the uncertainty in multi-hazard settings. These systems match predictive accuracy to decision schedules, allowing emergency managers to make sense of developing risk conditions and prioritize response efforts among interdependent hazard situations.

Adaptive forecasting models in real time enhance the decision-support abilities by continually updating models and reacting to the incoming streams of data. Forecasting systems have maintained a situational awareness in highly dynamic disaster environments by using adaptive learning processes and dynamic risk levels. This predictive modeling-decision-support integration enhances the emergency management practice by allowing evidence-based interventions aimed at reducing the cascading effects and promoting the coordinated disaster response activities in good time.

Translating Predictive Intelligence into Actionable Disaster Risk Mitigation

Hybrid, ensemble, and real-time adaptive forecasting systems allow transforming the outputs of predictive systems into operational knowledge that facilitates disaster risk reduction. With the combination of the results of complementary models, such structures can come up with risk intelligence that reveals changing hazard conditions, spatial exposure, and temporal escalation patterns. The situational awareness offered by predictive intelligence on the basis of these systems helps in identifying the pathways of emerging threats and emphasizes the areas with the highest risk such that plans in response to emergencies and preparedness are informed.

Figure 12. Translating Predictive Intelligence into Actionable Disaster Risk Mitigation

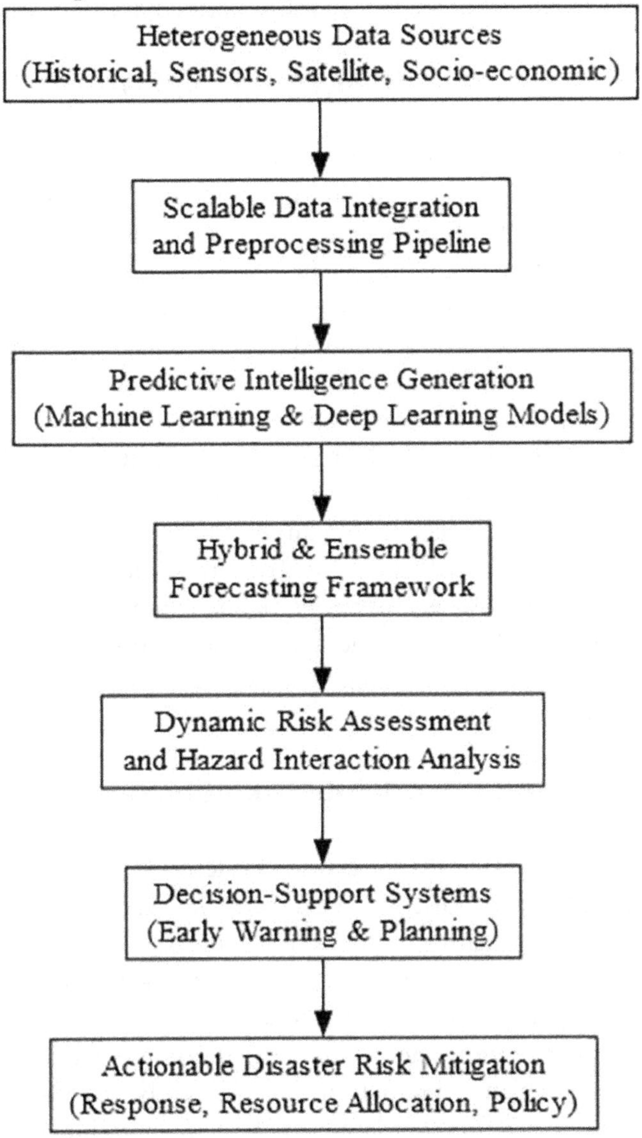

The timely disaster risk mitigation requires the correlation between the forecasting intelligence and the decision-support mechanisms and response mechanisms. Adaptive frameworks assist in constant optimization of risk indicators as new data streams are added to forecasting pipelines so as to maintain relevance during the rapidly evolving disaster circumstances. By providing early warning-enhancing communication of forecast-based information in time, these systems aid in the management of resources, coordination efforts within disaster management teams, and bolstering proactive risk reduction in multifaceted hazardous settings.

Operational Challenges in Real-World Deployment

Despite significant advancements in machine learning and data-driven forecasting, the real-world deployment of multi-hazard prediction systems presents several operational challenges. One of the primary issues is the availability and quality of data, as disaster forecasting often relies on heterogeneous data sources with varying spatial and temporal resolutions. In many regions, especially in developing countries, data scarcity and inconsistency limit the effectiveness of predictive models.

Another challenge lies in the integration of forecasting systems into real-time operational environments. High computational requirements, latency constraints, and the need for continuous model updating can hinder deployment in time-sensitive disaster scenarios. Additionally, the translation of complex model outputs into actionable insights for decision-makers remains a critical barrier.

Infrastructure limitations, lack of technical expertise, and institutional resistance to adopting advanced technologies further complicate implementation. Addressing these challenges requires the development of scalable, efficient, and user-friendly forecasting systems that can operate under real-world constraints while maintaining high predictive performance.

Model Validation and Uncertainty Quantification

The reliability of multi-hazard forecasting systems depends heavily on robust model validation and the quantification of uncertainty. Validation techniques such as cross-validation, back-testing with historical disaster data, and performance metrics including accuracy, precision, recall, and root mean square error (RMSE) are essential for evaluating model effectiveness.

In performance evaluation, understanding and communicating uncertainty is critical in disaster forecasting. Predictions are inherently probabilistic due to the complex and dynamic nature of hazard interactions. Incorporating uncertainty-aware models and probabilistic forecasting approaches can improve decision-making by providing confidence levels associated with predictions.

Recent advancements in explainable artificial intelligence (XAI) offer promising solutions for enhancing model transparency and interpretability. By enabling stakeholders to understand how predictions are generated, XAI techniques can increase trust in forecasting systems and support more informed policy and operational decisions.

Case Studies of Multi-Hazard and Cascading Disasters

To illustrate the practical relevance of multi-hazard disaster forecasting frameworks, it is essential to examine real-world cascading disaster events. One prominent example is the 2011 Japan disaster, where a massive earthquake triggered a tsunami, which subsequently led to the Fukushima nuclear crisis. This sequence of interconnected hazards highlights the complex dependencies between natural and technological systems. Traditional single-hazard models were insufficient to capture such cascading effects, whereas integrated multi-hazard forecasting approaches could significantly enhance early warning and risk mitigation strategies.

Another example can be observed in mountainous regions, where intense rainfall events often trigger floods that subsequently induce landslides. These cascading hazards create compounded risks for infrastructure, human settlements, and emergency response systems. Machine learning-based forecasting models, when combined with geospatial and meteorological data, can help identify such hazard interactions and improve predictive accuracy.

Similarly, tropical cyclones frequently result in compound disasters involving storm surges, inland flooding, and infrastructure failure. The integration of heterogeneous data sources, including satellite imagery, sensor networks, and historical disaster records, can support real-time forecasting and decision-making. These case studies demonstrate that multi-hazard forecasting is not only a theoretical construct but a critical requirement for modern disaster risk management systems.

The proposed model demonstrated consistent performance across different hazard scenarios, with improved prediction accuracy observed when multi-source data integration was applied. For instance, the model achieved an approximate accuracy improvement of 10–15% compared to baseline single-hazard models. The reduction in prediction error highlights the capability of the framework to capture complex interdependencies between hazards. These results confirm that the integration of spatial and temporal features significantly enhances forecasting reliability.

Table 1.

Model Type	Strength	Limitation
Traditional Models	Simple, interpretable	Poor multi-hazard handling
ML Models	High accuracy	Requires large data
Proposed Framework	Multi-hazard + adaptive	Computational complexity

The proposed system can be applied in real-world disaster management scenarios to support early warning systems, improve resource allocation, and enhance emergency response strategies.

CONCLUSION

Dynamic disaster environments require forecasting models that are able to capture interdependent hazard processes, time-escalating behaviors, and dynamic risk conditions. The traditional single-hazard and static forecasting models are no longer adequate in dealing with the complexity of cascading disasters, especially in the climate variability, urban sprawl, and growing connection between infrastructures. The problems discussed in this chapter of the book are the challenges that have been tackled by promoting a dynamic multi-hazard forecasting approach, which is based on machine learning, deep learning, and ensemble-based techniques. The chapter, with the integrated analytical frame, adds to the enhanced comprehension of how heterogeneous information and strategies of adaptive modeling could enhance predicting the cascading effects of disasters.

To be a foreteller of disaster, the socialization of multi-source data can be considered one of the key items. Through the integration of historical data, real-time sensor data, satellite-derived data, and socioeconomic data, forecasting systems acquire the ability to describe changing hazard conditions in a more faithful way. Continuous updating of the models due to scalable data pipelines and real-time data assimilation mechanisms ensures that predictions will be responsive in changing situations with rapid change. The result of such integration is the improved situational awareness and improvement of the early warning capabilities, which allows making more informed decisions when responding to disaster in its critical stages.

The machine learning and deep learning strategies are critical towards promoting the performance of multi-hazard forecasting. The ensemble-based algorithms add robustness and interpretability in feature-rich environments, whereas the temporal learning architecture adds sequential dependencies, which are very important to the interpretation of hazard propagation. Blended models combining these complementary features exhibit enhanced forecastability, computer efficiency and flexibility. These methodological improvements bring out the importance of integrating the spatial feature representation and temporal modeling to deal with the complexities of cascading disaster dynamics.

Besides methodological contributions, another significance of the chapter is the emphasis on the need to translate predictive intelligence to disaster risk mitigation actions. Prediction results that are compatible with operational decision support systems can be used to provide focused early warning andefficiency, rational resource distribution, as well as to coordinate the emergency response. Adaptive forecasting models can be used to constantly improve risk assessments and proactively plan, as opposed to reacting to them. This is a significant milestone in terms of increasing resilience in the disaster management systems because of this convergence of predictive modeling with the real-world application.

The challenges and opportunities identified in the chapter lead to the research directions in the future. Further development of scalable architectures, uncertainty-sensitive modeling methods, and explainable machine learning frameworks will enhance further operational deployment. Integration of new sources of data and improved cooperation between forecast and policy-making processes can and needs to be expanded to maximize benefit to society. The chapter has led to the changing discussion on disaster risk science by making contributions to the existing discourse on the topic of dynamic multi-hazard forecasting and offers a basis upon which more resilient, data-driven strategies towards controlling cascading disasters in complicated socio-environmental frameworks can be established.

In conclusion, this chapter highlights the growing importance of dynamic multi-hazard disaster forecasting in addressing the increasing complexity of cascading disaster events. The integration of machine learning, deep learning, and multi-source data provides a powerful framework for improving predictive accuracy and supporting proactive disaster risk management.

Including data limitations, model uncertainty, and the need for effective integration with policy and governance frameworks. Future research should focus on developing scalable and real-time forecasting architectures, incorporating uncertainty-aware and explainable models, and enhancing the interoperability of data systems.

Emerging technologies such as Internet of Things (IoT), edge computing, and real-time sensor networks offer new opportunities for advancing disaster forecasting capabilities. Additionally, interdisciplinary collaboration between data scientists, domain experts, and policymakers will be essential for translating technical innovations into practical solutions. By addressing these challenges and opportunities, multi-hazard forecasting systems can play a critical role in enhancing global disaster resilience and sustainability.

The discussion presented in this section provides a foundation for understanding the complexities of multi-hazard systems. Building upon these insights, the following section explores more advanced methodologies and practical considerations for improving disaster forecasting and risk management.

This study emphasizes the importance of integrating intelligent predictive models into disaster management systems to enhance resilience and support proactive decision-making in complex multi-hazard environments. Future work can focus on integrating real-time data streams and advanced deep learning techniques to further enhance the scalability and adaptability of the proposed framework.

REFERENCES

Afandi, G. E., Moustafa, A., Ibrahim, S., & Irfan, M. (2025). An in-depth analysis of approaches for forecasting extreme rainfall events leveraging artificial intelligence, machine learning, and deep learning techniques. *Earth Systems and Environment*, •••, 1–32. DOI: 10.1007/s41748-025-00813-w

Akiyama, M. (2025). Life-cycle approaches to sustainable and resilient infrastructure: Innovations in multi-hazard frameworks. *Structure and Infrastructure Engineering*, *21*(11-12), 1756–1781. DOI: 10.1080/15732479.2025.2524819

De Maio, F. V., De Stefano, R., Lanfranconi, C., Osmani, S., Fuggini, C., & Basso, P. (2024). Multi-hazard and risk informed system for Enhanced local and regional Disaster risk management.

Fearnley, C. (2024). Early warning of complex climate risk with integrated artificial intelligence. *Research Square* (*Research Square*).

Ferrario, D. M., Sanò, M., Maraschini, M., Critto, A., & Torresan, S. (2025). Harnessing Machine Learning methods for climate multi-hazard and multi-risk assessment. *EGUsphere*, *2025*, 1–72. DOI: 10.5194/egusphere-2025-670

Geiß, C. (2025). Collective Sensing and artificial intelligence techniques for natural hazard risk and impact assessment (Doctoral dissertation, Julius-Maximilians-University of Würzburg).

Hamid, H., & Abedlmajid, E. (2025). Intelligent Agents in Disaster Risk Management: A Systematic Review of Advances and Challenges. *International Journal of Advanced Computer Science and Applications*, *16*(6). Advance online publication. DOI: 10.14569/IJACSA.2025.01606100

He, X., Chen, Z., Yang, Q., & Xu, C. (2025). Advances in earthquake and cascading disasters. *Natural Hazards Review*, *5*(2), 421–431. DOI: 10.1016/j.nhres.2025.01.010

Jeddi, A. B., Shafieezadeh, A., Hur, J., Ha, J. G., Hahm, D., & Kim, M. K. (2022). Multi-hazard typhoon and earthquake collapse fragility models for transmission towers: An active learning reliability approach using gradient boosting classifiers. *Earthquake Engineering & Structural Dynamics*, *51*(15), 3552–3573. DOI: 10.1002/eqe.3735

Masud, N., Chowdhury, P., Al Bassam, A., Sourav, M. T. I., & Sobhan, A. (2025). A Machine Learning and IoT-Driven Framework for Real-Time Disaster Management. In Addressing Environmental Challenges With AI, Robotics, and Augmented Reality (pp. 247-288). IGI Global Scientific Publishing. DOI: 10.4018/979-8-3373-1892-9.ch010

Palanikkumar, D., Maashi, M., Alsamri, J., & Obayya, M. (2025). Machine learning driven multi-hazard risk framework for coastal resilience. *Journal of South American Earth Sciences*, *152*, 105331. DOI: 10.1016/j.jsames.2024.105331

Pugliese Viloria, A. D. J., Folini, A., Carrion, D., & Brovelli, M. A. (2024). Hazard susceptibility mapping with machine and deep learning: A literature review. *Remote Sensing (Basel)*, *16*(18), 3374. DOI: 10.3390/rs16183374

Razavi-Termeh, S. V., Sadeghi-Niaraki, A., Ali, F., Pradhan, B., & Choi, S. M. (2025). Optimizing ensemble learning for satellite-based multi-hazard monitoring and susceptibility assessment of landslides, land subsidence, floods, and wildfires. *Scientific Reports*, *15*(1), 30968. DOI: 10.1038/s41598-025-15381-2 PMID: 40846725

Sharma, A., Boora, A., & Kumar, Y. (2025). Artificial intelligence and machine learning for resilient transportation infrastructure. *Cureus J. Comput. Sci*, *2*, 1–16. DOI: 10.7759/s44389-025-09490-2

Shen, Z., & Zhou, H. (2025). Hazard-responsive digital twin for climate-driven urban resilience and equity. arXiv preprint arXiv:2510.22941.

Valarmathi, V., Ramkumar, J., Karthikeyan, R., & Doss, S. (2025). Harnessing AI for seismic hazard detection and prediction: Innovations and challenges. In *Emerging AI Applications in Earth Sciences: Challenges, Impact and Analysis* (pp. 97–110). Springer Nature Switzerland. DOI: 10.1007/978-3-031-84583-3_7

Wang, C. (2025). Multi-Modal Intelligence for Multi-Hazard Response and Urban Resilience (Doctoral dissertation, State University of New York at Stony Brook).

Wang, D. (2023). *Understanding Social Vulnerability to Natural Hazards Through Machine Learning and Dynamic Spatiotemporal Analysis*. Boise State University. DOI: 10.18122/td.2204.boisestate

Wang, J., Sun, P., Chen, L., Yang, J., Liu, Z., & Lian, H. (2023). Recent advances of deep learning in geological hazard forecasting. *Computer Modeling in Engineering & Sciences*, *137*(2), 1381–1418. DOI: 10.32604/cmes.2023.023693

Wu, X., & Zou, K. (2025, October). A machine learning-enhanced multi-agent framework for multi-timescale emergency resource allocation in natural disasters. In 2025 2nd International Symposium on AI and Cybersecurity (ISAICS) (pp. 1-6). IEEE. DOI: 10.1109/ISAICS66888.2025.11349790

Yang, Y., Wang, N., Wang, Z., Ouyang, M., & Wan, C. (2025). From Forecast to Action: A Deep Learning Model for Predicting Power Outages During Tropical Cyclones. arXiv preprint arXiv:2512.06644.

Yasmin, F., Hasan, M., Sinha, A., & Mondal, S. (2025). AI-driven hazard assessment: a comprehensive study of machine and deep learning approaches. Asian Journal of Civil Engineering, 1-16.

Zhang, T., Wang, D., & Lu, Y. (2023). Machine learning-enabled regional multi-hazards risk assessment considering social vulnerability. *Scientific Reports*, *13*(1), 13405. DOI: 10.1038/s41598-023-40159-9 PMID: 37591870

Zhao, Y., Lv, S., & Liu, P. (2024). Advances in earthquake prevention and reduction based on machine learning: A scoping review. *IEEE Access : Practical Innovations, Open Solutions*, *12*, 143908–143929. DOI: 10.1109/ACCESS.2024.3467149

Chapter 8
Land Suitability Analysis for Industrial Land Development in Pendik District of Istanbul, Türkiye

Eda Ustaoglu
http://orcid.org/0000-0001-6874-5162
Abdullah Gul University, Turkey

ABSTRACT

In many regions of Türkiye, land use development is chosen and managed without consideration to land capability, which results in disinvestment and lowers ecological capacity. In order to support the local authority's decision-making process, this article presents an analytical method that uses individual suitability maps to determine if a region is suitable for developing various economic sectors. Using a Multi-Criteria Decision Analysis (MCDA) integrated with Geographic Information Systems (GIS), the method is made up of the Analytic Hierarchy Process (AHP), a deterministic and fuzzy logic approach, the latter two were used for the standardisation of the criteria maps. The land-use suitability study took into account a number of criteria, such as geophysical characteristics, accessibility, built-up area and infrastructure, vegetation, and green and blue infrastructure. When compared to alternatives based on conventional approaches, the integration of fuzzy logic, AHP, and GIS is an advanced methodology that can enhance suitability evaluations.

DOI: 10.4018/979-8-2600-2383-9.ch008

© 2027, IGI Global Scientific Publishing. Copying or distributing in print or electronic forms without written permission of IGI Global Scientific Publishing is prohibited. Use of this chapter to ‹erative artificial intelligence (AI) technologies is expressly prohibited. The publisher reserves all rights to license its use for generative AI training and machine learning model development.

1. INTRODUCTION

Cities have developed rapidly in line with the rapidly expanding industries have been responsible for the improvements in the quality of life over the last century (Humbal et al., 2023). However, rapid growth of urban areas is now a treat for the environment as urban development has profoundly transformed natural landscapes in the worldwide (Tortora et al., 2015). Alongside growing urbanization, industrialization is another rising phenomenon in developing nations. Environmental quality and ecosystem services are now significantly impacted by urbanization and industrialization (Li et al., 2023). Terrestrial biological carbon cycles (Sleeter et al., 2018), hydrological processes (Zeng et al., 2020), and energy surface balances (Deng et al., 2015) have all been significantly impacted by the conversion of natural lands to urban and industrial regions. The accompanied urban and environmental problems such as urban heat island effects, environmental pollution, fragmentation of natural habitats, degradation of water resources, deterioration of historical centres, and unplanned or poorly planned urban developments constitute a significant threat to human well-being (Park et al., 2011; Varma et al., 2021; Miah et al., 2024). The adverse impacts on the environment can be reduced by utilisation of land more effectively. Therefore, with careful planning and policy making, land could be allocated efficiently with minimum destruction of urban ecosystems in the future.

Since the 1950s, Turkiye's population has grown rapidly, and migration from rural to urban regions has coincided with this development (Sertkaya Dogan, 2015). The mechanisation and commercialisation of agriculture and the resultant reduction in rural employment opportunities have resulted in a flow of rural population towards cities and urban regions (Keyder and Yenal, 2011). For example, the proportion of the population living in cities was 25% in the 1950s, but it rose to 36.6% in 1968, 44% in 1980, 65% in 2000, and 74.4% in 2017 (UN, 2014; World Bank, 2017). The national and local governments have taken an entrepreneurial position since the 1980s, implementing new urban rules to encourage private investment and development (Eraydın and Tasan-Kok, 2014). Therefore, from 1980s Turkish cities underwent a significant change from small-scale development to large-scale built-up areas through the intervention of large organisations and powerful stakeholders (Tekeli, 2009). Over the past few decades, there has been an unanticipated boom of construction. Profit-driven and speculative development activities have continued at an accelerating rate in many cities, and the government has encouraged construction by enacting a number of laws and amendments to control "deregulation" and promote "private sector investment and development" (Balaban, 2012; Gursoy and Edelman, 2017). The main forces for the rise of urban and industrial land in the post-1980 era can be attributed to GDP growth, capital investments, and rural-to-urban migration. However, because they are higher-level forces underlying the expansion of urban and industrial regions, national development plans and regional land-use restrictions have outweighed their effects. For example, Turkiye's amnesty policies have produced short-term economic benefits, such as increased tax revenues and the legalization of informal housing; however, due to lax regulatory oversight, they have also resulted in unchecked urban growth, environmental degradation, and strained infrastructure (Taktak, 2025).

Industrial areas are crucial in urban and regional planning particularly for developing nations where effective solutions for these areas may support economic and environmental sustainability. Pendik neighbourhood located in eastern Istanbul has seen economic growth over the last 30 years as a result of investor demands, and local authorities and government decisions. Pendik is among Türkiye's leading districts, boasting major companies that have made name for themselves in the food, textile, machinery and chemicals, forestry and mining industries, as well as in stone and soil-based industries. Pendik district was chosen as the study's case study area because, in contrast to many other Istanbul locations, it is one of the areas with many nature areas in the north that are interwoven with densely populated urban land and important industrial sites in the south. Thus, by linking the industrial area mostly found in the southern region with the forests and agricultural land found in the northern region, the region has the potential for sustainable industrial development. Despite Pendik's abundant natural resources, the district faces significant problems, which are exacerbated by climate change and related natural disasters. The rising temperature trend has serious implications for the region's environment, agriculture, water resources, and urban areas, emphasizing the need for a full assessment of the region's vulnerability to climate change. Due to its geographical location, Pendik is vulnerable to a variety of natural disasters, including particularly the earthquakes. The presence of many active fault lines, such as the North Anatolian Fault and the Marmara Sea Fault, highlights the region's significant seismic vulnerability. In order to support Pendik Municipality's decision-making process for deciding on potential sites for the new industrial developments that can promote socio-economic sustainability, seismic safety, and environmental preservation, this study presents an analytical method that integrated GIS and MCDA approaches that uses individual suitability maps to determine if a region is suitable, unsuitable, or conditionally appropriate for developing various economic sectors.

Land suitability mapping is one of the most prominent approaches for spatial planning and urban management (Malczewski, 2004). An integrated technique that measures and evaluates the degree of prospective capabilities of the land use by land requirement and attributes is land suitability analysis, which is based on GIS and multi-criteria assessment (MCA) methodology (Malczewski, 2004). In land-use planning, location and land suitability study, and spatial evaluation, GIS is a powerful and efficient tool for precisely and adaptably examining a variety of geographical data (Pramanik, 2016). Similarly, the land suitability analysis framework makes extensive use of MCA as a complementing tool. In order to evaluate various land-use issues with several options, the MCA technique takes into account a variety of factors, such as geophysical features and socioeconomic factors.

Using a MCA methodology integrated with geographic information systems (GIS), this study specifies the most suitable sites for industrial development in Pendik District, Istanbul. For this aim, principal criteria are defined and weighted following the Analytic Hierarchy Process (AHP) method. One of the most commonly used MCA methodology is the Analytic Hierarchy Process (AHP) which is utilised for deriving the relative weights among factors and the total values of each alternative based on the weights (Saaty, 1980). Saaty (1980) initially created AHP with the goal of creating a hierarchical model for resolving challenging land management issues. AHP has been applied extensively to a wide range of issues involving complicated parameters that interact at different levels. In order to determine the best sites for agricultural land usage, Saha et al. (2021) employed AHP in a GIS environment. Another example of using Boolean and AHP-based standardization in a GIS setting for the landfill site selection problem is Elkhrachy et al. (2023). Additionally, land suitability studies have specifically used fuzzy method (Akbari et al., 2019; Zarin et al., 2021), and GIS (Uy and Nakagoshi, 2008; Mendas and Delali, 2012).

Numerous studies have been carried out to precisely investigate the features of industrial spaces and possible locations that are appropriate for the development of industrial land in either a European or international context (Cerreta and de Toro, 2012; Johar et al., 2013; Vandermeer and Halleux, 2017; Wang et al., 2019). Additional research was carried out for Eastern nations, such as Hong Kong (Wang et al., 2015), China (Sun et al., 2016), and India (Midatana et al., 2018). Except for a few emerging nations like Bangladesh (Supto et al., 2025), Nigeria (Mohammed et al., 2025), and Indonesia (Septian and Saputra, 2022), there are fewer studies that examine the potential for manufacturing land growth in developing nations. Recently, there has been an increase in the body of research on the description and measurement of commercial and industrial changes in the Turkish setting. These studies have researched different case study areas mostly focusing on urban land use suitability as well as industrial/commercial land use suitability including the city of Antakya (Kılıç et al., 2005); Kemalpaşa District (Aktaş and Tahta, 2018); Mersin Metropolitan Area (Sahin et al., 2024); İnegöl District (Kuru et al., 2025) and Pendik District of Istanbul (Ustaoglu and Aydınoglu, 2020).

For the purpose of evaluating the suitability of industrial land development, a large number of studies rely on GIS analysis (Midatana et al., 2018; Septian and Saputra, 2022; Supto et al., 2025), Analytic Hierarchy Process (AHP) (Parry et al., 2018; Ramya and Devadas, 2019; Mohammed et al., 2025), or fuzzy set modeling (Ustaoglu and Aydınoglu, 2020). Rating and suitability classifications, which indicate a site's suitability for a particular land-use development, are represented numerically in traditional deterministic approaches. Expert comments and a survey of the literature were judged suitable for evaluating and standardizing the chosen criteria because there is no single standard for rating the criteria (Ustaoglu and Aydınoglu. 2020). By contrast to conventional methods, fuzzy logic permits objects to belong partially to multiple sets (Donevska et al., 2012). Zadeh (1965) invented fuzzy logic, defining a fuzzy set as 'a class of objects with a continuum grade of memberships'. According to Zadeh (1965), the objects do not have a fixed boundary, resulting in a gradual transition from zero to unity membership. In comparison to situations when the research either concentrate on GIS, fuzzy set, or AHP techniques alone, the fuzzy set modeling combined with the AHP technique can be seen as an advanced methodology that leads to more accurate judgments in the decision making process. To the best of the author's knowledge, very few research (Aydi et al., 2016; Ustaoglu and Aydınoglu, 2020) have used this integrated method to evaluate suitability for industrial land expansion.

The specification of relative weights of different criteria in the current study is performed by review of the relevant literature and expert opinions. The criteria have been classified under five groups: geo-physical factors, accessibility, water sources and recreational elements, built-up area and vegetation. There are two different approaches evaluated in the study. First, a deterministic approach is used to standardize variables and specify scales in map layers. Fuzzy logic, which permits the objects to partially belong to various sets, is the foundation of an alternative assessment. The results obtained from these two approaches are compared for the site suitability of industrial land uses. By combining the fuzzy set modeling and AHP with the GIS-based multi-criteria decision making process, the current study employs an integrated strategy for the suitability analysis of industrial land-uses. This study is beneficial for improving land use planning and decision support since it provides a better understanding of possible locations that are appropriate for industrial development.

2. STUDY AREA

The research area is Pendik District situated in Türkiye in Istanbul Metropolitan Region's eastern Anatolian region (Figure 1). The research area is roughly 179 km^2 and consists of 36 communities in the Municipality of Pendik. It is situated between the Tuzla District to the east and the Kartal District to the west. There are about 750,000 people living in the region as of 2024 (Pendik Governorship, 2025). The bulk of people live in the southern part of Pendik, and high-density projects are occurring close to the seaside, highways, metro, rail networks and Sebiha Gökçen

Airport. The region has a moderate climate, with daily highs and lows of 9.3 to 3.8°C in January and 29 to 18°C in July. There is between 24 and 120 mm of precipitation each year, most of which falls during the winter. Pendik District has an island area of 23,620 m^2 (Pavli Island), a lake area of 10 km^2, and a 13 km coastline. Bordering Aydos, Istanbul's highest peak, Pendik also hosts the Ömerli Dam, one of Istanbul's largest water resources, within its borders. Pendik's economy is comprised of industry, commerce, agriculture, livestock farming, tourism, and services sectors. Pendik Shipyard, a national leader in trade and industry, contributes to the district's economy through businesses such as Un Ro-Ro, as well as small and medium-sized industrial enterprises.

Figure 1. The study area of the Pendik District located in Istanbul Metropolitan Region (Türkiye)

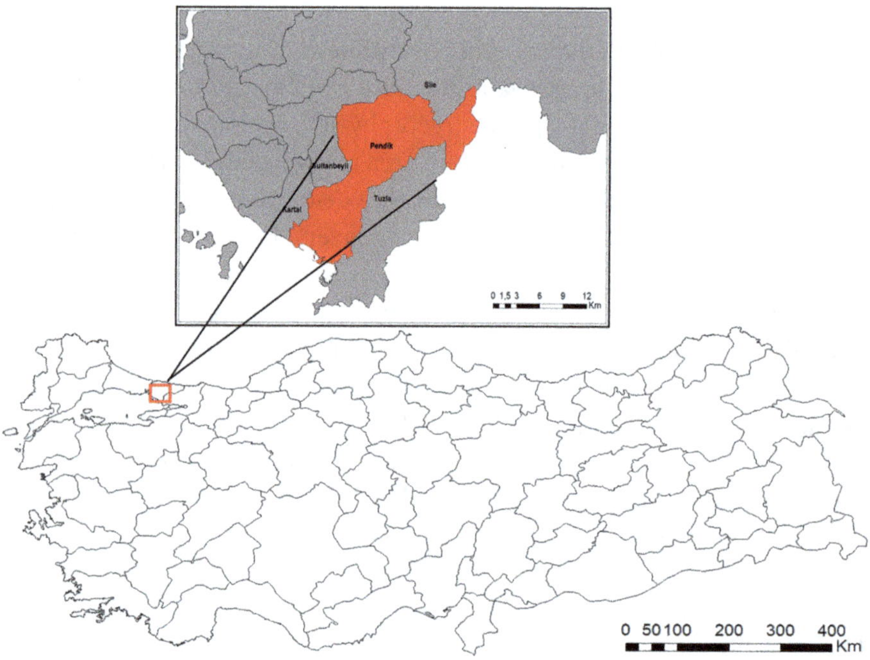

3. DATA AND METHODOLOGY

3.1. Data

ArcGIS 10.4 was used to do spatial analysis and map processing once pertinent data was gathered for each suitability criterion taken into account in the analysis. The ArcGIS 10.4 program was used to extract the land use data from the European Commission's (EC) 2018 Copernicus Urban Atlas project (see EEA, 2025). These maps were created using multispectral or pan-sharpened Earth Observation (EO) data with a spatial resolution of 2.5 m (EC, 2025). Ten rural classes with a minimum-mapping unit (MMU) of 1.0 ha and seventeen urban land cover classes with a MMU of 0.25 ha make up the data. The Municipality of Pendik's database was used to gather the digital elevation model (DEM) dataset with a resolution of 30–30 m, the transport network map, the land capability map, and the political boundary map. ArcGIS 10.4 was used to create the spatial data for elevation and slope from the DEM.

In Urban Atlas maps, industrial, commercial, public, military and private services are not separated and represented as a single land use class value. An ancillary data was used to separate industrial and commercial uses from the aggregated 'industrial, commercial, public, military and private units' class. The ancillary data is the map of 2018 showing the locations of public and private services buildings in Pendik District. The overlapping areas in the ancillary map and Urban Atlas map were re-classified as 'public, military and private units' and the remaining areas that do not cover public, military and private services were re-classified as 'industrial and commercial' uses. The resulting map is presented in Figure 2.

Figure 2. The separated land use classes of industrial and commercial uses, and public, military and private units

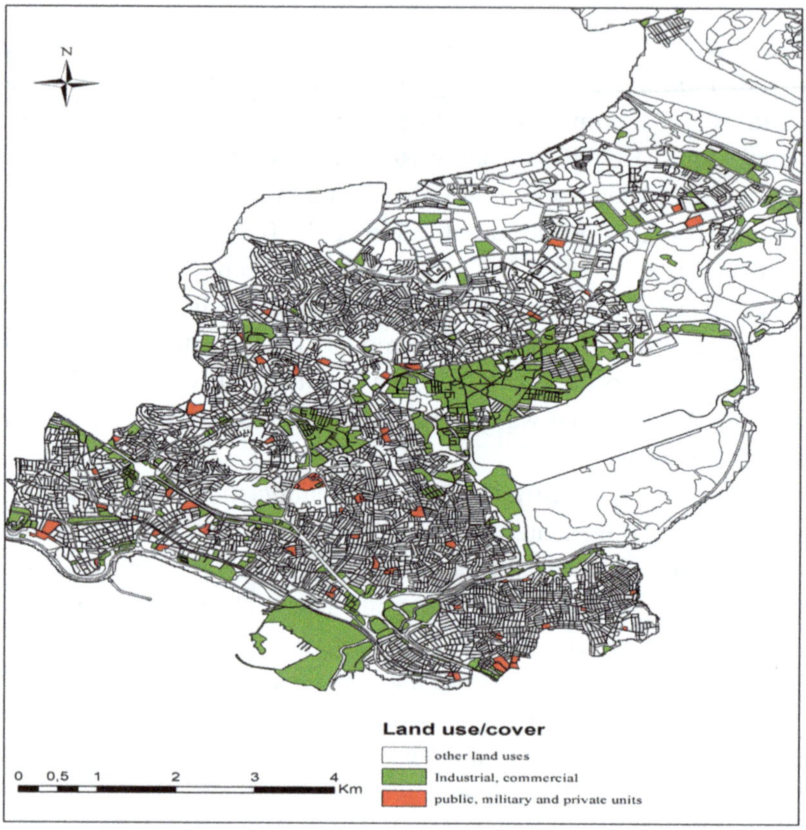

3.2. Land suitability criteria

In the current study, a variety of data are used to evaluate the suitability of industrial development. Considering the literature review (Park et al., 2011; Bathrellos et al., 2012; Zhang et al., 2013; Malmir et al., 2016), the study's goals, and the data at hand, industrial land use suitability criteria were categorized under six headings: developed area/infrastructure; landscape and geology; physical features; land use accessibility; and green and blue infrastructure (Figure 3). Physical features included elevation, slope, and aspect; accessibility-based factors included distance from highways, roads, ports, airports, and public transportation stops; blue and green infrastructure included distance from water sources, coastal areas, and urban green spaces; developed areas and infrastructure included distance from industrial/ commercial areas and population centers; vegetation was represented by current land use; and geological factors included soil capability (Ustaoglu and Aydınoglu, 2020) (Table 1).

Figure 3. Suitability value maps for each criterion used in the analysis

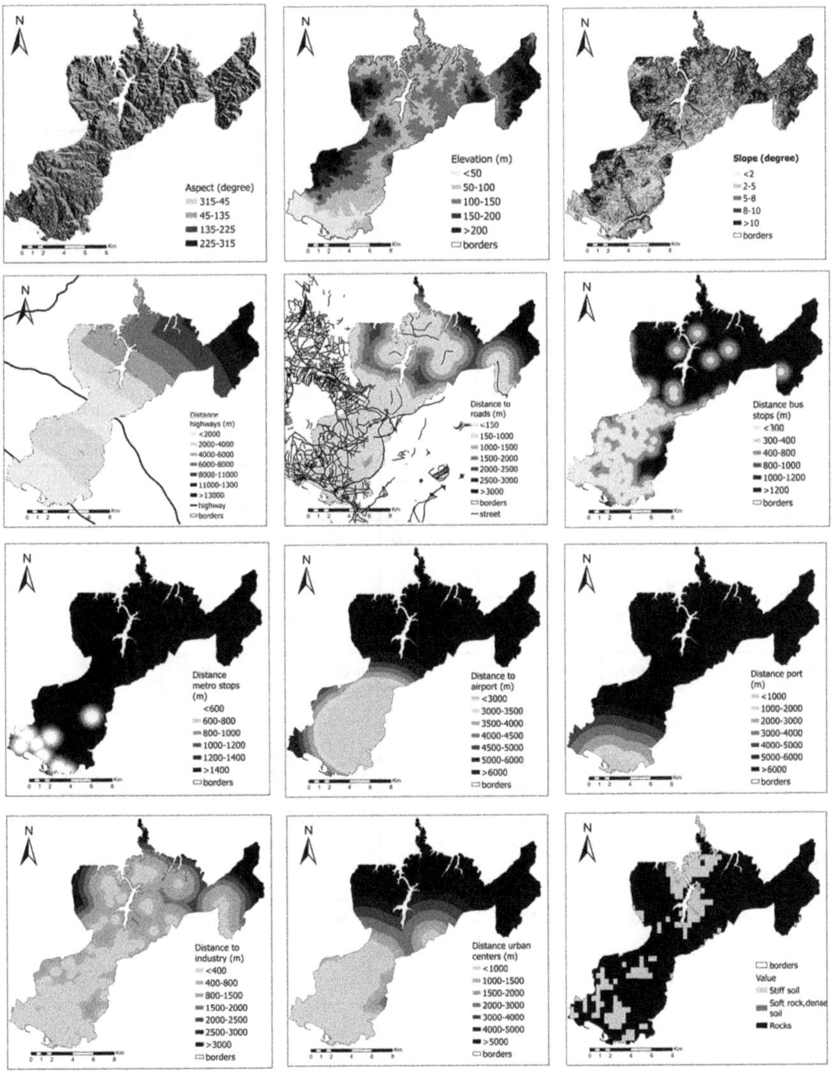

Table 1. Selected factors for industrial land suitability analysis

Sub-criteria	7 very high	6	5	4 medium	3	2	1 very low	References
Physical attributes								Bunruamkaew and Murayama (2011); Barthrellos et al. (2012); Zhang et al. (2013); Malmir et al. (2016); Li et al. (2018)
Aspect	135-225	-	45-135	-	225-315	-	315-45	
Elevation	0-100	-	-	100-300	-	-	>300	
Slope	0-2	2-5	5-15	15-25	>25	-	-	
Accessibility								
Distance from highways (m)	0-1000	1000-2000	2000-3000	3000-4000	4000-5000	5000-6000	>6000	Gül et al. (2006); Yigitcanlar et al. (2007); Bunruamkaew and Murayama (2011); Zhang et al. (2013); Tirachini (2014); Malmir et al. (2016); Yousefi et al. (2016); Zheng et al. (2017)
Distance from roads (m)	0-150	150-1000	1000-1500	1500-2000	2000-2500	2500-3000	>3000	Gül et al. (2006); Kienast et al. (2012); Barthrellos et al. (2012); Zhang et al. (2013); Morckel (2017); Dagistanlı et al. (2018); Li et al. (2018) Park et al. (2011); Zhang et al. (2013); Malmir et al. (2016); Yousefi et al. (2016); Dagistanli et al. (2018)
Distance from bus stops (m)	0-300	300-400	400-800	800-1000	1000-1200	>1200	-	
Distance from metro stops (m)	0-600	600-800	800-1000	1000-1200	1200-1400	>1400	-	
Distance from port (m)	1000-2000	2000-3000	3000-4000	4000-5000	5000-6000	6000-7000	<1000	
Distance from airport (m)	3000-3500	3500-4000	4000-4500	4500-5000	5000-6000	>6000	<3000	
Water sources and recreation								
Distance from reservoir (m)[a]	1000-2000	-	2000-4000	-	>4000	-	300-1000	
Distance from water courses (m)[b]	>600	600-500	500-400	400-300	300-200	200-100	50-100	
Distance from coastline (m)	>4000	3000-4000	2000-3000	1000-2000	500-1000	100-500	50-100	

continued on following page

Table 1. Continued

Sub-criteria	7 very high	6	5	4 medium	3	2	1 very low	References
Distance from urban green (m)	>2500	2000-2500	1500-2000	1000-1500	1000-500	500-250	<250	
Built-up area								
Distance from industry/ commerce (m)	0-400	400-800	800-1500	1500-2000	2000-2500	2500-3000	>3000	
Distance from high density residence (m)	0-1000	1000-1500	1500-2000	2000-3000	3000-4000	4000-5000	>5000	
Vegetation								
Land use	Urban land, Water bodies	Poorly vegetated areas	Agricultural land	Forest and natural vegetation	-	-	-	Dong et al. (2008); Park et al. (2011); Mahdavi and Niknejad (2014); Li et al. (2018)
Geology								
Soil capability[c]	Hard rocks	Rocks	Soft rock, very dense soil	Stiff soil	Soft soil		Other soils	Piran et al. (2013); Peng et al. (2016); Malmir et al. (2016); Dagistanli et al. (2018)

[a]0-300 m from the reservoir is not allowed for construction and 300-1000m is allowed for limited development (Regulation on Drinking Water Basins) (Ustaoglu and Aydınoglu, 2020), [b]0-50 m from the water course (and the coast) is not allowed and 50-100 m is allowed for limited development (Regulation on Coastal Zones) (Ustaoglu and Aydınoglu, 2020),[c]Soil capability is evaluated through shear-wave velocity (VS), which provides a single number for the average shear-wave velocity in the top 30m of the soil. If VS_{30} (m/sn) >1500: hard rocks (group A soils); VS_{30} (m/sn): 760-1500: rocks (group B soils); VS_{30} (m/sn): 360-760: soft rock, very dense soil (group C soils); VS_{30} (m/sn): 180-360: stiff soil (group D soils); VS_{30} (m/sn) <180: soft soil (group E soils); other soils: specific site evaluation required (group F soils) (Halaç, 2016; Ustaoglu and Aydınoglu, 2020)

Slope analysis is crucial for identifying possible locations for afforestation (Dubovyk et al., 2016), watershed management (Steiner et al., 2000), and land development (Bunruamkaew and Murayama, 2011). Slope is linked to soil thickness, fertility, erosion, irrigation and drainage (Ustaoglu and Aydınoglu, 2020). Deep soils are found on gentle slopes at the bottom of valleys and plain terrain, while thin soils are found on steep slopes with a higher danger of erosion (Zolekar and Bhagat, 2015). Small-scale climate variables as temperature, precipitation, wind, and solar radiation correspond with elevation and aspect (Zhang et al., 2015). The accessibility of various urban and non-urban land uses is significantly influenced by the current transportation network. In terms of transportation accessibility, land uses near ports, airports, road networks, and public transportation stops received the highest scores (Ustaoglu and Aydınoglu, 2020). This is because urban developments near existing transportation networks make land uses more accessible by lowering the cost of moving people, products, and services.

The viability of creating both urban and non-urban area is influenced by the current land use system, which is an important input that illustrates the spatial distribution and features of land. Numerous studies have taken land use/cover classification into account for land suitability assessments and land use planning (Thapa and Murayama, 2008; Rabia et al., 2013; Zolekar and Bhagat, 2015; Li et al., 2017). Water resources, urban land, agricultural land, forests, and natural vegetation are the categories into which the land use/cover in this study has been divided (Ustaoglu and Aydınoglu, 2020). Being in the vicinity of urban land or water bodies can increase the development potential of industrial land uses. Thus, one of the elements impacting Pendik's viability for industrial growth is its distance from urban centers and current water supplies. The water source is located in the northern half of the study area. Construction is prohibited within 300 m of the reservoir, while 300-1000 m allows for limited development (Drinking Water Basin Regulations) (Ustaoglu and Aydınoglu, 2020). Therefore, I selected zones that are less than 2000-1500 m but no less than 1000 m as ideal places for industrial growth. For watercourses and coastlines, construction is prohibited in zones less than 50 m, whereas 50-100 m zones allow for limited development (Regulation on Coastal Zones) (Ustaoglu and Aydınoglu, 2020). Zones less than 200-400 m but not less than 50 m were chosen for industrial development based on the defined criteria (Ustaoglu and Aydınoglu. 2020).

The remaining criteria used to evaluate a site's suitability for industrial areas are erosion risk and land capabilities. In hilly areas with greater rainfall, soil erosion is known to loose fertile topsoil (Shrestha et al., 2014). Because of the steep slopes and increased rainfall, higher elevation is associated with a higher risk of erosion (Chalise et al., 2019). The property with less erosion risk is highly valued since the eroded land is inappropriate for the development of industrial land uses. Another aspect of the strength of the earth's top layer that affects the growth of urban infrastructure is land capacity. Compared to stiff and soft soil, the ground with hard rocks is more suited for industrial development (Vondrackova et al., 2015).

3.3. Fuzzy modelling

A Boolean technique is used for modeling in a classical theory framework. As a result, an object's value either falls into a set or does not, suggesting that the method depicts reality in a discrete manner (Burrough, 1996). Examples of the Boolean technique are the conventional thematic maps that depict discrete features like polygons, lines, and points. These items on the maps either have a value or do not, and in the latter case, partial membership is not feasible (Fisher, 1996). Therefore, the Boolean technique cannot be used to simulate transitional memberships in the real world. Zadeh (1965) invented fuzzy set theory, which has since been refined by Kaufmann and Gupta (1991) and Zimmerman (1991). The partial membership problem has been resolved, enabling a more accurate depiction of reality.

A fuzzy subset A can be defined mathematically as follows: $X = \{x\}$ that is defined as a finite set of points; and a fuzzy subset A can be defined as: $A = \{x, \mu_A(x)\}$ for each $x \in X$, where $\mu_A(x)$ is the membership function that defines the grade of membership x in A. The $\mu_A(x)$ allows values between 0 and 1, including 1 and 0 for every A. A value of 1 indicates that the value fully belongs to the subset A, whilst a value of 0 indicates that the value of x absolutely does not belong to the subset A. On the other hand, a value between 0 and 1 denotes that x is somewhat related to A (Wang, 1990). If $X = \{x_1, x_2, \ldots, x_t\}$, the fuzzy subset can be written as: $A = x_1, \mu_A(x_1) + x_2, \mu_A(x_2) + \ldots + x_t, \mu_A(x_t)$. This equation shows that the membership function $\mu_A(x)$ explains the degree of ownership of x in A.

The suitability of land use for industrial development in relation to sub-factors like slope,, accessibility, distance to transport network etc. is one example of a statement that is connected to the membership function. The membership function's maximum suitability is represented by a value of 1, while its lowest suitability for industrial development is represented by a value of 0. The membership function might be symmetrical or asymmetrical, non-continuous or non-linear, and have any shape. The triangle form is the most basic function, although additional forms including Gaussian, sigmoid, inverted sigmoid, and linear functions are commonly used for the land suitability assessment, according to Torbert et al. (2008) and Zhang et al. (2015).

In the current work, various models were used to develop membership functions for the industrial land use suitability criteria. When the suitability value rises as the criterion's value increases, the Gaussian function is employed; once the maximum value is reached, the suitability value falls as the criterion's value increases. When a criterion's suitability declines as its value increases, inverted sigmoid and monotonically decreasing linear functions are employed. Lower and higher limits, their variable type, and the fuzzy membership function utilized for the topic criterion for the assessment of industrial land uses are shown in Table 2.

3.4. The AHP method

The AHP approach is utilized to offer the weights for each criterion taken into consideration in the suitability study of industrial land (Tables 1 and 2). Saaty (1980) created the multi-criteria AHP decision-making procedure. In order to create scales of preferences among a range of options, the model essentially focuses on a pairwise comparison procedure (Beynon, 2002). Pairwise comparisons produce a matrix in AHP, and the technique calculates both the final vector of weight coefficients for alternatives and their eigenvalues (Saaty, 1980). As a result, the matrix of pairwise comparisons yields a vector of weights that represent the relative significance of the different components (Park et al., 2011). The sub-criteria were rated using the AHP approach, with 1 denoting equal relevance between the two criteria and 9 denoting extreme significance (Table 3).

Table 2. The evaluation criteria for industrial land with fuzzy membership function

Main criteria	Sub-criteria	Upper limit	Lower limit	Type of fuzzy function
Physical	Aspect	359.5	-1	Gaussian
	Slope	39.5	0	Linear (decreasing)
	Elevation	536	0	Linear (decreasing)
Accessibility	Distance from highways	17166	0	Linear (decreasing)
	Distance from roads	6833.9	0	Linear (decreasing)
	Distance from bus stops	7800.3	0	Inverted sigmoid
	Distance from metro stops	21250	0	Inverted sigmoid
	Distance from airport			Gaussian
	Distance from port			Gaussian
	Distance from reservoir	14639	0	Inverted sigmoid
Water sources	Distance from water bodies	14844	0	Inverted sigmoid
	Distance from coastline	26977	0	Inverted sigmoid
Urban land	Distance from industry/commerce	6595.9	0	Gaussian
	Distance from high density residences	13653	0	Inverted sigmoid
	Distance from urban recreation	9257.9	0	Inverted sigmoid
Vegetation	Land use	-	-	Gaussian
Geology	Soil capability	6	4	Inverted sigmoid

Table 3. The rating scale for pairwise comparison matrix

Relative importance	Degree of preferences
1	Equal
3	Moderate
5	Strong
7	Very strong
9	Extreme
2, 4, 6, 8	Intermediate
Reciprocals	Less importance

1/9 1/7 1/5 1/3 1 3 5 7 9

← →

less importance more

Source: Saaty (1980)

Expert comments on the chosen criteria and information from the literature (Bathrellos et al., 2012, Zhang et al.,2013, Malmir et al., 2016, Yousefi et al., 2016) were taken into consideration when developing the pairwise comparison matrix. Any information found in the literature was used that is based on pairwise weighing of the chosen criterion. In the absence of information, expert opinions were used to evaluate the remaining criteria. The expert group was made up of six people with backgrounds in urban development and planning, environmental policy, and geomatics engineering. On a scale of 1 to 9, they were asked to rate the importance of criteria and sub-criteria. In this sense, they provided relative weights on the subject criteria according to what they knew. The results were then averaged to provide a single value on each pair of criteria, which was then represented as a single number in the pairwise comparison matrix (Table 4). The study used paired comparison matrices with the required criteria and sub-criteria listed in the first column of Table 1. Every sub-criterion was compared to the others and given a weight based on how important it was in relation to the other criteria that were evaluated (Table 4).

After the pairwise matrix was produced, relative weights were calculated using Saaty's (1980) method, and the consistency ratio index (CR) was used to measure the degree of consistency employed in creating the ratings.

$$CR = \frac{\lambda_{max} - n}{(n-1) \times RI} \tag{1}$$

Where λ_{max} is the principle eigenvalue of matrix, n is the order of the matrix and RI represents the average of resulting consistency index depending on the order n. The likelihood that the matrix ratings were produced at random is displayed by the CR index. The matrix shows contradictions if CR ≥ 0.1, and the AHP approach

Table 4. Criteria pairwise comparison matrix for industrial land suitability evaluation

	aspect	elevation	slope	highways	roads	bus stop	metro stop	port	airport	reservoir	water sources	coastline	urban green	commercial centres	residential centres	land use/ cover	soil capability
aspect	1	2	1/3	1/5	1/6	1/5	1/4	1/3	1/3	1/3	1/4	1/5	1/4	1/2	1/4	1/5	1/3
elevation		1	1/3	1/7	1/8	1/6	1/5	1/3	1/4	1/4	1/6	1/5	4	1/3	1/5	1/7	1
slope			1	1/3	1/6	1/4	1/3	1/4	1/5	1/3	1/5	1/4	4	1/7	1/6	1/6	1
highways				1	1	1	1	2	2	2	2	3	4	2	1/2	4	2
roads					1	3	4	2	3	2	2	2	4	1	1/2	3	2
bus stop						1	1	2	3	4	3	4	3	2	1	2	3
metro stop							1	3	2	4	3	4	3	1	1	3	3
port								1	1/3	2	1/3	2	2	1/2	1/3	1	2
airport									1	3	2	2	4	1/2	1	3	2
reservoir										1	1/2	1	1	1/4	1/5	1/2	1/3
water source											1	1	1	1/4	1/5	1	1/3
coastline												1	1	1/3	1/4	1	1/3
urban green													1	1/4	1/5	1/2	1/3
commerce cent.														1	1/3	3	2
residential cent.															1	3	2
land use/cover																1	1
soil capability																	1

Note: [1]Relative weights assigned to pairs of criteria were adapted from Barthrellos et al. (2012); Romana et al. (2015); Malmir et al. (2016).
[2] Principal Eigen value: 19.440

3.5. Land suitability modelling

The chosen sub-criteria taken into account in the analysis were used to calculate the overall suitability index. As indicated by equation (2), the suitability index was calculated by weighted linear summing of the layers.

$$S = \sum_{j=1}^{n} W_j MF_j \tag{2}$$

Where S is the suitability score; n is the number of factors considered in the analysis; W_j is the weight of criterion j which was computed by using the AHP method; and MF_j is the membership function for criterion j (see Ustaoglu and Aydınoglu, 2020). The suitability score, or S, has a value between 0 and 1, with a number of 0 denoting unsuitable land and a value of 1 denoting exceptionally acceptable land. A model of the land use suitability was created using the Arc GIS 10.4 model builder tool. The overall evaluation methodology used in the study is presented in Figure 4.

Figure 4. The overall industrial land suitability evaluation methodology used in the study

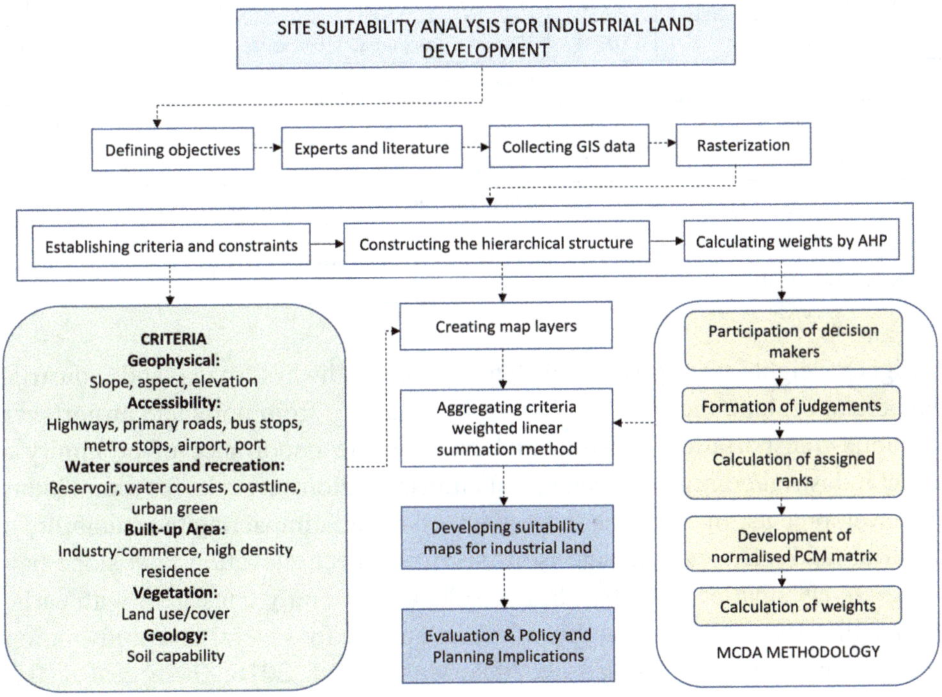

4. RESULTS

4.1. The weights obtained from AHP

The analysis considered geophysical characteristics, accessibility, green and blue infrastructure, urban infrastructure/developed area, and vegetation from the assessment of 17 sub-criteria. Pairwise comparison matrices that were computed using the AHP technique were used to determine the weighting of each criterion. The results of the criteria weights obtained from the AHP can be seen in Table 5.

Table 5. Weights of sub-criteria for industrial land suitability evaluation

Sub-criteria	Weights	Sub-criteria	Weights
Physical attributes		*Water resources*	
Aspect	0.020	Distance from reservoirs	0.037
Elevation	0.019	Distance from coastline	0.035
Slope	0.025	Distance from water bodies	0.037
Accessibility		*Urban land*	
Distance from highways	0.085	Distance from industry/commerce	0.126
Distance from roads	0.122	Distance from high density residential areas	0.080
Distance from bus stops	0.070	Distance from urban green areas	0.024
Distance from metro stops	0.071	*Vegetation*	
Distance from port	0.084	Current land use	0.040
Distance from airport	0.076	*Geology*	
		Soil capability	0.049
Principal Eigen value: 18.68			
Consistency ratio (CR): 0.065			

Higher weights were given to distances from highways, roads, and industrial/ commercial centers; these were followed by distances from ports and airports and the public transportation network, underscoring the importance of proximity to current industrial/commercial centers and transportation networks for future industrial developments. In comparison to other sub-criteria impacting the suitability of industrial land, physical characteristics such as aspect, elevation, and slope have comparatively smaller weights. These results are generally consistent with earlier research that assessed the suitability of urban land use for several case study regions in Europe and elsewhere (Dong et al., 2008; Malmir et al., 2016; Zheng et al., 2017).

4.2. Industrial land suitability in Pendik

Figure 5 shows the outcomes of the land suitability assessment for industrial construction, representing the conclusions from deterministic approach and fuzzy logic model. The most suitable locations, with the greatest values, are shown in the Figure in dark green, while the least suitable locations, with the lowest values, are shown in brown. The northern section of the study region is severely limited by the resulting suitability maps (Figure 5). Given that the northern part of the research region is primarily made up of forests, natural vegetation, and agricultural uses, and that industrial growth is more constrained there due to the area's protection status, this result is in line with the current situation. The area associated with the highest suitability class for industrial developments is primarily found near highways, close to existing residential and commercial centers, and along the public transportation network, as shown in Figure 5. The northern section of Pendik has a small amount of land with development potential for commercial and industrial applications. Table 6 provides a quantitative analysis of the variations in suitability. Figure 6 displays the percentage distributions of several suitability classes.

Figure 5. Land suitability map for industrial development from deterministic model and fuzzy logic model

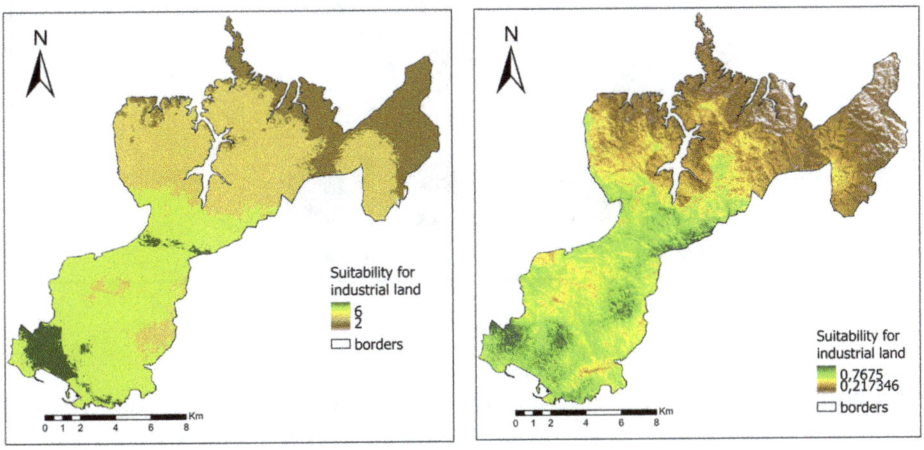

Table 6. Comparison of suitability differences in different models

	Areas by suitability levels			
	Deterministic approach		**Fuzzy logic method**	
Very low suitability	0-1	-	0.35-0.45	-
	1-2	0.76 km	0.45-0.55	7.60 km
↓	2-3	38.85 km	0.55-0.65	52.39 km
	3-4	48.74 km	0.65-0.75	49.82 km
	4-5	73.26 km	0.75-0.85	59.20 km
Very high suitability	5-6	16.39 km	>0.85	14.01 km

Figure 6. Percentage distribution of land use according to different suitability categories

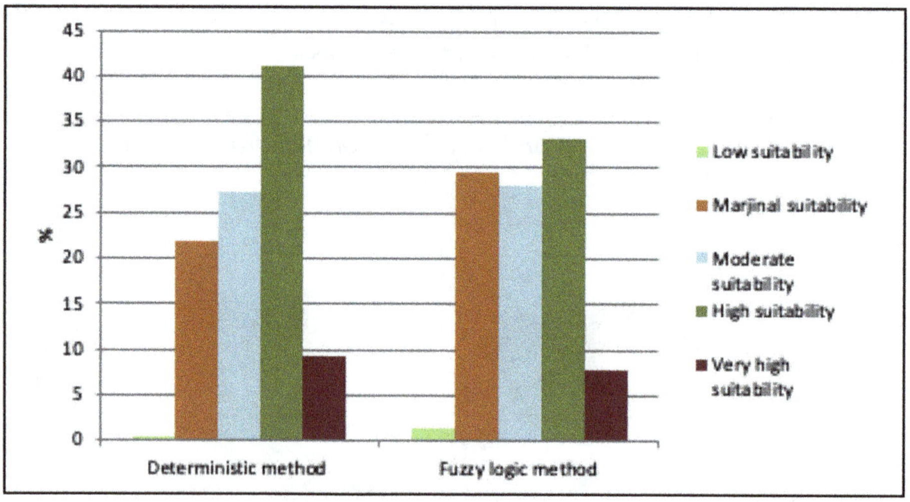

According to Table 6's results from deterministic approach, comparatively lesser percentage of the region is categorized as very favourable for industrial development. Regarding fuzzy logic model results, a lower proportion of highly suitable land is linked to industrial purposes. When it comes to the fuzzy logic technique, the low suitable class makes up less than 5% of the research area's total land, and when it comes to both deterministic and fuzzy logic methodologies, the very high suitable class makes up more than 10%. The distribution of suitable land among the low, moderate, marginal, and high suitability classes is more equitable, according to the fuzzy logic model's findings. Since various techniques were used to standardize maps and land-use classes, a direct comparison of suitable land areas between these two models may produce unreliable results.

The results from both models showed that the research area's central and south-eastern regions include extremely suitable terrain, with only a major fraction found in the southern region close to the airport and the coast. These urban centers are densely populated residential areas encircled by commercial and industrial constructions. In the future, a third growth center might be constructed in the center of the southern section due to the area's tremendous potential for residential and industrial/commercial development as well as its first-rate public transit infrastructure.

4.3. Sensitivity analysis

A sensitivity study was conducted on the overall suitability analysis, which was shown in Table 6. Sensitivity testing was done to the highest weight (i.e. distance to industrial/commercial areas) in accordance with Romano et al. (2015), and the findings are presented appropriately. A series of suitability assessments were conducted using a method similar to that used by Chen et al. (2010), in which each criterion weight was altered by a quarter percent (e.g., ± 25%, ± 50%, and ± 75%), and the weights of the remaining criteria were adjusted proportionately so that all criterion weights added up to one (see Table 7). The suitability differences for the selected criterion—distance to industrial/commercial areas are shown in Figure 7. The distribution of suitability classes (in km^2) for the criteria "industrial/commercial areas" is shown in Table 8 for the fuzzy logic method and deterministic method.

The findings show that the most sensitive criterion is the one with the largest weight, which is "industrial/commercial areas". In terms of the fuzzy logic model, when there is ± 75% change in the weight value, low suitability class (S3) results in +16.35% and -44.26% changes whereas the changes within the high suitability class (S6) ranges between +88.18% and −50.24%. According to the deterministic model, when there is ± 75% change in the weight value, very low suitability class (S2) have changes about +190.23% and -78.42% whereas the changes in the high suitability class vary between +97.68% and -38.77%. The findings show that the percentage change of the criterion weights has less of an impact on the high and moderate suitability classes (S3 and S5).

Table 7. Criteria for sensitivity analysis

Factors/parameters subject to sensitivity testing	Sensitivity Test	Explanation
Industrial Suitability Analysis		
Distance to industrial/ commercial areas	±25%, ±50%, ±75% change in the value of the subject parameter	Sensitivity analysis was performed on the largest weights (e.g. Distance to industrial/commercial regions). To meet the requirement that all factor weights add up to one, the factor weight was altered in percentage increments (for example, from ±25% to ±75), and the weights of the other criteria were adjusted proportionately.

Note: See Ustaoglu and Aydınoglu (2020)

Figure 7. Summary of results from sensitivity analysis across different models

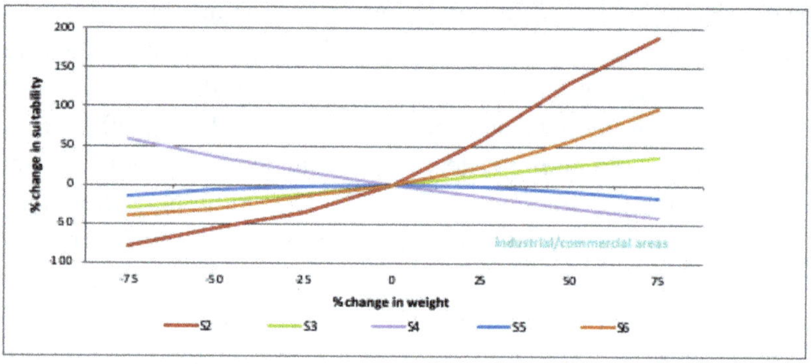

(a) Deterministic model

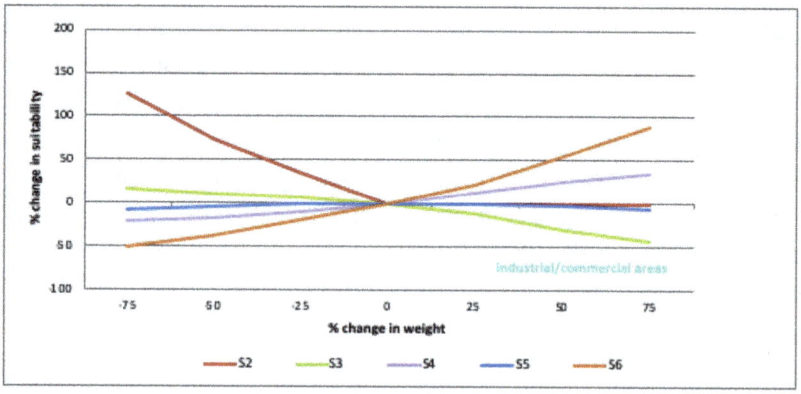

(b) Fuzzy logic model

Table 8. Results from sensitivity analysis from fuzzy logic and deterministic methods for the criterion 'industrial/commercial areas' regarding industrial suitability analysis

	Suitability (km2)					Changes in suitability (%)				
Change (%)	**S2**	**S3**	**S4**	**S5**	**S6**	**S2**	**S3**	**S4**	**S5**	**S6**
	Fuzzy logic method									
-75	17.187	60.953	38.965	53.927	6.973	126.09	16.35	-21.78	-8.89	-50.24
-50	13.264	58.266	41.049	56.714	8.712	74.49	11.22	-17.60	-4.18	-37.83
-25	8.612	55.705	44.656	58.667	11.366	35.14	6.33	-10.36	-0.88	-18.89
0	7.602	52.387	49.815	59.189	14.013	0	0	0	0	0
25	-	46.388	55.851	58.690	17.077	-	-11.45	12.12	-0.84	21.86
50	-	36.423	62.676	57.213	21.694	-	-30.47	25.82	-3.34	54.81
75	-	29.200	67.217	55.220	26.370	-	-44.26	34.93	-6.71	88.18
Deterministic method										
-75	0.165	27.833	76.820	63.150	10.038	-78.42	-28.36	57.61	-13.80	-38.77
-50	0.337	31.018	66.267	68.993	11.391	-55.90	-20.17	35.96	-5.82	-30.51
-25	0.499	34.229	56.968	72.316	13.994	-34.67	-11.90	16.88	-1.28	-14.63
0	0.763	38.854	48.740	73.256	16.393	0	0	0	0	0
25	1.211	43.677	41.441	71.605	20.072	58.61	12.41	-14.97	-2.25	22.44
50	1.768	48.420	34.821	67.236	25.761	131.60	24.62	-28.56	-8.22	57.15
75	2.008	52.711	28.504	61.728	32.405	190.23	35.67	-41.52	-15.74	97.68

5. DISCUSSION

5.1. The effectiveness of integration of GIS, AHP and fuzzy logic

This study employed an integrated Multi-Criteria Decision Analysis (MCDA) approach, combining Geographic Information Systems (GIS), the Analytic Hierarchy Process (AHP), and fuzzy logic, to assess land suitability for industrial development in the Pendik district of Istanbul, Türkiye. The findings reveal a significant concentration of suitable land for industrial expansion within the central and southern regions of the case study area. This spatial distribution aligns with the inherent advantages these areas possess, likely due to existing infrastructure and accessibility, factors that are consistently prioritized in industrial siting.

Compared to fuzzy approaches, the deterministic technique combined with AHP is more straightforward. It is simple to understand how weights are calculated and how the criteria are standardized, and it is easy to evaluate the suitability maps that are produced. The literature suggests that because AHP's inconsistency can be assessed and managed in the presence of conflicting criteria, it is a more effective method for determining the weights than its alternatives (Munier and Hontoria, 2022). The parametric approach (Bagherzadeh and Gholizadeh, 2016), boolean logic in GIS (Cheng and Thompson, 2016), Entropy and TOPSIS (Li et al., 2022), PROMETHEE (Benabbou. 2024) regression analysis (Park et al., 2011), grey relation analysis (Ni et al., 2024), fuzzy-AHP method (Sarkar et al., 2022), and principal component analysis (Ustaoglu et al., 2023) are some of the alternative methods used to determine factor weights. Numerous studies in the literature that successfully applied the subject technique in land suitability assessments have validated the efficiacy of the AHP approach (Akıncı et al., 2013; Bozdağ et al., 2016; Parry et al., 2018; Tolche et al., 2022; Choudhary et al., 2023; AbdelRahman et al., 2025). Since the AHP approach necessitates a questionnaire survey to identify relative weights and subject methodology is based on the subjective assessment of relative importance of the two criteria, interrelationships between different elements may be neglected. Additional disadvantages of the AHP technique include a lack of knowledge about the field of interest, reproducibility of the results, and subjectivity of the relative weights applied to the variables (Ustaoglu and Aydınoglu. 2020).

The application of an integrated methodology, particularly the incorporation of fuzzy logic for standardizing criteria maps alongside the AHP for weight derivation, represents a notable advancement over conventional, deterministic approaches. The suitability of industrial land usage was categorized using Boolean logic in the first model. This model is simple to comprehend and use; the quality of the input maps and the restrictions used determine how well it performs. This model has been widely used in the literature for the site suitability assessment of landfill site selection (Mahmood et al., 2021), PV solar plant location (Noorollahi et al., 2022), wind farm (Wimhurst et al., 2023), water aquaculture and livestock development (Qiu et al., 2017), agricultural land (Sengupta et al., 2022), and urban and industrial development (Ramya and Devedas, 2019). As highlighted in the research, fuzzy logic's capacity to allow objects to belong partially to multiple sets provides a more nuanced and realistic representation of complex environmental and socio-economic factors. This is particularly crucial in land suitability analysis where boundaries between suitable and unsuitable conditions are often soft transitional rather than hard boundaries. The comparison of suitability differences between the deterministic and fuzzy methods, as presented in Table 6, would further highlight the value of this enhanced methodological precision. GIS and fuzzy logic have been used in different studies including suitability assessment of landfill site (Zarin et al., 2021), agricultural land (Tennakoon et al., 2023), groundwater (Mallik et al., 2021), wind power plant location (Aghaloo et al., 2023), and urban development (Aliani et al.,

2021). There is hardly any study that integrates fuzzy logic, GIS and AHP for the site suitability analysis of industrial development; therefore, this study aims to contribute to the literature on this subject.

Similar studies based on industrial/commercial land suitability evaluations in other Turkish cities and regions may be found in the literature. For example, Aktas and Tahta (2018) employed GIS and AHP techniques to evaluate potential industrial growth locations for the Kemalpaşa District. The main elements to be evaluated for site suitability of industrial development included ecosystem development areas, protected areas, forests, olive groves, pastures, agricultural lands, water bodies, streams, land capability areas, distance to residential areas, modes of transportation, slopes, and erosion. Unlike the current study, they did not integrate fuzzy logic and AHP approach for the land suitability analysis. Their suitability of industrial land use was categorized using Boolean logic. A more recent study by Sahin et al. (2024) developed an urban growth suitability map for the current period and simulated it to 2027 to evaluate the potential development areas in Mersin Metropolitan Area against their specified suitability criteria. They included land cover, lithology, transport network, water resources, soil capability, and geo-physical factors as the main determinants of land suitability of urban development. A further study by Kuru et al. (2025) focused on location suitability modelling for industrial areas in İnegöl District. Their aim was to develop economic and ecological sustainability for the İnegöl District, which is Türkiye's one of the leading settlements in the furniture industry. They used GIS-based multicriteria decision analysis to assess thirty-seven criteria. Similar to Aktas and Tahta (2018), both studies did not consider fuzzy logic approach in the standardisation of their criteria maps. Therefore, the current study contributes to the subject articles through providing a comparison of deterministic and fuzzy methods in the development of suitability maps for industrial land use and has shown that fuzzy method resulted in more precise suitability mapping for industrial development.

The sensitivity analysis proved to be a critical component of this study, evaluating the stability and reliability of the suitability models. By systematically varying the weights of the most influential criteria, specifically "distance to industrial/commercial areas" (which held the highest weight), I was able to quantify the impact of such variations on the final suitability classifications. As shown in Table 8 and Figure 6, a significant ±75% change in the weight of this criterion led to substantial shifts in the suitability classes. For the fuzzy logic model, the low suitability class (S3) exhibited changes of +16.35% to -44.26%, while the high suitability class (S6) ranged from +88.18% to -50.24%. Similarly, the deterministic model showed even more pronounced fluctuations, with the very low suitability class (S2) experiencing changes from +190.23% to -78.42%, and the high suitability class varying between +97.68% and -38.77%.

Furthermore, the sensitivity analysis revealed that the moderate suitability classes (S3 and S5 in the fuzzy logic model, and S4 and S5 in the deterministic model) exhibited less variability compared to the very high suitability classes. This suggests that these transitional zones have some inbuilt resilience. This implies to planners that regions that fall into categories of moderate suitability may present more stable development alternatives, less likely to undergo drastic reclassification if criterion weights are slightly changed. This stability could be advantageous for long-term planning, as these areas might represent adaptable buffers or areas suitable for diverse industrial types that can prosper even in the absence of ideal conditions.

5.2. Implications for land use policies

The study's emphasis on Pendik District is pertinent given its significant economic growth over the past three decades, driven by investor demands and its strategic location within the Istanbul Metropolitan Region. Identifying potential sites for industrial expansion directly supports the Pendik Municipality's decision-making process, contributing to more effective land use planning and policy-making. This is especially vital in rapidly urbanizing regions like Pendik, where uncontrolled urban growth poses threats to the environment and challenges sustainable development goals. By providing a spatially explicit understanding of appropriate industrial locations, the study contributes to minimizing adverse impacts on urban ecosystems, such as traffic congestion, environmental pollution, and the degradation of natural habitats, which are key concerns identified in the introduction.

Through the implementation of New Urban Agenda, the Turkish government has been advocating for a new urbanization strategy to create efficient, sustainable, and livable cities since 2016. The results of this study on industrial land suitability may be a useful tool for its application. In particular, the creation of appropriate industrial development strategies and policies toward suitable lands can better preserve the study area's environmental resources and limit the expansion of industrial developments in unsuitable locations. Different suitability levels can be allocated for various land development reasons, as the study specifies. For example, land that isn't suitable for urban and industrial development can be used for agriculture, forests, or natural vegetation. Low-suitable areas for industrial/commercial uses may be better suited for recreational developments. Before making a decision on the alternative use of land, further evaluation is required.

On August 6, 2019, the Ministry of Industry and Technology ("Ministry") of the Turkish Republic released the Regulation on the Industrial Zones ("Regulation") in the Official Gazette. The Regulation includes comprehensive guidelines for the creation, administration, and functioning of private industrial zones, exclusive investment sites, and mixed/specialized industrial zones. In compliance with the regulations, a zoning plan for the industrial zone will be prepared by the Ministry or the management business. The zoning plan will be posted on the Ministry's official website for a week following approval, at which time stakeholders will have seven days to object. Subdivision plans for the arrangement of the lands and properties will thereafter be prepared by the Ministry or the management firm. Plans for infrastructure and superstructure, zoning and subdivisions, and the necessary licenses and permits will all need Ministry approval and administration. The current study provides the potential sites that can be evaluated for industrial zoning and therefore can guide the Ministry in its approval of industrial zoning and subdivisions.

5.3. Implications for future research

The growing demands of a fast-changing global landscape, where the supply of suitable land is increasingly limited by environmental and demographic shifts, must be anticipated by effective long-term industrial planning (Metternicht, 2017). The area of resilient developable sites is getting smaller due to accelerating climate change threats like rising sea levels, heat stress, and more frequent floods, especially in the vicinity of major coastal logistical centers (Lee et al., 2018). The urgent requirement for industrial land and the growing demand for residential housing and commercial services are at odds with each other as a result of anticipated population growth and economic expansion (Wu et al., 2016). The footprint needed for housing and retail places tremendous strain on limited land resources as urban areas grow denser and spread to accommodate new residents, frequently encroaching upon or pricing out critical industrial operations (Turok, 2016). In order to ensure that industrial infrastructure continues to function effectively in the face of these compounding physical factors, future zoning and site selection procedures must go beyond conventional metrics to prioritize hazard reduction, sustainable density, adaptive reuse and urban regeneration.

The integrity and effectiveness of Multi-Criteria Decision Analysis (MCDA) depend on the active integration of various stakeholders, from local communities and municipal authorities to business actors. MCDA need a participatory framework to guarantee that technical outputs are in line with social and political reality because it is intrinsically value-driven and depends on assigning weights to competing criteria (Beutler et al., 2024). Local communities promote social welfare and environmental preservation, industrial stakeholders contribute insights into economic viability and financial limits, and municipalities provide a well-defined regulatory background (Onu et al., 2023). The decision-making process may gain the necessary transparency by involving these groups in the definition of criteria and the calibration of preferences. This will ultimately reduce conflict and guarantee that the final strategic choices are not only analytically sound but also socially acceptable and feasible.

The omission of certain social and environmental impact parameters, such as air quality issues, noise pollution, and waste-disposal limits, is a major drawback of the existing evaluation approach. The approach ignores the negative externalities inherent in industrial operations that have a direct impact on ecological integrity and community well-being by concentrating mainly on geo-spatial and physical suitability. The absence of these factors creates conflict such that theoretically high-scoring locations might be found in places where noise production or industrial pollution would cause serious land-use disputes or public health issues. In order to fill these crucial gaps, the ensuing suitability rankings should be viewed as a measure of geo-spatial and physical feasibility rather than total suitability. This means that environmental impact evaluations must be systematically included into following evaluation stages as part of a future research.

Future research could build upon this foundation by integrating additional dynamic factors, such as predicted population shifts, climate change impacts, impacts of air/noise pollution or economic growth scenarios, to project future land suitability. Furthermore, while the current study provides macro-level suitability, a micro-level analysis incorporating more granular local regulations and community preferences could refine the findings. Although the Urban Atlas dataset offers a consistent and high-resolution spatial framework, the intrinsic restrictions of the data's temporal validity and resolution ultimately limit the analysis's validity. The Minimum Mapping Unit (MMU) of the dataset results in a generalization of land-use features, which may mask complexities of mixed-use or small-scale fragmentation that are essential for accurate site selection. Additionally, model uncertainty is introduced by the temporal lag between the dataset's capture and the current operational setting, since the categorization layers may not yet reflect recent zoning initiatives or rapid urbanization._Consequently, the resulting suitability maps must be interpreted as strategic approximations rather than definitive site plans, necessitating rigorous ground-truthing and cross-referencing with real-time cadastral data to mitigate potential misclassification errors. Investigating the specific types of industrial development best suited for the identified regions, rather than a generic "industrial" classification, could also provide more insights for local authorities. Finally, expanding the application of this integrated GIS-AHP-fuzzy logic framework to other rapidly developing urban districts in Türkiye and beyond would validate its generalizability and continue to support sustainable urban planning globally.

6. CONCLUSION

This study successfully developed and applied an integrated GIS-MCDA framework, combining the Analytic Hierarchy Process (AHP) with both deterministic and fuzzy logic approaches, to assess land suitability for industrial development in the Pendik District of Istanbul, Türkiye. The research addressed a critical need for structured decision support in urban planning, aiming to optimize industrial land allocation while considering a multifaceted array of environmental, social, and economic factors.

The core findings highlight that the central and southern regions of Pendik exhibit the highest suitability for industrial expansion derived from the two approaches employed in this investigation. A total area of 73.26 km^2 and 16.39 km^2 were determined to be suitable for industrial developments, respectively representing the high and very high suitability classes, based on the deterministic model findings. In contrast, fuzzy logic model's results showed that 59.2 and 14.01 km^2 of land are ideal for industrial developments, respectively representing high and very high suitability classes. Larger land areas were assigned a marginal or moderate suitability class for the industrial land developments: the fuzzy logic framework found that more than 100 km^2 of land were marginally and moderately suitable, while the deterministic approach found that nearly 85 km^2 of land corresponded to marginal and moderate suitability classes. Industrial development had an equal proportion of low, moderate, marginal, and high suitability classifications, according to fuzzy logic approach results. This suitability is largely driven by proximity to existing industrial and commercial centers, well-developed transportation networks (highways, roads, port, airport, public transport), and urban infrastructure. Conversely, the northern areas, characterized by significant natural vegetation and agricultural land, were identified as largely unsuitable, highlighting the importance of environmental protection and sustainable land management practices. Existing industrial areas appear to be more connected in the southern region, while there is limited or no connectivity to industrial land in the northern region. As a result, I advocate conducting future research on the environmental characteristics of land uses in the northern region, as well as developing an industrial zoning map in the southern region by connecting existing industrial spaces with the newly proposed industrial land to highlight the benefits of agglomeration economies and potential networking between old and new industrial areas. In the southern region, the high and very high suitability classes can correspond to industrial land zonings of degree 1 and 2 according to the emergent need of industrial development; moderate and marginal classes can be used for required infrastructure development and low suitability classes can be allocated for urban green space development. These spatial patterns provide clear, data-driven guidance for urban planners in Pendik Municipality.

The sensitivity analysis based on the highest weighted criterion used in the suitability assessment of industrial development demonstrated the impact of variations in the model results. In the case of the highest weighted criterion (i.e., distance from industrial/commercial centers), low and extremely high suitability classes are vulnerable to changes, based on industrial suitability evaluation. The fuzzy logic technique produced the least sensitive results for the industrial/commercial categories, while the traditional approach appears to be more sensitive to changes in the criteria weight. The stability of fuzzy logic model outputs in this case indicates that the fuzzy approach produces more reliable results than the deterministic method, and that this technique will increase the stability of outcomes when the model integrates additional possibilities or alternatives. This finding is crucial for policy-making, indicating that strategic investments in infrastructure and connectivity to existing economic hubs will have the most profound impact on enhancing industrial land suitability. The relative stability observed by the fuzzy logic model developing suitable areas also offers flexible options for planners, providing adaptable zones less susceptible to minor changes in criterion importance.

To sum up, this work offers a strong, multi-criteria spatial decision support system that can greatly improve Pendik's industrial growth land-use planning. A clear and convincing foundation for determining ideal locations, encouraging sustainable development, and guiding strategic infrastructure investments is provided by the integrated methodology and the accompanying suitability maps. In order to improve the decision-making process, future studies may investigate dynamic modeling of land-use change based on these suitability assessments or add other socioeconomic factors. In the end, this study offers a useful guide for other quickly urbanizing areas facing comparable land-use issues, promoting a sensible strategy that promotes economic growth while preserving ecological integrity.

REFERENCES

AbdelRahman, M. A. E., Yossif, T. M. H., & Metwaly, M. M. (2025). Enhancing land suitability assessment through integration of AHP and GIS-based for efficient agricultural planning in arid regions. *Scientific Reports, 15*(1), 31370. DOI: 10.1038/s41598-025-14051-7

Aghaloo, K., Ali, T., Chiu, Y.-R., & Sharifi, A. (2023). Optimal site selection for the solar-wind hybrid renewable energy systems in Bangladesh using an integrated GIS-based BWM-fuzzy logic method. *Energy Conversion and Management, 283*, 116899. DOI: 10.1016/j.enconman.2023.116899

Akbari, M., Neamatollahi, E., & Neamatollahi, P. (2019). Evaluating land suitability for spatial planning in arid regions of eastern Iran using fuzzy logic and multi-criteria analysis. *Ecological Indicators, 98*, 587–598. DOI: 10.1016/j.ecolind.2018.11.035

Akıncı, H., Yavuz Özalp, A., & Turgut, B. (2013). Agricultural land use suitability analysis using GIS and AHP technique. *Computers and Electronics in Agriculture, 97*, 71–82. DOI: 10.1016/j.compag.2013.07.006

Aktaş, E., & Tahta, B. T. (2018). Investigation of suitable land use potential for industrial sites: The case of Kemalpaşa. *Environmental Monitoring and Assessment, 190*(11), 654. DOI: 10.1007/s10661-018-7007-6

Aydi, A., Abichou, T., Nasr, I. H., Louati, M., & Zairi, M. (2016). Assessment of land suitability for olive mill wastewater disposal site selection by integrating fuzzy logic, AHP, and WLC in a GIS. *Environmental Monitoring and Assessment, 188*(1), 59. DOI: 10.1007/s10661-015-5076-3

Bagherzadeh, A., & Gholizadeh, A. (2016). Modelling land suitability evaluation for wheat production by parametric and TOPSIS approaches using GIS, northeast of Iran. *Modeling Earth Systems and Environment, 2*(3), 126. DOI: 10.1007/s40808-016-0177-8

Balaban, O. (2012). The negative effects of construction boom on urban planning and environment in Turkey: Unraveling the role of the public sector. *Habitat International, 36*(1), 26–35. DOI: 10.1016/j.habitatint.2011.05.003

Bathrellos, G. D., Gaki-Papanastassiou, K., Skilodimou, H. D., Papanastassiou, D., & Chousianitis, K. G. (2012). Potential suitability for urban planning and industry development using natural hazards maps and geological-geomorphological parameters. *Environmental Earth Sciences, 66*(2), 537–548. DOI: 10.1007/s12665-011-1263-x

Benabbou, A. (2024). Land use suitability analysis using an approach combining GIS, regionalization, and PROMETHEE II. *Annals of GIS*, *30*(1), 121–136. DOI: 10.1080/19475683.2024.2304183

Beutler, P., Larsen, T. A., Maurer, M., Staufer, P., & Lienert, J. (2024). A participatory multi-criteria decision analysis framework reveals transition potential towards non-grid wastewater management. *Journal of Environmental Management*, *367*, 121962. DOI: 10.1016/j.jenvman.2024.121962

Bozdağ, A., Yavuz, F., & Günay, A. S. (2016). AHP and GIS based land suitability analysis for Cihanbeyli (Turkey) County. *Environmental Earth Sciences*, *75*(9), 813. DOI: 10.1007/s12665-016-5558-9

Bunruamkaew, K., & Murayama, Y. (2011). Site suitability evaluation for ecotourism using GIS & AHP: A case study of Surat Thani province, Thailand. *Procedia: Social and Behavioral Sciences*, *21*, 269–288. DOI: 10.1016/j.sbspro.2011.07.024

Burrough, P. A. (1996). Natural objects with indeterminate boundaries. In Burrough, P. A., & Frank, A. (Eds.), *Geographic Objects with Indeterminate Boundaries*. CRC Press.

Cerreta, M., & de Toro, P. (2012). Urbanization suitability maps: A dynamic spatial decision support system for sustainable land use. *Earth System Dynamics: ESD*, *3*(2), 157–171. DOI: 10.5194/esd-3-157-2012

Chalise, D., Kumar, L., & Kristiansen, P. (2019). Land degradation by soil erosion in Nepal: A review. *Soil Systems*, *3*(1), 12. DOI: 10.3390/soilsystems3010012

Chen, Y., Yu, J., & Khan, S. (2010). Spatial sensitivity analysis of multi-criteria weights in GIS-based land suitability evaluation. *Environmental Modelling & Software*, *25*(12), 1582–1591. DOI: 10.1016/j.envsoft.2010.06.001

Cheng, C., & Thompson, R. G. (2016). Application of boolean logic and GIS for determining suitable locations for Temporary Disaster Waste Management Sites. *International Journal of Disaster Risk Reduction*, *20*, 78–92. DOI: 10.1016/j.ijdrr.2016.10.011

Choudhary, K., Boori, M. S., Shi, W., Valiev, A., & Kupriyanov, A. (2023). Agricultural land suitability assessment for sustainable development using remote sensing techniques with analytic hierarchy process. *Remote Sensing Applications: Society and Environment*, *32*, 101051. DOI: 10.1016/j.rsase.2023.101051

Dagıstanlı, C., Demirag Turan, I., & Dengiz, O. (2018). Evaluation of the suitability of sites for outdoor recreation using a multi-criteria assessment method. *Arabian Journal of Geosciences*, *11*(17), 492. DOI: 10.1007/s12517-018-3856-0

Deng, X., Shi, Q., Zhang, Q., Shi, C., & Yin, F. (2015). Impacts of land use and land cover changes on surface energy and water balance in the Heihe River Basin of China, 2000-2010. *Physics and Chemistry of the Earth Parts A/B/C*, *79-82*, 2–10. DOI: 10.1016/j.pce.2015.01.002

Donevska, K. R., Gorsevski, P. V., Jovanovski, M., & Pesevski, I. (2012). Regional non-hazardous landfill site selection by integrating fuzzy logic, AHP and geographic information systems. *Environmental Earth Sciences*, *67*(1), 121–131. DOI: 10.1007/s12665-011-1485-y

Dong, J., Zhuang, D., Xu, X., & Ying, L. (2008). Integrated evaluation of urban development suitability based on remote sensing and GIS techniques-a case study in Jingjinji Area, China. *Sensors (Basel)*, *8*(9), 5975–5986. DOI: 10.3390/s8095975

Dubovyk, O., Menz, G., & Khamzina, A. (2016). Land suitability assessment for afforestation with Elaeagnus Angustifolia L. in degraded agricultural areas of the Lower Amudarya River Basin. *Land Degradation & Development*, *27*(8), 1831–1839. DOI: 10.1002/ldr.2329

EC. (2025). Mapping guide-Urban Atlas Land Cover/Land Use and street tree layer 2012 and 2018. Retrieved from: https://land.copernicus.eu/en/technical-library/urban_atlas_2012_2018_mapping _guide Accessed: 18 November 2025

EEA. (2025). Online data and maps. Retrieved from: https://www.eea.europa.eu/data-and-maps Accessed: 18 November 2024

Elkhrachy, I., Alhamami, A., & Alyami, S. H. (2023). Landfill site selection using multi-criteria decision analysis, remote sensing data, and Geographic Information System tools in Najran City, Saudi Arabia. *Remote Sensing (Basel)*, *15*(15), 3754. DOI: 10.3390/rs15153754

Eraydın, A., & Tasan-Kok, T. (2014). State response to contemporary urban movements in Turkey: A critical overview of state entrepreneurialism and authoritarian interventions. *Antipode*, *46*(1), 110–129. DOI: 10.1111/anti.12042

Fisher, P. (1996). Boolean and fuzzy regions. In Burrough, P. A., & Frank, A. (Eds.), *Geographic Objects with Indeterminate Boundaries*. CRC Press.

Gül, A., Örücü, K. M., & Karaca, Ö. (2006). An approach for recreation suitability analysis to recreation planning in Gölcük Nature Park. *Environmental Management*, *37*(5), 606–625. DOI: 10.1007/s00267-004-0322-4

Gursoy, N., & Edelman, D. J. (2017). Regional development planning in Istanbul: Recent issues and challenges. *Current Urban Studies*, *5*(2), 146–163. DOI: 10.4236/cus.2017.52009

Humbal, A., Chaudhary, N., & Pathak, B. (2023). Urbanization Trends, Climate Change, and Environmental Sustainability. In Pathak, B., & Dubey, R. S. (Eds.), *Climate Change and Urban Environment Sustainability. Disaster Resilience and Green Growth.* Springer. DOI: 10.1007/978-981-19-7618-6_9

Johar, A., Jain, S. S., & Garg, P. K. (2013). Land suitability analysis for industrial development using GIS. *Journal of Genomics*, *7*(2), 101–106.

Kaufmann, A., & Gupta, M. M. (1991). *Fuzzy mathematical models in engineering and management science*. North-Holland.

Keyder, C., & Yenal, Z. (2011). Agrarian change under globalization: Markets and insecurity in Turkish agriculture. *Journal of Agrarian Change*, *11*(1), 60–86. DOI: 10.1111/j.1471-0366.2010.00294.x

Kienast, F., Degenhardt, B., Weilenmann, B., Wager, Y., & Buchecker, M. (2012). GIS-assisted mapping of landscape suitability for nearby recreation. *Landscape and Urban Planning*, *105*(4), 385–399. DOI: 10.1016/j.landurbplan.2012.01.015

Kılıç, Ş., Evrendilek, F., Şenol, S., & Çelik, I. (2005). Developing a suitability index for land uses and agricultural land covers: A case study in Turkey. *Environmental Monitoring and Assessment*, *102*(1-3), 323–335. DOI: 10.1007/s10661-005-6030-6

Kuru, A., Yüzer, M. A., Yüzer, A. Ş., Güney, B. G., & Yüzer, M. E. (2025). Integrated site selection model for industrial areas: Case study for İnegöl furniture industry. *Environmental Science and Pollution Research International*, *32*(8), 4771–4793. DOI: 10.1007/s11356-025-35965-9

Lee, K., Chun, H., & Song, J. (2018). New strategies for resilient planning in response to climate change for urban development. *Procedia Engineering*, *212*, 840–846. DOI: 10.1016/j.proeng.2018.01.108

Li, G., Messina, J. P., Peter, B. G., & Snapp, S. S. (2017). Mapping land suitability for agriculture in Malawi. *Land Degradation & Development*, *28*(7), 2001–2016. DOI: 10.1002/ldr.2723

Li, J., Xie, B., Dong, H., Zhou, K., & Zhang, X. (2023). The impact of urbanization on ecosystem services: Both time and space are important to identify driving forces. *Journal of Environmental Management*, *347*, 119161. DOI: 10.1016/j.jenvman.2023.119161

Li, Z., Fan, Z., & Shen, S. (2018). Urban green space suitability evaluation based on the AHP-CV combined weight method: A case study of Fuping County, China. *Sustainability (Basel)*, *10*(8), 2656. DOI: 10.3390/su10082656

Li, Z., Luo, Z., Wang, Y., Fan, G., & Zhang, J. (2022). Suitability evaluation system for the shallow geothermal energy implementation in region by Entropy Weight Method and TOPSIS method. *Renewable Energy*, *184*, 564–576. DOI: 10.1016/j.renene.2021.11.112

Mahdavi, A., & Niknejad, M. (2014). Site suitability evaluation for ecotourism using MCDM methods and GIS: Case study-Lorestan province, Iran. *Journal of Biodiversity and Environmental Sciences*, *4*(6), 425–437.

Mahmood, K. W., Khzr, B. O., Othman, R. M., Rasul, A., Ali, S. A., & Ibrahim, G. R. F. (2021). Optimal site selection for landfill using the Boolean-analytic hierarchy process. *Environmental Earth Sciences*, *80*(5), 173. DOI: 10.1007/s12665-021-09501-0

Malczewski, J. (2004). GIS-based land-use suitability analysis: A critical overview. *Progress in Planning*, *62*(1), 3–65. DOI: 10.1016/j.progress.2003.09.002

Mallik, S., Mishra, U., & Paul, N. (2021). Groundwater suitability analysis for drinking using GIS based fuzzy logic. *Ecological Indicators*, *121*, 107179. DOI: 10.1016/j.ecolind.2020.107179

Malmir, M., Zarkesh, M. M. K., Monavari, S. M., Jozi, S. A., & Sharifi, E. (2016). Analysis of land suitability for urban development in Ahwaz County in southwestern Iran using fuzzy logic and analytic network process (ANP). *Environmental Monitoring and Assessment*, *188*(8), 447. DOI: 10.1007/s10661-016-5401-5

Mendas, A., & Delali, A. (2012). Integration of multicriteria decision analysis in GIS to develop land suitability for agriculture: Application to drum wheat cultivation in the region of Mleta in Algeria. *Computers and Electronics in Agriculture*, *83*, 117–126. DOI: 10.1016/j.compag.2012.02.003

Metternicht, G. (2017). Land use planning. Global Land Outlook Working Paper. United Nations Convention to Combat Desertification.

Miah, M. T., Fariha, J. N., Jodder, P. K., Kafy, A. A., Raiyan, R., Usha, S. A., Hossan, J., & Rahaman, K. R. (2024). Urban heat island and environmental degradation analysis utilizing a remote sensing technique in rapidly urbanizing South Asian cities. *WORLD (Oakland, Calif.)*, *5*(4), 1023–1053. DOI: 10.3390/world5040052

Midatana, S., Saran, S., & Ramana, K. M. (2018). Site suitability analysis for industries using GIS and multi criteria decision making. *ISPRS Annals of the Photogrammetry, Remote Sensing and Spatial Information Sciences*, *IV-5*, 447–454. DOI: 10.5194/isprs-annals-IV-5-447-2018

Mohammed, J. K., Saidu, U. A., & Sheriff, I. H. (2025). GIS-AHP based suitability model for commercial property investment in Bida, Nigeria. *Journal of Inclusive Cities and Built Environment, 5*(9), 36–44. DOI: 10.54030/2788-564X/2025/is9v5a3

Morckel, V. (2017). Using suitability analysis to select and prioritize naturalization efforts in legacy cities: An example from Flint, Michigan. *Urban Forestry & Urban Greening, 27*, 343–351. DOI: 10.1016/j.ufug.2017.09.006

Munier, N., & Hontoria, E. (2022). *Uses and limitations of the AHP method: A non-mathematical and rational analysis*. Springer Cham.

Ni, X., Li, J., Xu, J., Shen, Y., & Liu, X. (2024). Grey relation analysis and multiple criteria decision analysis method model for suitability evaluation of underground space development. *Engineering Geology, 338*, 107608. DOI: 10.1016/j.enggeo.2024.107608

Noorollahi, Y., Senani, A. G., Fadaei, A., Simaee, M., & Moltames, R. (2022). A framework for GIS-based site selection and technical potential evaluation of PV solar farm using Fuzzy-Boolean logic and AHP multi-criteria decision-making approach. *Renewable Energy, 186*, 89–104. DOI: 10.1016/j.renene.2021.12.124

Onu, P., Pradhan, A., & Mbohwa, C. (2023). The potential of industry 4.0 for renewable energy and materials development-The case of multinational energy companies. *Heliyon, 9*(10), e20547. DOI: 10.1016/j.heliyon.2023.e20547

Park, S., Jeon, S., Kim, S., & Choi, C. (2011). Prediction and comparison of urban growth by land suitability index mapping using GIS and RS in South Korea. *Landscape and Urban Planning, 99*(2), 104–114. DOI: 10.1016/j.landurbplan.2010.09.001

Parry, J. A., Ganaie, S. A., & Bhat, M. S. (2018). GIS based land suitability analysis using AHP model for urban services planning in Srinagar and Jammu urban centres of J&K, India. *Journal of Urban Management, 7*(2), 46–56. DOI: 10.1016/j.jum.2018.05.002

Pendik Governorship. (2025). Online Data. Retrieved from: http://www.pendik.gov.tr Accessed: 18 November 2025

Peng, J., Ma, J., Du, Y., Zhang, L., & Hu, X. (2016). Ecological suitability evaluation for mountainous area development based on conceptual model of landscape structure, function, and dynamics. *Ecological Indicators, 61*(2), 500–511. DOI: 10.1016/j.ecolind.2015.10.002

Piran, H., Maleknia, R., Akbari, H., Soosani, J., & Karami, O. (2013). Site selection for local forest park using analytic hierarchy process and geographic information system (case study: Badreh County). *International Research Journal of Applied and Basic Sciences*, *6*(7), 930–935.

Pramanik, M. K. (2016). Site suitability analysis for agricultural land use of Darjeeling district using AHP and GIS techniques. *Modeling Earth Systems and Environment*, *2*(2), 56. DOI: 10.1007/s40808-016-0116-8

Rabia, A. H., Figueredo, H., Huong, T. L., Lopez, B. A. A., Solomon, H. W., & Alessandro, V. (2013). Land suitability analysis for policy making assistance: A GIS based land suitability comparison between surface and drip irrigation system. *International Journal of Environmental Sciences and Development*, *4*(1), 1–6. DOI: 10.7763/IJESD.2013.V4.292

Ramya, S., & Devadas, V. (2019). Integration of GIS, AHP and TOPSIS in evaluating suitable locations for industrial development: A case of Tehri Garhwal district, Uttarakhand, India. *Journal of Cleaner Production*, *238*, 117872. DOI: 10.1016/j.jclepro.2019.117872

Romano, G., Dal Sasso, P., Trisorio Liuzzi, G., & Gentile, F. (2015). Multi-criteria decision analysis for land suitability mapping in a rural area of Southern Italy. *Land Use Policy*, *48*, 131–143. DOI: 10.1016/j.landusepol.2015.05.013

Saaty, T. L. (1980). *The Analytical Hierarchy Process*. McGraw Hill.

Saha, S., Sarkar, D., Mondal, P., & Goswami, S. (2021). GIS and multi-criteria decision-making assessment of sites suitability for agriculture in an anabranching site of sooin river, India. *Modeling Earth Systems and Environment*, *7*(1), 571–588. DOI: 10.1007/s40808-020-00936-1

Sahin, E., Iban, M. C., & Bilgilioglu, S. S. (2024). Remote sensing-enabled urban growth simulation overlaid with AHP-GIS-based urban land suitability for potential development in Mersin Metropolitan Area, Türkiye. *Applied Sciences (Basel, Switzerland)*, *14*(8), 3484. DOI: 10.3390/app14083484

Sengupta, S., Mohinuddin, S., Arif, M., Sengupta, B., & Zhang, W. (2022). Assessment of agricultural land suitability using GIS and fuzzy analytical hierarchy process approach in Ranchi District, India. *Geocarto International*, *37*(26), 13337–13368. DOI: 10.1080/10106049.2022.2076925

Septian, S. A., & Saputra, R. A. (2022). Spatial model of industrial area suitability using spatial multi criteria evaluation: A case study in Kendari City. *Sustinere Journal of Environment & Sustainability*, *6*(3), 174–270.

Sertkaya Dogan, Ö. (2015). Growth of population and population policies implemented in Turkey between 1965 and 2014. *International Journal of Scientific Research*, *4*(4), 1907–1911.

Shrestha, D. P., Suriyaprasit, M., & Prachansri, S. (2014). Assessing soil erosion in inaccessible mountainous areas in the tropics: The use of land cover and topographic parameters in a case study in Thailand. *Catena*, *121*, 40–52. DOI: 10.1016/j.catena.2014.04.016

Sleeter, B. M., Liu, J., Daniel, C., Rayfield, B., Sherba, J., Hawbaker, T. J., Zhu, Z., Selmants, P. C., & Loveland, T. R. (2018). Effects of contemporary land-use and land-cover change on the carbon balance of terrestrial ecosystems in the United States. *Environmental Research Letters*, *13*(4), 045006. DOI: 10.1088/1748-9326/aab540

Steiner, F., McSherry, L., & Cohen, J. (2000). Land suitability analysis for the upper Gila River watershed. *Landscape and Urban Planning*, *50*(4), 199–214. DOI: 10.1016/S0169-2046(00)00093-1

Sun, C., Sun, C., Yang, Z., Zhang, J., & Deng, Y. (2016). Urban land development for industrial and commercial use: A case study of Beijing. *Sustainability (Basel)*, *8*(12), 1323. DOI: 10.3390/su8121323

Supto, S. T. J., Karzai, D. R. H., & Adib, E. I. (2025). Assessing industrial land suitability for sustainable urban planning in Dhaka Region using geospatial techniques. *Environmental and Earth Sciences Proceedings*, *36*(1), 5. DOI: 10.3390/eesp2025036005

Taktak, F. (2025). Comparative building amnesty and land use policies: Insights from ten countries with a focus on Turkey. *Land Use Policy*, *158*, 107763. DOI: 10.1016/j.landusepol.2025.107763

Tekeli, I. (2009). An exploratory approach to urban historiography through a new paradigm: The case of Turkey. In Sarkis, H., & Turan, N. (Eds.), *A Turkish Triangle: Ankara, Istanbul and Izmir at the gates of Europe*.

Tennakoon, S., Apan, A., Maraseni, T., & Altarez, R. D. D. (2023). Decoding the impacts of space and time on honey bees: GIS based fuzzy AHP and fuzzy overlay to assess land suitability for apiary sites in Queensland, Australia. *Applied Geography (Sevenoaks, England)*, *155*, 102951. DOI: 10.1016/j.apgeog.2023.102951

Thapa, R. B., & Murayama, Y. (2008). Land evaluation for peri-urban agriculture using analytical hierarchical process and geographic information system techniques: A case study of Hanoi. *Land Use Policy*, *25*(2), 225–239. DOI: 10.1016/j.landusepol.2007.06.004

Tirachini, A. (2014). The economics and engineering of bus stops: Spacing, design and congestion. *Transportation Research Part A, Policy and Practice*, *59*, 37–57. DOI: 10.1016/j.tra.2013.10.010

Tolche, A. D., Gurara, M. A., Pham, Q. B., Ditthakit, P., & Anh, D. T. (2022). Agricultural land use suitability analysis using AHP and GIS techniques at basin scale. *Arabian Journal of Geosciences*, *15*(18), 1522. DOI: 10.1007/s12517-022-10799-8

Torbert, H. A., Krueger, E., & Kurtener, D. (2008). Soil quality assessment using fuzzy modelling. *International Agrophysics*, *22*, 365–370.

Tortora, A., Statuto, D., & Picuno, P. (2015). Rural landscape planning through spatial modelling and image processing of historical maps. *Land Use Policy*, *42*, 71–82. DOI: 10.1016/j.landusepol.2014.06.027

Turok, I. (2016). Getting urbanization to work in Africa: The role of the urban land-infrastructure-finance nexus. *Area Development and Policy*, *1*(1), 30–47. DOI: 10.1080/23792949.2016.1166444

United Nations (UN). (2014). *World urbanization prospects, the 2014 revision*. UN Department of Economic and Social Affairs-Population Division.

Ustaoglu, E., & Aydınoglu, A. C. (2020). Suitability evaluation of urban construction land in Pendik district of Istanbul, Turkey. *Land Use Policy*, *99*, 104783. DOI: 10.1016/j.landusepol.2020.104783

Ustaoglu, E., Ortega Lopez, G., & Gutierrez-Alcoba, A. (2023). Building composite indicators for the territorial quality of life assessment in European regions. Combining data reduction and alternative weighting techniques. *Environment, Development and Sustainability*, *27*(3), 6025–6063. Advance online publication. DOI: 10.1007/s10668-023-04116-w

Uy, P. D., & Nakagoshi, N. (2008). Application of land suitability analysis and landscape ecology to urban greenspace planning in Hanoi, Vietnam. *Urban Forestry & Urban Greening*, *7*(1), 25–40. DOI: 10.1016/j.ufug.2007.09.002

Vandermeer, M. C., & Halleux, J. M. (2017). Evaluation of the spatial and economic effectiveness of industrial land policies in northwest Europe. *European Planning Studies*, *25*(8), 1454–1475. DOI: 10.1080/09654313.2017.1322042

Varma, K., Srivastava, V., Singhal, A., & Jha, P. K. (2021). Urban and environmental hazards. In Rai, P. K., Singh, P., & Mishra, V. N. (Eds.), *Recent Technologies for Disaster Management and Risk Reduction. Earth and Environmental Sciences Library*. Springer. DOI: 10.1007/978-3-030-76116-5_19

Vondrackova, T., Musilek, J., & Kais, L. (2015). The issue of soft rocks causing problems in foundation engineering. *Proedia Earth and Planetary Science, 15*, 54–59. DOI: 10.1016/j.proeps.2015.08.015

Wang, F. (1990). Fuzzy supervised classification of remote sensing maps. *IEEE Transactions on Geoscience and Remote Sensing*, *28*(2), 194–201. DOI: 10.1109/36.46698

Wang, H., Shen, Q., & Tang, B.-S. (2015). GIS-based framework for supporting land use planning in urban renewal: Case study in Hong Kong. *Journal of Urban Planning and Development*, *141*(3), 05014015. Advance online publication. DOI: 10.1061/(ASCE)UP.1943-5444.0000216

Wang, T., Kazak, J., Han, Q., & de Vries, B. (2019). A framework for path-dependent industrial land transition analysis using vector data. *European Planning Studies*, *27*(7), 1391–1412. DOI: 10.1080/09654313.2019.1588852

Wimhurst, J. J., Nsude, C. C., & Greene, J. S. (2023). Standardizing the factors used in wind farm site suitability models: A review. *Heliyon*, *9*(5), e15903. DOI: 10.1016/j.heliyon.2023.e15903

World Bank. (2017). Population estimates and projections, https://datacatalog.worldbank.org/dataset/population-estimates-and-projections

Wu, Y., Luo, J., Zhang, X., & Skitmore, M. (2016). Urban growth dilemmas and solutions in China: Looking forward to 2030. *Habitat International*, *56*, 42–51. DOI: 10.1016/j.habitatint.2016.04.004

Yigitcanlar, T., Sipe, N., Evans, R., & Pitot, M. (2007). A GIS-based land use and public transport accessibility indexing model. *Australian Plants*, *44*(3), 30–37. DOI: 10.1080/07293682.2007.9982586

Yousefi, E., Salehi, E., Zahiri, S. H., & Yavari, A. (2016). Green space suitability analysis using evolutionary algorithm and Weighted Linear Combination (WLC) method. *Space Ontology International Journal*, *5*(4), 51–60.

Zadeh, L. A. (1965). Fuzzy sets. *Information and Control*, *8*(3), 338–353. DOI: 10.1016/S0019-9958(65)90241-X

Zarin, R., Azmat, M., Naqvi, S. R., Saddique, Q., & Ullah, S. (2021). Landfill site selection by integrating fuzzy logic, AHP and WLC method based on multi-criteria decision analysis. *Environmental Science and Pollution Research International*, *28*(16), 19726–19741. DOI: 10.1007/s11356-020-11975-7

Zeng, S., Xia, J., Chen, X., Zou, L., Du, H., & She, D. (2020). Integrated land-surface hydrological and biogeochemical processes in simulating water, energy and carbon fluxes over two different ecosystems. *Journal of Hydrology (Amsterdam)*, *582*, 124390. DOI: 10.1016/j.jhydrol.2019.124390

Zhang, J., Su, Y., Wu, J., & Liang, H. (2015). GIS based land suitability assessment for tobacco production using AHP and fuzzy set in Shandong province of China. *Computers and Electronics in Agriculture*, *114*, 202–211. DOI: 10.1016/j.compag.2015.04.004

Zhang, X., Fang, C., Wang, Z., & Ma, H. (2013). Urban construction land suitability evaluation based on improved multi-criteria evaluation based on GIS (MCE-GIS): Case of New Hefei City, China. *Chinese Geographical Science*, *23*(6), 740–753. DOI: 10.1007/s11769-013-0609-6

Zheng, Q., Yang, X., Wang, K., Huang, L., Shahtahmassebi, A. R., Gan, M., & Weston, M. V. (2017). Deliminating urban growth boundary through combining land suitability evaluation and cellular automata. *Sustainability (Basel)*, *9*(12), 2213. DOI: 10.3390/su9122213

Zimmerman, H. J. (1991). *Fuzzy set theory and its applications*. Kluwer Academic Publishers. DOI: 10.1007/978-94-015-7949-0

Zolekar, B. R., & Bhagat, V. S. (2015). Multi-criteria land suitability analysis for agriculture in hilly zone: Remote sensing and GIS approach. *Computers and Electronics in Agriculture*, *118*, 300–321. DOI: 10.1016/j.compag.2015.09.016

Compilation of References

AbdelRahman, M. A. E., Yossif, T. M. H., & Metwaly, M. M. (2025). Enhancing land suitability assessment through integration of AHP and GIS-based for efficient agricultural planning in arid regions. *Scientific Reports*, *15*(1), 31370. DOI: 10.1038/s41598-025-14051-7

Abdelwahab, S. A. M., El-Rifaie, A. M., Hegazy, H. Y., Tolba, M. A., Mohamed, W. I., & Mohamed, M. (2024). Optimal control and optimization of grid-connected PV and wind turbine hybrid systems using electric eel foraging optimization algorithms. *Sensors (Basel)*, *24*(7), 2354. DOI: 10.3390/s24072354 PMID: 38610565

Adeniyi, L. A. (2023). Assessment of the effects of location of disposal-sites on waste disposal practices of urban residents in Nigeria. *GeoJournal*, *88*(4), 3889–3903. DOI: 10.1007/s10708-023-10835-4

Adli, H. K., Remli, M. A., Wan Salihin Wong, K. N. S., Ismail, N. A., González-Briones, A., Corchado, J. M., & Mohamad, M. S. (2023). Recent advancements and challenges of AIoT application in smart agriculture: A review. *Sensors (Basel)*, *23*(7), 3752. DOI: 10.3390/s23073752 PMID: 37050812

Afandi, G. E., Moustafa, A., Ibrahim, S., & Irfan, M. (2025). An in-depth analysis of approaches for forecasting extreme rainfall events leveraging artificial intelligence, machine learning, and deep learning techniques. *Earth Systems and Environment*, ●●●, 1–32. DOI: 10.1007/s41748-025-00813-w

Aghaloo, K., Ali, T., Chiu, Y.-R., & Sharifi, A. (2023). Optimal site selection for the solar-wind hybrid renewable energy systems in Bangladesh using an integrated GIS-based BWM-fuzzy logic method. *Energy Conversion and Management*, *283*, 116899. DOI: 10.1016/j.enconman.2023.116899

Ahmadi-Gh, Z., & Bello-Pintado, A. (2024). Towards sustainable manufacturing: How does digitalization and development affect sustainability barriers? *Journal of Cleaner Production*, *476*(1), 1–13. DOI: 10.1016/j.jclepro.2024.143792

Ahmed, F., Ali, I., Kousar, S., & Ahmed, S. (2022). The environmental impact of industrialization and foreign direct investment: Empirical evidence from Asia-Pacific region. *Environmental Science and Pollution Research International, 29*(20), 29778–29792. DOI: 10.1007/s11356-021-17560-w PMID: 34993824

Akbari, M., Neamatollahi, E., & Neamatollahi, P. (2019). Evaluating land suitability for spatial planning in arid regions of eastern Iran using fuzzy logic and multi-criteria analysis. *Ecological Indicators, 98*, 587–598. DOI: 10.1016/j.ecolind.2018.11.035

Akhavan, M., Alivirdi, M., Jamalpour, A., Kheradranjbar, M., Mafi, A., Jamalpour, R., & Ravanshadnia, M. (2022). Impact of Industry 5.0 on the Construction Industry (Construction 5.0): Systematic Literature Review and Bibliometric Analysis. Buildings 2025, 15, 1491. Artificial Intelligence.

Akıncı, H., Yavuz Özalp, A., & Turgut, B. (2013). Agricultural land use suitability analysis using GIS and AHP technique. *Computers and Electronics in Agriculture, 97*, 71–82. DOI: 10.1016/j.compag.2013.07.006

Akiyama, M. (2025). Life-cycle approaches to sustainable and resilient infrastructure: Innovations in multi-hazard frameworks. *Structure and Infrastructure Engineering, 21*(11-12), 1756–1781. DOI: 10.1080/15732479.2025.2524819

Aktaş, E., & Tahta, B. T. (2018). Investigation of suitable land use potential for industrial sites: The case of Kemalpaşa. *Environmental Monitoring and Assessment, 190*(11), 654. DOI: 10.1007/s10661-018-7007-6

Al-Ani, M. A., Zdiri, M. A., Salem, F. B., & Derbel, N. (2024). Optimized grid-connected hybrid renewable energy power generation: A comprehensive analysis of photovoltaic, wind, and fuel cell systems. Engineering, Technology &. *Applied Scientific Research, 14*(3), 13929–13936.

Almeida, F., & Okon, E. (2025). Impact of digitalization on SMEs performance: The mediating role of IOT. *Digital Transformation and Society, 4*(4), 407–422. DOI: 10.1108/DTS-11-2024-0240

Almtireen, N., Reddy, V., Sutton, M., Nedvidek, A., Karn, C., Ryalat, M., Elmoaqet, H., & Rawashdeh, N. (2025). PLC-Controlled Intelligent Conveyor System with AI-Enhanced Vision for Efficient Waste Sorting. *Applied Sciences (Basel, Switzerland), 15*(3), 1550. DOI: 10.3390/app15031550

Anuradha, P., Vasanth, K., Renuka, G., & Rao, A. R. (2023). IoT based enabling home automation system for individuals with diverse disabilities. e-Prime - Advances in Electrical Engineering. *Electronics and Energy, 6*, 100366.

Arif, M. E., & Herawan, T. (2025). Driving performance in SMEs: Exploring entrepreneurial ecosystems, leadership, and dynamic capability in emerging economies. *Journal of the International Council for Small Business*, *1*(1), 1–21. DOI: 10.1080/26437015.2025.2531873

Arindam, R., Apurba, M., Jungmin, K., & Ilkyeong, M. (2022). IoT-based smart bin allocation and vehicle routing in solid waste management: A case study in South Korea. *Computers & Industrial Engineering*, *171*, 108457. DOI: 10.1016/j.cie.2022.108457

Arora, J., Joshi, A., & Ray, R. C. (2024). *Transforming agriculture residues for sustainable development*. Springer., DOI: 10.1007/978-3-031-61133-9

Astuti, S. D., Tamimi, M. H., Pradhana, A. A. S., Alamsyah, K. A., Purnobasuki, H., Khasanah, M., Susilo, Y., Triyana, K., Kashif, M., & Syahro, A. (2021). Gas sensor array to classify the chicken meat with E. coli contaminant by using random forest and support vector machine. *Biosensors & Bioelectronics: X*, *9*, 100083. DOI: 10.1016/j.biosx.2021.100083

Autio, E., Komlósi, E., Szerb, L., Tiszberger, M. G., Park, D., & Jinjarak, Y. (2025). Digital entrepreneurship landscapes in developing Asia: Insights from the global index of digital entrepreneurship systems. *European Journal of Innovation Management*, *28*(7), 2845–2872. DOI: 10.1108/EJIM-02-2024-0121

Aydi, A., Abichou, T., Nasr, I. H., Louati, M., & Zairi, M. (2016). Assessment of land suitability for olive mill wastewater disposal site selection by integrating fuzzy logic, AHP, and WLC in a GIS. *Environmental Monitoring and Assessment*, *188*(1), 59. DOI: 10.1007/s10661-015-5076-3

Babu, T. M., Chenchireddy, K., Kumar, K. K., Nehal, V., Srihitha, S., & Vikas, M. R. (2024). Intelligent control strategies for grid-connected photovoltaic wind hybrid energy systems using ANFIS. *International Journal of Advances in Applied Sciences*, *13*(3), 497–506. DOI: 10.11591/ijaas.v13.i3.pp497-506

Bagherzadeh, A., & Gholizadeh, A. (2016). Modelling land suitability evaluation for wheat production by parametric and TOPSIS approaches using GIS, northeast of Iran. *Modeling Earth Systems and Environment*, *2*(3), 126. DOI: 10.1007/s40808-016-0177-8

Balaban, O. (2012). The negative effects of construction boom on urban planning and environment in Turkey: Unraveling the role of the public sector. *Habitat International*, *36*(1), 26–35. DOI: 10.1016/j.habitatint.2011.05.003

Bathrellos, G. D., Gaki-Papanastassiou, K., Skilodimou, H. D., Papanastassiou, D., & Chousianitis, K. G. (2012). Potential suitability for urban planning and industry development using natural hazards maps and geological-geomorphological parameters. *Environmental Earth Sciences*, *66*(2), 537–548. DOI: 10.1007/s12665-011-1263-x

Benabbou, A. (2024). Land use suitability analysis using an approach combining GIS, regionalization, and PROMETHEE II. *Annals of GIS*, *30*(1), 121–136. DOI: 10.1080/19475683.2024.2304183

Beutler, P., Larsen, T. A., Maurer, M., Staufer, P., & Lienert, J. (2024). A participatory multi-criteria decision analysis framework reveals transition potential towards non-grid wastewater management. *Journal of Environmental Management*, *367*, 121962. DOI: 10.1016/j.jenvman.2024.121962

Bhatia, V., & Jaglan, V. (2024). Integrating Artificial Intelligence and IoT for Smart Cities: Applications and Challenges. Reshaping Intelligent Business and Industry: Convergence of AI and IoT at the Cutting Edge, 267-278.

Boffa, E., & Maffei, A. (2024). Investigating the impact of digital transformation on manufacturers' business model: Insights from Swedish industry. *Journal of Open Innovation*, *10*(2), 100312. DOI: 10.1016/j.joitmc.2024.100312

Borni, A., Bessous, N., Zaghba, L., Bouchakour, A., Agmas, M. S., Ali, E., Ghoneim, S. S. M., & Hoballah, A. (2025). Enhancing grid connected wind energy conversion systems through fuzzy logic control optimization with PSO and GA techniques. *Scientific Reports*, *15*(1), 27678. DOI: 10.1038/s41598-025-12593-4 PMID: 40730602

Boulouard, Z., Ouaissa, M., Ouaissa, M., Siddiqui, F., Almutiq, M., & Krichen, M. (2022). An integrated artificial intelligence of things environment for river flood prevention. *Sensors (Basel)*, *22*(23), 9485. DOI: 10.3390/s22239485 PMID: 36502187

Boutracheh, H., Mejjad, N., & El Bouhadioui, M. (2024a, August). for Trends and Opportunities. In GIS, Applied Computing and Data Science for Water Management

Boutracheh, H., Mejjad, N., El Bouhadioui, M., & Moumen, A. (2024b, August). Water Research in the Age of AI: A Bibliometric Heuristic Analysis for Trends and Opportunities. In International Conference GIRE3D Participatory and Integrated Management of Water Resources in Arid Zones (pp. 3-45). Cham: Springer International Publishing. DOI: 10.1007/978-3-031-63038-5_1

Bozdağ, A., Yavuz, F., & Günay, A. S. (2016). AHP and GIS based land suitability analysis for Cihanbeyli (Turkey) County. *Environmental Earth Sciences*, *75*(9), 813. DOI: 10.1007/s12665-016-5558-9

Bozma, H. I., & Yalçın, H. (2002). Visual processing and classification of items on a moving conveyor: A selective perception approach. *Robotics and Computer-integrated Manufacturing*, *18*(2), 125–133. DOI: 10.1016/S0736-5845(01)00035-7

Bunruamkaew, K., & Murayama, Y. (2011). Site suitability evaluation for ecotourism using GIS & AHP: A case study of Surat Thani province, Thailand. *Procedia: Social and Behavioral Sciences*, *21*, 269–288. DOI: 10.1016/j.sbspro.2011.07.024

Burrough, P. A. (1996). Natural objects with indeterminate boundaries. In Burrough, P. A., & Frank, A. (Eds.), *Geographic Objects with Indeterminate Boundaries*. CRC Press.

Buteau, S. (2021). Roadmap for digital technology to foster India's MSME ecosystem—Opportunities and challenges. *CSI Transaction on ICT*, *9*(1), 233–244. DOI: 10.1007/s40012-021-00345-4

Carvalho, T. P., Soares, F. A. A. M. N., Vita, R., Francisco, R. P., Basto, J. P., & Alcalá, S. G. (2019). A systematic literature review of machine learning methods applied to predictive maintenance. *Computers & Industrial Engineering*, *137*, 106024. Advance online publication. DOI: 10.1016/j.cie.2019.106024

Cerreta, M., & de Toro, P. (2012). Urbanization suitability maps: A dynamic spatial decision support system for sustainable land use. *Earth System Dynamics: ESD*, *3*(2), 157–171. DOI: 10.5194/esd-3-157-2012

Chalise, D., Kumar, L., & Kristiansen, P. (2019). Land degradation by soil erosion in Nepal: A review. *Soil Systems*, *3*(1), 12. DOI: 10.3390/soilsystems3010012

Chang, X., Fu, K., & Conteh, M. (2025). How digitalization impacts the high-qulality development of manufacturing industry: Evidence from China. *Sustainability (Basel)*, *17*(12), 5586. DOI: 10.3390/su17125586

Cheng, C., & Thompson, R. G. (2016). Application of boolean logic and GIS for determining suitable locations for Temporary Disaster Waste Management Sites. *International Journal of Disaster Risk Reduction*, *20*, 78–92. DOI: 10.1016/j.ijdrr.2016.10.011

Cheng, T., Kojima, D., Hu, H., Onoda, H., & Pandyaswargo, A. H. (2024). Optimizing Waste Sorting for Sustainability: An AI-Powered Robotic Solution for Beverage Container Recycling. *Sustainability (Basel)*, *16*(23), 10155. DOI: 10.3390/su162310155

Chen, H. S., Tam, K. I., Zhao, Y. L., Yuan, L., Wang, W., Lin, M., & Chiang, P. C. (2023a). Development of environmental action plans for adaptation to climate change: A perspective of air quality management. *Aerosol and Air Quality Research, 23*(10), 220377. DOI: 10.4209/aaqr.220377

Chen, L. B., Huang, X. R., Chen, W. H., Pai, W. Y., Huang, G. Z., & Wang, W. C. (2023b). Design and implementation of an artificial intelligence of things-based autonomous mobile robot system for cleaning garbage. *IEEE Sensors Journal, 23*(8), 8909–8922. DOI: 10.1109/JSEN.2023.3254902

Chen, X., Despeisse, M., & Johnasson, B. (2020). Environment sustainability of digitalization in manufacturing: A review. *Sustainability (Basel), 12*(24), 1–31. DOI: 10.3390/su122410298

Chen, Y., Yu, J., & Khan, S. (2010). Spatial sensitivity analysis of multi-criteria weights in GIS-based land suitability evaluation. *Environmental Modelling & Software, 25*(12), 1582–1591. DOI: 10.1016/j.envsoft.2010.06.001

Chhotaray, P., Behera, B. C., Moharana, B. R., Muduli, K., & Sephyrin, F. T. R. (2023). Enhancement of manufacturing sector performance with the application of Industrial Internet of Things (IIoT). In *Smart Technologies for Improved Performance of Manufacturing Systems and Services* (pp. 1–19). CRC Press. DOI: 10.1201/9781003346623-1

Choudhary, K., Boori, M. S., Shi, W., Valiev, A., & Kupriyanov, A. (2023). Agricultural land suitability assessment for sustainable development using remote sensing techniques with analytic hierarchy process. *Remote Sensing Applications: Society and Environment, 32*, 101051. DOI: 10.1016/j.rsase.2023.101051

Chowdhury, E. K. (2025). Drivers of entrepreneurship development in South Asia: Empirical insights and policy recommendations. *Journal of Innovation and Entrepreneurship, 14*(46), 1–30. DOI: 10.1186/s13731-025-00512-w

Cunningham, J. A., Damij, N., Modic, D., & Olan, F. (2023). MSME technology adoption, entrepreneurial mindset and value creation: A configurational approach. *The Journal of Technology Transfer, 48*(1), 1574–1598. DOI: 10.1007/s10961-023-10022-0

Dagıstanlı, C., Demirag Turan, I., & Dengiz, O. (2018). Evaluation of the suitability of sites for outdoor recreation using a multi-criteria assessment method. *Arabian Journal of Geosciences, 11*(17), 492. DOI: 10.1007/s12517-018-3856-0

De Maio, F. V., De Stefano, R., Lanfranconi, C., Osmani, S., Fuggini, C., & Basso, P. (2024). Multi-hazard and risk informed system for Enhanced local and regional Disaster risk management.

Deng, X., Shi, Q., Zhang, Q., Shi, C., & Yin, F. (2015). Impacts of land use and land cover changes on surface energy and water balance in the Heihe River Basin of China, 2000-2010. *Physics and Chemistry of the Earth Parts A/B/C*, *79-82*, 2–10. DOI: 10.1016/j.pce.2015.01.002

Dhaya, R., & Kanthavel, R. (2022). IoT based urban flooding high definition surveillance using concurrent multipath wireless system. *Earth Science Informatics*, *15*(3), 1407–1416. DOI: 10.1007/s12145-022-00817-4

Díaz-Pelaez, A., Chura-Quispe, G., Clemente-Almendros, A. A., & Velarde-Molina, J. F. (2024). Analysing the influence of digital strategy and innovation on MSMEs' performance in emerging countries. *Creativity and Innovation Management*, *34*(2), 351–363. DOI: 10.1111/caim.12642

Donevska, K. R., Gorsevski, P. V., Jovanovski, M., & Pesevski, I. (2012). Regional non-hazardous landfill site selection by integrating fuzzy logic, AHP and geographic information systems. *Environmental Earth Sciences*, *67*(1), 121–131. DOI: 10.1007/s12665-011-1485-y

Dong, J., Zhuang, D., Xu, X., & Ying, L. (2008). Integrated evaluation of urban development suitability based on remote sensing and GIS techniques-a case study in Jingjinji Area, China. *Sensors (Basel)*, *8*(9), 5975–5986. DOI: 10.3390/s8095975

Dubovyk, O., Menz, G., & Khamzina, A. (2016). Land suitability assessment for afforestation with Elaeagnus Angustifolia L. in degraded agricultural areas of the Lower Amudarya River Basin. *Land Degradation & Development*, *27*(8), 1831–1839. DOI: 10.1002/ldr.2329

Dumitrescu, E. I., & Hurlin, C. (2012). Testing for Granger non-causality in heterogeneous panels. *Economic Modelling*, *29*(4), 1450–1460. DOI: 10.1016/j.econmod.2012.02.014

EC. (2025). Mapping guide-Urban Atlas Land Cover/Land Use and street tree layer 2012 and 2018. Retrieved from: https://land.copernicus.eu/en/technical-library/urban_atlas_2012_2018_mapping _guide Accessed: 18 November 2025

EEA. (2025). Online data and maps. Retrieved from: https://www.eea.europa.eu/data-and-maps Accessed: 18 November 2024

Egere, O. M., Mass, G., & Jones, P. (2022). A critical analysis of the Nigerian entrepreneurial ecosystem on transformational entrepreneurship. *Journal of Small Business Management, 62*(3), 1187–1218. DOI: 10.1080/00472778.2022.2123109

Elkhrachy, I., Alhamami, A., & Alyami, S. H. (2023). Landfill site selection using multi-criteria decision analysis, remote sensing data, and Geographic Information System tools in Najran City, Saudi Arabia. *Remote Sensing (Basel), 15*(15), 3754. DOI: 10.3390/rs15153754

Elmorshedy, M. F., Habib, H. U. R., Ali, M. M., Sathik, M. J., & Almakhles, D. J. (2022). Improved performance of hybrid pv and wind generating system connected to the grid using finite-set model predictive control. *IEEE Access : Practical Innovations, Open Solutions, 10*, 110344–110361. DOI: 10.1109/ACCESS.2022.3214996

ElMougy, S., Alrahmawy, M., El-dosuky, A. D. M., & Hamed, A. E. (2023) COMPREHENSIVE SURVEY OF AIOT APPROACHES FOR WATERLOGGING CRISIS MONITORING 19.

Embia, G., Moharana, B. R., Mohamed, A., Muduli, K., & Muhammad, N. B. (2023). 3D Printing Pathways for Sustainable Manufacturing. In Nayyar, A., Naved, M., & Rameshwar, R. (Eds.), *New Horizons for Industry 4.0 in Modern Business. Contributions to Environmental Sciences & Innovative Business Technology* (pp. 253–272). Springer International Publishing., DOI: 10.1007/978-3-031-20443-2_12

Eraydın, A., & Tasan-Kok, T. (2014). State response to contemporary urban movements in Turkey: A critical overview of state entrepreneurialism and authoritarian interventions. *Antipode, 46*(1), 110–129. DOI: 10.1111/anti.12042

Erumban, A. A., & de Vriers, G. J. (2024). Structural change and poverty reduction in developing economies. *World Development, 181*(1), 106674. DOI: 10.1016/j.worlddev.2024.106674

Faiz, F., Le, V., & Masli, E. K. (2024). Determinants of digital technology adoption in innovative SMEs. *Journal of Innovation and Knowledge, 9*(4), 1–20. DOI: 10.1016/j.jik.2024.100610

Fearnley, C. (2024). Early warning of complex climate risk with integrated artificial intelligence. *Research Square* (*Research Square*).

Feng, Z., Yang, J., Chen, L., Chen, Z., & Li, L. (2022). An intelligent waste-sorting and recycling device based on improved EfficientNet. *International Journal of Environmental Research and Public Health, 19*(23), 15987. DOI: 10.3390/ijerph192315987 PMID: 36498058

Fernberg, P. J. (2024). *Artificial Intelligence in Landscape Architecture: A Survey of Theory*. Culture, and Practice.

Ferrario, D. M., Sanò, M., Maraschini, M., Critto, A., & Torresan, S. (2025). Harnessing Machine Learning methods for climate multi-hazard and multi-risk assessment. *EGUsphere*, *2025*, 1–72. DOI: 10.5194/egusphere-2025-670

Ferronato, N., & Torretta, V. (2019). Waste mismanagement in developing countries: A review of global issues. *International Journal of Environmental Research and Public Health*, *16*(6), 1060. DOI: 10.3390/ijerph16061060 PMID: 30909625

Galvani, S., Carloni, E., Bocconcelli, R., & Pagano, A. (2025). When digitalization meets sustainability: Exploring interactions within a manufacturing firm. *Journal of Business and Industrial Marketing*, *40*(13), 30–46. DOI: 10.1108/JBIM-03-2024-0132

Gbadega, P. A., & Balogun, O. A. (2022). Modeling and control of grid-connected solar-wind hybrid micro-grid system with multiple-input ćuk DC-dc converter for household & high power applications. *International Journal of Engineering Research in Africa*, *58*, 191–224. DOI: 10.4028/www.scientific.net/JERA.58.191

Geiß, C. (2025). Collective Sensing and artificial intelligence techniques for natural hazard risk and impact assessment (Doctoral dissertation, Julius-Maximilians-University of Würzburg).

Goel, A., Raut, G., Sharma, A., & Taneja, U. (2024). Artificial Intelligence and Sustainable Business: A Review. *South Asian Journal of Business and Management Cases*, *13*(3), 340–365. DOI: 10.1177/22779779241302146

Gül, A., Örücü, K. M., & Karaca, Ö. (2006). An approach for recreation suitability analysis to recreation planning in Gölcük Nature Park. *Environmental Management*, *37*(5), 606–625. DOI: 10.1007/s00267-004-0322-4

Gulzar, M. M., Iqbal, A., Sibtain, D., & Khalid, M. (2023). An innovative converterless solar PV control strategy for a grid connected hybrid PV/wind/fuel-cell system coupled with battery energy storage. *IEEE Access : Practical Innovations, Open Solutions*, *11*, 23245–23259. DOI: 10.1109/ACCESS.2023.3252891

Gupta, M., Tiwari, P. M., Viral, R. K., Shrivastava, A., Zneid, B. A., & Hunko, I. (2025). Grid-connected PV inverter system control optimization using Grey Wolf optimized PID controller. *Scientific Reports*, *15*(1), 28869. DOI: 10.1038/s41598-025-10617-7 PMID: 40774975

Gursoy, N., & Edelman, D. J. (2017). Regional development planning in Istanbul: Recent issues and challenges. *Current Urban Studies*, *5*(2), 146–163. DOI: 10.4236/cus.2017.52009

Haitao, Ch. (2021). Optimization of an Intelligent Sorting and Recycling System for Solid Waste Based on Image Recognition Technology. *Advances in Mathematical Physics, ID, 4094684*, 12.

Halkos, G., de Alba, J. M., & Todorov, V. (2021). Analyzing manufacturing sector and selected development challenges: A panel data analysis. *Energy, 235*(1), 121253. DOI: 10.1016/j.energy.2021.121253

Halubanza, B. (2024). A framework for an early warning system for the management of the spread of locust invasion based on artificial intelligence technologies (Doctoral dissertation, The University of Zambia).

Hamid, B., Hussain, I., Iqbal, S. J., Singh, B., Das, S., & Kumar, N. (2022). Optimal MPPT and BES control for grid-tied DFIG-based wind energy conversion system. *IEEE Transactions on Industry Applications, 58*(6), 7966–7977. DOI: 10.1109/TIA.2022.3202757

Hamid, H., & Abedlmajid, E. (2025). Intelligent Agents in Disaster Risk Management: A Systematic Review of Advances and Challenges. *International Journal of Advanced Computer Science and Applications, 16*(6). Advance online publication. DOI: 10.14569/IJACSA.2025.01606100

Hassan, G., Chemori, A., El Rafei, M., Gouttefarde, M., Francis, C., Hervé, P.E., Sallé, D. (2022). Ultra-Local Model-Based Intelligent Robust Control of PKMs: Theory and Simulations. IFAC-PapersOnLine, 55, 12, 2022, 634-640.

He, X., Chen, Z., Yang, Q., & Xu, C. (2025). Advances in earthquake and cascading disasters. *Natural Hazards Review, 5*(2), 421–431. DOI: 10.1016/j.nhres.2025.01.010

Humbal, A., Chaudhary, N., & Pathak, B. (2023). Urbanization Trends, Climate Change, and Environmental Sustainability. In Pathak, B., & Dubey, R. S. (Eds.), *Climate Change and Urban Environment Sustainability. Disaster Resilience and Green Growth.* Springer. DOI: 10.1007/978-981-19-7618-6_9

Hussein, N.M., Mohialden, Y.M., Salman S.A. (2024). Impact of IoT-Based Environmental Monitoring on Lab Safety and Sustainability. Babylonian Journal of Internet of Things, 16-26.

Jardine, A. K. S., Lin, D., & Banjevic, D. (2006). A review on machinery diagnostics and prognostics implementing condition-based maintenance. *Mechanical Systems and Signal Processing, 20*(7), 1483–1510. DOI: 10.1016/j.ymssp.2005.09.012

Jayadatta, S. (2024). A Study on AI-Driven Agricultural Innovations for Rural and Industrial Development in the Indian Context. Journal of Rural and Industrial Development, 12(1), 01-16.

Jeddi, A. B., Shafieezadeh, A., Hur, J., Ha, J. G., Hahm, D., & Kim, M. K. (2022). Multi-hazard typhoon and earthquake collapse fragility models for transmission towers: An active learning reliability approach using gradient boosting classifiers. *Earthquake Engineering & Structural Dynamics, 51*(15), 3552–3573. DOI: 10.1002/eqe.3735

Jena, J. R., Biswal, S. K., Panigrahi, R. R., & Shrivastava, A. K. (2024). Investigating the potential areas in artificial intelligence and financial innovation: A bibliometric analysis. *Journal of Scientometric Research, 13*(1), 71–80. DOI: 10.5530/jscires.13.1.6

Johar, A., Jain, S. S., & Garg, P. K. (2013). Land suitability analysis for industrial development using GIS. *Journal of Genomics, 7*(2), 101–106.

Jyoti, B., & Singh, A. K. (2023). Does digitalization have a causal relationship with economic development? an experience from a country-wise panel data statistical analysis. In R. Sine Nazlı & G. Sari (Eds.), *Handbook of Research on Perspectives on Society and Technology Addiction* (pp. 34-61). IGI Global. DOI: 10.4018/978-1-6684-8397-8.ch003

Karyakarte, Y. (2024). Enhancing manufacturing efficiency with advanced GIS solutions. https://edstechnologies.com/newsletter/2024/june/Manufacturing-GIS.pdf

Kaufmann, A., & Gupta, M. M. (1991). *Fuzzy mathematical models in engineering and management science*. North-Holland.

Kaza, S., Yao, L., Bhada-Tata, P., & Van Woerden, F. (2018). *What a Waste 2.0: A Global Snapshot of Solid Waste Management to 2050*. World Bank Publications. DOI: 10.1596/978-1-4648-1329-0

Kellow, P., Joel, J. P. C. R., Ousmane, D., Ashok, K. D., Victor, H. C. A., & Sergei, A. K. (2020). A Smart Waste Management Solution Geared towards Citizens. *Sensors (Basel), 20*(8), 2380. DOI: 10.3390/s20082380 PMID: 32331464

Kesavan, P. K., Subramaniam, U., Almakhles, D. J., & Selvam, S. (2024). Modelling and coordinated control of grid connected photovoltaic, wind turbine driven PMSG, and energy storage device for a hybrid DC/AC microgrid. *Protection and Control of Modern Power Systems, 9*(1), 154–167. DOI: 10.23919/PCMP.2023.000272

Keyder, C., & Yenal, Z. (2011). Agrarian change under globalization: Markets and insecurity in Turkish agriculture. *Journal of Agrarian Change, 11*(1), 60–86. DOI: 10.1111/j.1471-0366.2010.00294.x

Khemakhem, S., Rekik, M., & Krichen, L. (2019). Impact of Electric Vehicles integration on residential demand response system to peak load minimizing in smart grid. *19th International Conference on Sciences and Techniques of Automatic Control and Computer Engineering (STA)*. DOI: 10.1109/STA.2019.8717304

Khemakhem, S., Rekik, M., & Krichen, L. (2022). A Smart Home Power Supervision Strategy Based on Optimal Plug-in Electric Vehicles Integration into Smart City Context for Peak Loads Profiles Flattening. *Arabian Journal for Science and Engineering*, *47*(3), 3383–3396. DOI: 10.1007/s13369-021-06315-x

Kienast, F., Degenhardt, B., Weilenmann, B., Wager, Y., & Buchecker, M. (2012). GIS-assisted mapping of landscape suitability for nearby recreation. *Landscape and Urban Planning*, *105*(4), 385–399. DOI: 10.1016/j.landurbplan.2012.01.015

Kılıç, Ş., Evrendilek, F., Şenol, S., & Çelik, I. (2005). Developing a suitability index for land uses and agricultural land covers: A case study in Turkey. *Environmental Monitoring and Assessment*, *102*(1-3), 323–335. DOI: 10.1007/s10661-005-6030-6

Kirchherr, J., Reike, D., & Hekkert, M. (2017). Conceptualizing the circular economy: An analysis of 114 definitions. *Resources, Conservation and Recycling*, *127*, 221–232. DOI: 10.1016/j.resconrec.2017.09.005

Kong, S. M., Yoo, C., Park, J., Park, J. H., & Lee, S. W. (2024). AIOT monitoring technology for optimal fill dam installation and operation. *Applied Sciences (Basel, Switzerland)*, *14*(3), 1024. DOI: 10.3390/app14031024

Konstantia, M., Freiderikos, R., George, L., Stavros, O., Maria, P., Michalis, Z., & Michail, M. (2022). Intelligent robotic system for urban waste recycling. IEEE International Conference on Imaging Systems and Techniques, Proceedings, ISBN (Electronic) 9781665481021.

Kumar, K., Lakshmi Devi, V., Dhanamjayulu, C., Kotb, H., & ELrashidi, A. (2024). Evaluation and deployment of a unified MPPT controller for hybrid Luo converter in combined PV and wind energy systems. *Scientific Reports*, *14*(1), 3248. DOI: 10.1038/s41598-024-53605-z PMID: 38332236

Kumar, M., Singh, A. K., & Lal, S. (2024). Design And Simulation Of Grid-Connected Solar Wind Hybrid Power System With MPPT Techniques. *Nanotechnology Perceptions*, *20*(S15), 4049–4064.

Kupriyanova, M., Evdokimova, E., Soloviova, I., & Simikova, I. (2023). The impact of digitalization on the manufacturing industry. E3S Web of Conferences, 431, 05030. DOI: 10.1051/e3sconf/202343105030

Kuru, A., Yüzer, M. A., Yüzer, A. Ş., Güney, B. G., & Yüzer, M. E. (2025). Integrated site selection model for industrial areas: Case study for İnegöl furniture industry. *Environmental Science and Pollution Research International*, *32*(8), 4771–4793. DOI: 10.1007/s11356-025-35965-9

Laiwang, X., Abbas, A., Kunpeng, A., & We, X. (2025). Quantifying the role of digitalization, social factors and energy transition to achieve sustainability in East Asia. *Humanities & Social Sciences Communications*, *12*(1), 1613. DOI: 10.1057/s41599-025-05893-7

Larios-Francia, R. P., & Ferasso, M. (2023). The relationship between innovation and performance in MSMEs: The case of wearing apparel sector in emerging countries. *Journal of Open Innovation*, *9*(1), 100018. DOI: 10.1016/j.joitmc.2023.100018

Laski, P. A., Takosoglu, J. E., & Blasiak, S. (2015). Design of a 3-DOF tripod electro-pneumatic parallel manipulator. *Robotics and Autonomous Systems*, *72*, 59–70. DOI: 10.1016/j.robot.2015.04.009

Lee, J., Bagheri, B., & Kao, H. A. (2015). A cyber-physical systems architecture for Industry 4.0-based manufacturing systems. *Manufacturing Letters*, *3*, 18–23. DOI: 10.1016/j.mfglet.2014.12.001

Lee, K., Chun, H., & Song, J. (2018). New strategies for resilient planning in response to climate change for urban development. *Procedia Engineering*, *212*, 840–846. DOI: 10.1016/j.proeng.2018.01.108

Lei, Y., Li, N., Guo, L., Li, N., Yan, T., & Lin, J. (2018). Machinery health prognostics: A systematic review from data acquisition to RUL prediction. *Mechanical Systems and Signal Processing*, *104*, 799–834. DOI: 10.1016/j.ymssp.2017.11.016

Li, G., Messina, J. P., Peter, B. G., & Snapp, S. S. (2017). Mapping land suitability for agriculture in Malawi. *Land Degradation & Development*, *28*(7), 2001–2016. DOI: 10.1002/ldr.2723

Li, J., Xie, B., Dong, H., Zhou, K., & Zhang, X. (2023). The impact of urbanization on ecosystem services: Both time and space are important to identify driving forces. *Journal of Environmental Management*, *347*, 119161. DOI: 10.1016/j.jenvman.2023.119161

Li, W., Xue, D., & Huang, X. (2018). The role of manufacturing in sustainable economic development: A case of Guangzhou, China. *Sustainability (Basel)*, *10*(9), 3039. DOI: 10.3390/su10093039

Li, Z., Fan, Z., & Shen, S. (2018). Urban green space suitability evaluation based on the AHP-CV combined weight method: A case study of Fuping County, China. *Sustainability (Basel)*, *10*(8), 2656. DOI: 10.3390/su10082656

Li, Z., Luo, Z., Wang, Y., Fan, G., & Zhang, J. (2022). Suitability evaluation system for the shallow geothermal energy implementation in region by Entropy Weight Method and TOPSIS method. *Renewable Energy*, *184*, 564–576. DOI: 10.1016/j.renene.2021.11.112

Lokeshwari, M., Jha, G. K., Praveen, K. V., & Bharadwaj, A. (2024). Artificial intelligence for crop yield prediction: a bibliometric analysis. Current Science (00113891), 126(10).

Lubongo, C., Bin Daej, M. A., & Alexandridis, P. (2024). Recent developments in technology for sorting plastic for recycling: The emergence of artificial intelligence and the rise of the robots. *Recycling*, *9*(4), 59. DOI: 10.3390/recycling9040059

Lu, S., Lu, J., An, K., Wang, X., & He, Q. (2023). Edge computing on IoT for machine signal processing and fault diagnosis: A review. *IEEE Internet of Things Journal*, *10*(13), 11093–11116. DOI: 10.1109/JIOT.2023.3239944

Maalouf, A., & Mavropoulos, A. (2022). Re-assessing global municipal solid waste generation. *Waste Management & Research*, *41*(4), 936–947. DOI: 10.1177/0734242X221074116 PMID: 35075952

Maaruf, M., Khan, K., & Khalid, M. (2022). Robust control for optimized islanded and grid-connected operation of solar/wind/battery hybrid energy. *Sustainability (Basel)*, *14*(9), 5673. DOI: 10.3390/su14095673

MacArthur, E. (2013). Towards the circular economy. *Journal of Industrial Ecology*, *2*, 23–44.

Maffei, A., Grahn, S., & Nuur, C. (2019). Characterization of the impact of digitalization on the adoption of sustainable business models in manufacturing. *Procedia CIRP*, *81*(1), 765–770. DOI: 10.1016/j.procir.2019.03.191

Mahdavi, A., & Niknejad, M. (2014). Site suitability evaluation for ecotourism using MCDM methods and GIS: Case study-Lorestan province, Iran. *Journal of Biodiversity and Environmental Sciences*, *4*(6), 425–437.

Mahmood, K. W., Khzr, B. O., Othman, R. M., Rasul, A., Ali, S. A., & Ibrahim, G. R. F. (2021). Optimal site selection for landfill using the Boolean-analytic hierarchy process. *Environmental Earth Sciences*, *80*(5), 173. DOI: 10.1007/s12665-021-09501-0

Malczewski, J. (2004). GIS-based land-use suitability analysis: A critical overview. *Progress in Planning*, *62*(1), 3–65. DOI: 10.1016/j.progress.2003.09.002

Mallik, S., Mishra, U., & Paul, N. (2021). Groundwater suitability analysis for drinking using GIS based fuzzy logic. *Ecological Indicators*, *121*, 107179. DOI: 10.1016/j.ecolind.2020.107179

Malmir, M., Zarkesh, M. M. K., Monavari, S. M., Jozi, S. A., & Sharifi, E. (2016). Analysis of land suitability for urban development in Ahwaz County in southwestern Iran using fuzzy logic and analytic network process (ANP). *Environmental Monitoring and Assessment*, *188*(8), 447. DOI: 10.1007/s10661-016-5401-5

Mangla, S. K., Luthra, S., Garza-Reyes, J. A., Jose, C., Jabbour, C., & Brem, A. (2024). Sustainability and Industry 4.0: The role of social, environmental and technological factors in the development of digital manufacturing. *Technological Forecasting and Social Change*, *201*(1), 123223. DOI: 10.1016/j.techfore.2024.123223

Maraveas, C. (2022). Incorporating artificial intelligence technology in smart greenhouses: Current State of the Art. *Applied Sciences (Basel, Switzerland)*, *13*(1), 14. DOI: 10.3390/app13010014

Maria, K., Fredy, R., George, P., Nikitas, M., & Michail, M. (2021). Robotic Waste Sorting Technology: Toward a Vision-Based Categorization System for the Industrial Robotic Separation of Recyclable Waste. *IEEE Robotics & Automation Magazine*, *28*(2), 50–60. DOI: 10.1109/MRA.2021.3066040

Maria, V., Priyanka, P., Oluwaseyi, F., Katherine, H. I., Liang, Z., Hossain, S., & Sheikh, I. A. (2022). Development of a Noninvasive Blood Glucose Monitoring System Prototype: Pilot Study. *JMIR Formative Research*, *6*, 8. PMID: 36018623

Masud, N., Chowdhury, P., Al Bassam, A., Sourav, M. T. I., & Sobhan, A. (2025). A Machine Learning and IoT-Driven Framework for Real-Time Disaster Management. In Addressing Environmental Challenges With AI, Robotics, and Augmented Reality (pp. 247-288). IGI Global Scientific Publishing. DOI: 10.4018/979-8-3373-1892-9.ch010

Material Economics. (2018). *The Circular Economy—A Powerful Force for Climate Mitigation*. Material Economics Sverige AB.

Mayouf, O. M., Rekik, M., & Krichen, L. (2022). Designing a hybrid system using solar cells and batteries to power a clinic in Tajarhi desert village in the Libya-Niger border off-grid. 2022 IEEE 21st International Conference on Sciences and Techniques of Automatic Control and Computer Engineering (STA).

Mendas, A., & Delali, A. (2012). Integration of multicriteria decision analysis in GIS to develop land suitability for agriculture: Application to drum wheat cultivation in the region of Mleta in Algeria. *Computers and Electronics in Agriculture, 83*, 117–126. DOI: 10.1016/j.compag.2012.02.003

Menzri, F., Boutabba, T., Benlaloui, I., Chrifi-Alaoui, L., Alkuhayli, A., Khaled, U., & Mahmoud, M. M. (2024). Applications of novel combined controllers for optimizing grid-connected hybrid renewable energy systems. *Sustainability (Basel), 16*(16), 6825. DOI: 10.3390/su16166825

Metternicht, G. (2017). Land use planning. Global Land Outlook Working Paper. United Nations Convention to Combat Desertification.

Miah, M. T., Fariha, J. N., Jodder, P. K., Kafy, A. A., Raiyan, R., Usha, S. A., Hossan, J., & Rahaman, K. R. (2024). Urban heat island and environmental degradation analysis utilizing a remote sensing technique in rapidly urbanizing South Asian cities. *WORLD (Oakland, Calif.), 5*(4), 1023–1053. DOI: 10.3390/world5040052

Midatana, S., Saran, S., & Ramana, K. M. (2018). Site suitability analysis for industries using GIS and multi criteria decision making. *ISPRS Annals of the Photogrammetry, Remote Sensing and Spatial Information Sciences, IV-5*, 447–454. DOI: 10.5194/isprs-annals-IV-5-447-2018

Mishra, D., Muduli, K., Raut, R., Narkhede, B. E., Shee, H., & Jana, S. K. (2023). Challenges facing artificial intelligence adoption during COVID-19 pandemic: An investigation into the agriculture and agri-food supply chain in India. *Sustainability (Basel), 15*(8), 6377. DOI: 10.3390/su15086377

Mitra, A., Vangipuram, S. L., Bapatla, A. K., Bathalapalli, V. K., Mohanty, S. P., Kougianos, E., & Ray, C. (2022). Everything you wanted to know about smart agriculture. arXiv preprint arXiv:2201.04754.

Mittal, R., Sheetal, , & Soriya, S. (2025). Fostering competitiveness of Indian MSMEs through IT and digitalization. *Information Technology for Development, 31*(4), 1066–1089. DOI: 10.1080/02681102.2025.2453211

Mobley, R. K. (2002). *An introduction to predictive maintenance* (2nd ed.). Butterworth-Heinemann.

Mohammed, J. K., Saidu, U. A., & Sheriff, I. H. (2025). GIS-AHP based suitability model for commercial property investment in Bida, Nigeria. *Journal of Inclusive Cities and Built Environment, 5*(9), 36–44. DOI: 10.54030/2788-564X/2025/is9v5a3

Morckel, V. (2017). Using suitability analysis to select and prioritize naturalization efforts in legacy cities: An example from Flint, Michigan. *Urban Forestry & Urban Greening*, *27*, 343–351. DOI: 10.1016/j.ufug.2017.09.006

Muduli, K., Luthra, S., Garza-Reyes, J. A., & Huisingh, D. (2023). Application of blockchain technology for addressing reverse logistics challenges: current status and future opportunities. *Supply Chain Forum: An International Journal*, 24(1), 1-6. DOI: 10.1080/16258312.2023.2165279

Mukherjee, S. (2018). Challenges to Indian micro small scale and medium enterprises in the era of globalization. *Journal of Global Entrepreneurship Research*, *8*(28), 1–14. DOI: 10.1186/s40497-018-0115-5

Munier, N., & Hontoria, E. (2022). *Uses and limitations of the AHP method: A non-mathematical and rational analysis*. Springer Cham.

Mu, X., Chen, D., & Zhang, H. (2024). The impact of digital transformation on the sustainable development of enterprises: An analysis from the perspective of digital finance. *Journal of Applied Econometrics*, *28*(1), 1–13. DOI: 10.1080/15140326.2025.2464488

Naeem, S., Nabeel, N., Beg, W., Ali, S., Kanojiya, R. N., Mandawade, S. S., & Patil, A. V. (2024). A Review on Exploring Artificial Intelligence Applications, Advancements, Issues, and Future Challenges. *POLYTECHNIC JOURNAL*, *14*(2), 96–113.

Naima, B., Belkacem, B., Ahmed, T., Benbouhenni, H., Riyadh, B., Samira, H., Sarra, Z., Elbarbary, Z. M. S., & Mohammed, S. A. (2025). Enhancing MPPT optimization with hybrid predictive control and adaptive P&O for better efficiency and power quality in PV systems. *Scientific Reports*, *15*(1), 24559. DOI: 10.1038/s41598-025-10335-0 PMID: 40629005

Nasef, S. A., Hassan, A. A., ElMadany, H. T., Zahran, M. B., El-Shaer, M. K., & Abdelaziz, A. Y. (2022). Optimal power management and control of hybrid photovoltaic-battery for grid-connected doubly-fed induction generator based wind energy conversion system. *International Journal of Renewable Energy Research*, *12*(1), 408–421.

Nerella, O., & Ahmed, S. M. (2023, March). Advance warning and alert system for detecting lightning risk to reduce human disaster using AIoT Platform—A proposed model to support rural India. In International Conference on Communications and Cyber Physical Engineering 2018 (pp. 925-935). Singapore: Springer Nature Singapore.

Nidhya, R., Manish, K., Renjith, V. R., & Deepak, V. (2020). Enhanced Route Selection (ERS) algorithm for IoT enabled smart waste management system. *Environmental Technology & Innovation*, *20*, 101116. DOI: 10.1016/j.eti.2020.101116

Ni, X., Li, J., Xu, J., Shen, Y., & Liu, X. (2024). Grey relation analysis and multiple criteria decision analysis method model for suitability evaluation of underground space development. *Engineering Geology*, *338*, 107608. DOI: 10.1016/j.enggeo.2024.107608

Noorollahi, Y., Senani, A. G., Fadaei, A., Simaee, M., & Moltames, R. (2022). A framework for GIS-based site selection and technical potential evaluation of PV solar farm using Fuzzy-Boolean logic and AHP multi-criteria decision-making approach. *Renewable Energy*, *186*, 89–104. DOI: 10.1016/j.renene.2021.12.124

Ochoa, M., & Nonnecke, B. M. (2019). Developing a digital inclusion index at the state level: The case of Mexico. TPRC47: The 47th Research Conference on Communication, Information and Internet Policy 2019. DOI: 10.2139/ssrn.3432615

Onu, P., Pradhan, A., & Mbohwa, C. (2023). The potential of industry 4.0 for renewable energy and materials development-The case of multinational energy companies. *Heliyon*, *9*(10), e20547. DOI: 10.1016/j.heliyon.2023.e20547

Palanikkumar, D., Maashi, M., Alsamri, J., & Obayya, M. (2025). Machine learning driven multi-hazard risk framework for coastal resilience. *Journal of South American Earth Sciences*, *152*, 105331. DOI: 10.1016/j.jsames.2024.105331

Pandey, P. K., Kumar, R., Gupta, V., Kanungo, A., & Diwania, S. (2024). An innovative hybrid controller-based combined grid-connected hybrid renewable energy system. *Electrical Engineering*, *106*(6), 7055–7072. DOI: 10.1007/s00202-024-02363-2

Park, S., Jeon, S., Kim, S., & Choi, C. (2011). Prediction and comparison of urban growth by land suitability index mapping using GIS and RS in South Korea. *Landscape and Urban Planning*, *99*(2), 104–114. DOI: 10.1016/j.landurbplan.2010.09.001

Parry, J. A., Ganaie, S. A., & Bhat, M. S. (2018). GIS based land suitability analysis using AHP model for urban services planning in Srinagar and Jammu urban centres of J&K, India. *Journal of Urban Management*, *7*(2), 46–56. DOI: 10.1016/j.jum.2018.05.002

Patric, M., Diogo, M., Eduardo, D., Jonatan, C., Rodrigo, C., Juergen, R., Edison, P., & Rafael, K. (2019). An IoT-based smart cities infrastructure architecture applied to a waste management scenario. *Ad Hoc Networks*, *87*(1), 200–208.

Pech, M., Vrchota, J., & Bednář, J. (2021). Predictive maintenance and intelligent sensors in smart factory. *Sensors (Basel), 21*(4), 1470. Advance online publication. DOI: 10.3390/s21041470 PMID: 33672479

Pendik Governorship. (2025). Online Data. Retrieved from: http://www.pendik.gov.tr Accessed: 18 November 2025

Peng, J., Ma, J., Du, Y., Zhang, L., & Hu, X. (2016). Ecological suitability evaluation for mountainous area development based on conceptual model of landscape structure, function, and dynamics. *Ecological Indicators, 61*(2), 500–511. DOI: 10.1016/j.ecolind.2015.10.002

Piran, H., Maleknia, R., Akbari, H., Soosani, J., & Karami, O. (2013). Site selection for local forest park using analytic hierarchy process and geographic information system (case study: Badreh County). *International Research Journal of Applied and Basic Sciences, 6*(7), 930–935.

Pourroostaei Ardakani, S., & Cheshmehzangi, A. (2023). Big data-enabled time series analysis for climate change analysis in brazil: an artificial neural network machine learning model. In *Big Data Analytics for Smart Transport and Healthcare Systems* (pp. 141–156). Springer Nature Singapore. DOI: 10.1007/978-981-99-6620-2_8

Prah, K., & Gajšek, B. (2024). Geographical information system modeling for planning internal transportation in a manufacturing plant's outdoor area. *Processes (Basel, Switzerland), 12*(1), 1–6. DOI: 10.3390/pr12010180

Pramanik, M. K. (2016). Site suitability analysis for agricultural land use of Darjeeling district using AHP and GIS techniques. *Modeling Earth Systems and Environment, 2*(2), 56. DOI: 10.1007/s40808-016-0116-8

Pugliese Viloria, A. D. J., Folini, A., Carrion, D., & Brovelli, M. A. (2024). Hazard susceptibility mapping with machine and deep learning: A literature review. *Remote Sensing (Basel), 16*(18), 3374. DOI: 10.3390/rs16183374

Qiao, N., & Niu, L. (2024). The impact of the industrial innovation ecosystem on innovation performance – using the equipment manufacturing industry as an example. *Systems, 12*(12), 1–21. DOI: 10.3390/systems12120578

Qing, C. H. G., Oktaviani, E., & Heripracoyo, S. (2019). The literature review of manufacturing industry 4.0 using mapping design with GIS approach. *2019 International Conference on Information Management and Technology (ICIMTech)*, Jakarta/Bali, Indonesia, 2019, pp. 438-443, DOI: DOI: 10.1109/ICIMTech.2019.8843824

Rabia, A. H., Figueredo, H., Huong, T. L., Lopez, B. A. A., Solomon, H. W., & Alessandro, V. (2013). Land suitability analysis for policy making assistance: A GIS based land suitability comparison between surface and drip irrigation system. *International Journal of Environmental Sciences and Development*, *4*(1), 1–6. DOI: 10.7763/IJESD.2013.V4.292

Rakholia, R., Suárez-Cetrulo, A. L., Singh, M., & Carbajo, R. S. (2025). Integrating AI and IoT for predictive maintenance in Industry 4.0 manufacturing environments: A practical approach. *Information (Basel)*, *16*(9), 737. Advance online publication. DOI: 10.3390/info16090737

Raman, R., Pattnaik, D., Lathabai, H. H., Kumar, C., Govindan, K., & Nedungadi, P. (2024). Green and sustainable AI research: An integrated thematic and topic modeling analysis. *Journal of Big Data*, *11*(1), 1–28. DOI: 10.1186/s40537-024-00920-x

Ramson, S. R. J., & Moni, D. J. (2017). Wireless sensor networks based smart bin. *Computers & Electrical Engineering*, *64*, 337–353. DOI: 10.1016/j.compeleceng.2016.11.030

Ramya, S., & Devadas, V. (2019). Integration of GIS, AHP and TOPSIS in evaluating suitable locations for industrial development: A case of Tehri Garhwal district, Uttarakhand, India. *Journal of Cleaner Production*, *238*, 117872. DOI: 10.1016/j.jclepro.2019.117872

Rao, T. E., Sundaram, E., Chenniappan, S., Almakhles, D., Subramaniam, U., & Bhaskar, M. S. (2022). Performance improvement of grid interfaced hybrid system using distributed power flow controller optimization techniques. *IEEE Access : Practical Innovations, Open Solutions*, *10*, 12742–12752. DOI: 10.1109/ACCESS.2022.3146412

Razavi-Termeh, S. V., Sadeghi-Niaraki, A., Ali, F., Pradhan, B., & Choi, S. M. (2025). Optimizing ensemble learning for satellite-based multi-hazard monitoring and susceptibility assessment of landslides, land subsidence, floods, and wildfires. *Scientific Reports*, *15*(1), 30968. DOI: 10.1038/s41598-025-15381-2 PMID: 40846725

Rekioua, D., Mezzai, N., Mokrani, Z., Oubelaid, A., Kakouche, K., Logerais, P. O., Alshareef, M., Bajaj, M., Tuka, M. B., & Ghoneim, S. S. (2024). Effective optimal control of a wind turbine system with hybrid energy storage and hybrid MPPT approach. *Scientific Reports*, *14*(1), 30013. DOI: 10.1038/s41598-024-78847-9 PMID: 39622862

Rekioua, D., Mokrani, Z., Kakouche, K., Oubelaid, A., Rekioua, T., Alhazmi, M., Ali, E., Bajaj, M., Dost Mohammadi, S. A., & Ghoneim, S. S. (2024). Coordinated power management strategy for reliable hybridization of multi-source systems using hybrid MPPT algorithms. *Scientific Reports*, *14*(1), 10267. DOI: 10.1038/s41598-024-60116-4 PMID: 38704399

Romano, G., Dal Sasso, P., Trisorio Liuzzi, G., & Gentile, F. (2015). Multi-criteria decision analysis for land suitability mapping in a rural area of Southern Italy. *Land Use Policy*, *48*, 131–143. DOI: 10.1016/j.landusepol.2015.05.013

Rosyidiana, R. N., & Narsa, I. M. (2024). Micro, small, and medium-sized enterprises (MSMEs) during the post-pandemic economic recovery period: Digitalization, literation, innovation, and its impact on financial performance. *Cogent Business &. Cogent Business & Management*, *11*(1), 1–14. DOI: 10.1080/23311975.2024.2342488

Rui, H., Aaron, U., Dian, J., Jiang, Z., & Jae-Won, C. (2022). Inductive proximity sensors within a ceramic package manufactured by material extrusion of binder-coated zirconia. *Sensors and Actuators. A, Physical*, *338*, 113497. DOI: 10.1016/j.sna.2022.113497

Saaty, T. L. (1980). *The Analytical Hierarchy Process*. McGraw Hill.

Sabato, A., Niezrecki, C., & Fortino, G. (2017). Wireless MEMS-based accelerometer sensor boards for structural vibration monitoring: A review. *IEEE Sensors Journal*, *17*(2), 226–235. DOI: 10.1109/JSEN.2016.2630008

Saha, S., Sarkar, D., Mondal, P., & Goswami, S. (2021). GIS and multi-criteria decision-making assessment of sites suitability for agriculture in an anabranching site of sooin river, India. *Modeling Earth Systems and Environment*, *7*(1), 571–588. DOI: 10.1007/s40808-020-00936-1

Sahin, E., Iban, M. C., & Bilgilioglu, S. S. (2024). Remote sensing-enabled urban growth simulation overlaid with AHP-GIS-based urban land suitability for potential development in Mersin Metropolitan Area, Türkiye. *Applied Sciences (Basel, Switzerland)*, *14*(8), 3484. DOI: 10.3390/app14083484

Sánchez, R.-V., Lucero, P., Vásquez, R.-E., Cerrada, M., Macancela, J.-C., & Cabrera, D. (2018). Feature ranking for multi-fault diagnosis of rotating machinery using random forest and k-nearest neighbors. *Journal of Intelligent & Fuzzy Systems*, *34*(6), 3463–3473. DOI: 10.3233/JIFS-169526

Sankar, M. S. K., Gupta, S., Luthra, S., Jagtap, S., Singh, S., & Kumar, A. (2024). The emergence of digitalization to the manufacturing sector in the sustainability context: A multi-stakeholder perspective analysis. *Journal of Cleaner Production, 468*(1), 142983. DOI: 10.1016/j.jclepro.2024.142983

Sathish, P. G., Subrata, H., & Atul, T. (2017). A review on automated sorting of source-separated municipal solid waste for recycling. *Waste Management (New York, N.Y.), 60*, 56–74. DOI: 10.1016/j.wasman.2016.09.015 PMID: 27663707

Saxena, R., Pant, D. K., Pant, S. C., Joshi, L., Paul, R. K., & Singh, R. (2023). Artificial intelligence-led innovations for agricultural transformation: A scoping study. *Agricultural Economics Research Review, 36*, 1–18.

Selcuk, S. (2017). Predictive maintenance, its implementation and latest trends. *Proceedings of the Institution of Mechanical Engineers. Part B, Journal of Engineering Manufacture, 231*(9), 1670–1679. DOI: 10.1177/0954405415601640

Sengupta, S., Mohinuddin, S., Arif, M., Sengupta, B., & Zhang, W. (2022). Assessment of agricultural land suitability using GIS and fuzzy analytical hierarchy process approach in Ranchi District, India. *Geocarto International, 37*(26), 13337–13368. DOI: 10.1080/10106049.2022.2076925

Septian, S. A., & Saputra, R. A. (2022). Spatial model of industrial area suitability using spatial multi criteria evaluation: A case study in Kendari City. *Sustinere Journal of Environment & Sustainability, 6*(3), 174–270.

Serradilla, O., Zugasti, E., Rodriguez, J., & Zurutuza, U. (2022). Deep learning models for predictive maintenance: A survey, comparison, challenges, and prospects. *Applied Intelligence, 52*(10), 10934–10964. DOI: 10.1007/s10489-021-03004-y

Sertkaya Dogan, Ö. (2015). Growth of population and population policies implemented in Turkey between 1965 and 2014. *International Journal of Scientific Research, 4*(4), 1907–1911.

Shah, S., Menon, S., Ojo, O. O., & Naghi, G. E. (2020). Digitalisation in sustainable manufacturing – a literature review. *2020 IEEE International Conference on Technology Management, Operations and Decisions (ICTMOD)*, Marrakech, Morocco, pp. 1-6. DOI: DOI: 10.1109/ICTMOD49425.2020.9380605

Shang, Z., & Zhang, L. (2022). The sustainable digitalization in the manufacturing industry: A bibliometric analysis and research trend. *Mobile Information Systems, 2022*(1), 1–14. DOI: 10.1155/2022/1451705

Shan-Huen, H., & Pei-Chun, L. (2015). Vehicle routing–scheduling for municipal waste collection system under the "Keep Trash off the Ground" policy. *Omega, 55*, 24–37. DOI: 10.1016/j.omega.2015.02.004

Sharabati, A. A. A., Ali, A. A. A., Allahham, M. I., Hussein, A. A., Alhhet, A. F., & Mohammad, A. S. (2024). The impact of digital marketing on the performance of SMEs: An analytical study in light of modern digital transformations. *Sustainability (Basel), 16*(19), 1–25. DOI: 10.3390/su16198667

Sharma, S., Chauhan, B. K., & Saxena, N. K. (2023). Artificial neural network grid-connected MPPT-based techniques for hybrid PV-WIND with battery energy storage system. Journal of The Institution of Engineers (India): Series B, 104(6), 1217-1226.

Sharma, A., Boora, A., & Kumar, Y. (2025). Artificial intelligence and machine learning for resilient transportation infrastructure. *Cureus J. Comput. Sci, 2*, 1–16. DOI: 10.7759/s44389-025-09490-2

Sharma, S., Chauhan, B. K., & Saxena, N. K. (2023). Grid connected fuzzy logic control-based MPPT techniques for hybrid photovoltaic wind with battery system. *International Journal of Power and Energy Conversion, 14*(2-3), 280–309. DOI: 10.1504/IJPEC.2023.134873

Shen, Z., & Zhou, H. (2025). Hazard-responsive digital twin for climate-driven urban resilience and equity. arXiv preprint arXiv:2510.22941.

Shraddha, M. K., Sameer, U., Amogh, R. M., Shreyas, S. A., & Meenaxi, M. R. (2025). Grading and Classification of Tomatoes Using ESP32-CAM and Impulse Cloud based IoT Platform for Sustainability. Seventh International Conference on Recent Trends in Image Processing and Pattern Recognition (RTIP2R-2024). *Procedia Computer Science, 260*, 938–946. DOI: 10.1016/j.procs.2025.03.277

Shrestha, D. P., Suriyaprasit, M., & Prachansri, S. (2014). Assessing soil erosion in inaccessible mountainous areas in the tropics: The use of land cover and topographic parameters in a case study in Thailand. *Catena, 121*, 40–52. DOI: 10.1016/j.catena.2014.04.016

Singh, A. K. (2024). Causality of digitalization with social sustainability in selected countries. In Siyal, S. (Ed.), *Impact of Digitalization on Education and Social Sustainability* (pp. 39–78). IGI Global., DOI: 10.4018/979-8-3693-1854-6.ch002

Singh, A. K., & Jyoti, B. (2023). Does digitalization have a bi-directional causal relationship with the entrepreneurship ecosystem? evidence from a country-wise panel data analysis. In Yanamandra, R., & Indiran, L. (Eds.), *Handbook of Research on Designing Sustainable Strategies to Develop Entrepreneurial Intention* (pp. 178–172). IGI Global., DOI: 10.4018/978-1-6684-8781-5.ch007

Singh, A. K., & Kumar, S. (2021). Assessing the performance and factors affecting industrial development in Indian states: An empirical analysis. *Journal of Social Economics Research*, *8*(2), 135–154. DOI: 10.18488/journal.35.2021.82.135.154

Singh, A. K., Singh, B., & Ashraf, S. N. (2020). Implications of intellectual property protection, and science and technological development in the manufacturing sector in selected economies. *Journal of Advocacy. Research in Education*, *7*(1), 16–35. https://oaji.net/articles/2021/1704-1615668997.pdf

Singh, A. K., & Singh, R. (2025). Implications of digital infrastructure on social sustainability in global countries. In Jaswal, A., Mukherjee, D., Gunasekaran, A., & Kandpal, V. (Eds.), *Green Futures. Approaches to Global Sustainability, Markets, and Governance* (pp. 381–403). Springer., DOI: 10.1007/978-981-96-0843-0_19

Singh, A. K., Singh, S., & Gupta, M. (2024). A literature review for the association of social media with micro and small enterprises in India. In Hossain, S. (Ed.), *Cases on Social Media and Entrepreneurship* (pp. 38–62). IGI Global., DOI: 10.4018/979-8-3693-1781-5.ch003

Singh, M. (2019). Solid waste management in urban India: Imperatives for improvement. *Journal of Contemporary Issues in Business and Government*, *25*(1), 87–92.

Si, X.-S., Wang, W., Hu, C.-H., & Zhou, D.-H. (2011). Remaining useful life estimation: A review on the statistical data driven approaches. *European Journal of Operational Research*, *213*(1), 1–14. DOI: 10.1016/j.ejor.2010.11.018

Skubis, I., Wolniak, R., & Grebski, W. W. (2024). AI and Human-Centric Approach in Smart Cities Management: Case Studies from Silesian and Lesser Poland Voivodships. *Sustainability (Basel)*, *2024*(16), 8279. DOI: 10.3390/su16188279

Sleeter, B. M., Liu, J., Daniel, C., Rayfield, B., Sherba, J., Hawbaker, T. J., Zhu, Z., Selmants, P. C., & Loveland, T. R. (2018). Effects of contemporary land-use and land-cover change on the carbon balance of terrestrial ecosystems in the United States. *Environmental Research Letters*, *13*(4), 045006. DOI: 10.1088/1748-9326/aab540

Somashekhar Gaddanakeri, V. S., Singh, S., Samal, I., Dubey, V. K., & Bhoi, T. K. (2024). An In-Depth Analysis of Artificial Intelligence-Based Crop Pest Management and Water Supply. Artificial Intelligence Techniques in Smart Agriculture, 151.

Soumyabrata, S., Rituparna C. (2023). IoT based smart waste management system in aspect of COVID-19. Journal of Open Innovation: Technology, Market, and Complexity, 9, 2, 100048.

Steiner, F., McSherry, L., & Cohen, J. (2000). Land suitability analysis for the upper Gila River watershed. *Landscape and Urban Planning*, *50*(4), 199–214. DOI: 10.1016/S0169-2046(00)00093-1

Subasi, A., Qaisar, S. M., & Nisar, H. (Eds.). (2024). *Artificial Intelligence and Multimodal Signal Processing in Human-Machine Interaction*. Elsevier.

Sun, C., Sun, C., Yang, Z., Zhang, J., & Deng, Y. (2016). Urban land development for industrial and commercial use: A case study of Beijing. *Sustainability (Basel)*, *8*(12), 1323. DOI: 10.3390/su8121323

Supto, S. T. J., Karzai, D. R. H., & Adib, E. I. (2025). Assessing industrial land suitability for sustainable urban planning in Dhaka Region using geospatial techniques. *Environmental and Earth Sciences Proceedings*, *36*(1), 5. DOI: 10.3390/eesp2025036005

Surya, B., Menne, F., Sabhan, H., Suriani, S., Abubakar, H., & Idris, M. (2021). Economic growth, increasing productivity of SMEs, and open innovation. *Journal of Open Innovation*, *7*(1), 1–37. DOI: 10.3390/joitmc7010020

Swain, S., Muduli, K., Biswal, J. N., Tripathy, S., & Panda, T. K. (2017). Evaluation of barriers of health care waste management in India—a gray relational analysis approach. *Proceedings of the 5th International Conference on Frontiers in Intelligent Computing: Theory and Applications (FICTA 2016)*, Volume 1, 181–188. Springer, Singapore. DOI: 10.1007/978-981-10-3153-3_18

Swain, S., Muduli, K., Kommula, V. P., & Sahoo, K. K. (2022). Innovations in internet of medical things, artificial intelligence, and readiness of the healthcare sector towards health 4.0 adoption. *International Journal of Social Ecology and Sustainable Development*, *13*(1), 1–14. DOI: 10.4018/IJSESD.292078

Swain, S., Muduli, K., Kumar, A., & Luthra, S. (2024). Analysis of barriers of mHealth adoption in the context of sustainable operational practices in health care supply chains. *International Journal of Industrial Engineering and Operations Management*, *6*(2), 85–116. DOI: 10.1108/IJIEOM-12-2022-0067

Taktak, F. (2025). Comparative building amnesty and land use policies: Insights from ten countries with a focus on Turkey. *Land Use Policy*, *158*, 107763. DOI: 10.1016/j.landusepol.2025.107763

Tan, Y.-M., Amorós, E., Autio, E., Fu, K., & Park, D. (2025). The impact of digitalization on entrepreneurial innovation: a comparative analysis between developing economies in asia and in other regions. ADB Economics Working Paper Series No. 776, DOI: 10.2139/ssrn.5233001

Tao, F., Qi, Q., Wang, L., & Nee, A. Y. C. (2019). Digital twins and cyber-physical systems toward smart manufacturing and Industry 4.0. *Engineering (Beijing)*, *5*(4), 653–661. DOI: 10.1016/j.eng.2019.01.014

Tekeli, I. (2009). An exploratory approach to urban historiography through a new paradigm: The case of Turkey. In Sarkis, H., & Turan, N. (Eds.), *A Turkish Triangle: Ankara, Istanbul and Izmir at the gates of Europe*.

Tennakoon, S., Apan, A., Maraseni, T., & Altarez, R. D. D. (2023). Decoding the impacts of space and time on honey bees: GIS based fuzzy AHP and fuzzy overlay to assess land suitability for apiary sites in Queensland, Australia. *Applied Geography (Sevenoaks, England)*, *155*, 102951. DOI: 10.1016/j.apgeog.2023.102951

Texas Instruments Incorporated. (2021). Sensor Design for Inductive Sensing Applications Using LDC, SNOA930C – March 2015 – Revised May 2021.

Tham, M. L., Wong, Y. J., Kwan, B. H., Ng, X. H., & Owada, Y. (2023, January). Artificial intelligence of things (AIoT) for disaster monitoring using wireless mesh network. In *Proceedings of the 2023 6th International Conference on Software Engineering and Information Management* (pp. 234-239). DOI: 10.1145/3584871.3584905

Thapa, R. B., & Murayama, Y. (2008). Land evaluation for peri-urban agriculture using analytical hierarchical process and geographic information system techniques: A case study of Hanoi. *Land Use Policy*, *25*(2), 225–239. DOI: 10.1016/j.landusepol.2007.06.004

Tianchen, J., Huaiying, F., Rencheng, Z., Jianhong, Y., Lulu, F., Yangyang, H., & Zhengxing, C. (2023). Low-value recyclable waste identification based on NIR feature analysis and RGB-NIR fusion. *Infrared Physics & Technology*, *131*, 104693. DOI: 10.1016/j.infrared.2023.104693

Tirachini, A. (2014). The economics and engineering of bus stops: Spacing, design and congestion. *Transportation Research Part A, Policy and Practice*, *59*, 37–57. DOI: 10.1016/j.tra.2013.10.010

Tolche, A. D., Gurara, M. A., Pham, Q. B., Ditthakit, P., & Anh, D. T. (2022). Agricultural land use suitability analysis using AHP and GIS techniques at basin scale. *Arabian Journal of Geosciences*, *15*(18), 1522. DOI: 10.1007/s12517-022-10799-8

Torbert, H. A., Krueger, E., & Kurtener, D. (2008). Soil quality assessment using fuzzy modelling. *International Agrophysics*, *22*, 365–370.

Tortora, A., Statuto, D., & Picuno, P. (2015). Rural landscape planning through spatial modelling and image processing of historical maps. *Land Use Policy*, *42*, 71–82. DOI: 10.1016/j.landusepol.2014.06.027

Tsui, K. L., Chen, N., Zhou, Q., Hai, Y., & Wang, W. (2015). Prognostics and health management: A review on data driven approaches. *Mathematical Problems in Engineering*, *2015*, 793161. Advance online publication. DOI: 10.1155/2015/793161

Turok, I. (2016). Getting urbanization to work in Africa: The role of the urban land-infrastructure-finance nexus. *Area Development and Policy*, *1*(1), 30–47. DOI: 10.1080/23792949.2016.1166444

UN Environment Programme. International SWA. (2024). Global Waste Management Outlook 2024—Beyond an Age of Waste: Turning Rubbish into a Resource. Technical Report, Available online: https://wedocs.unep.org/20.500.11822/44939

United Nation Development Programmes (UNDP), Human Development Reports. (2026). [Online]. https://hdr.undp.org/data-center/documentation-and-downloads

United Nations (UN). (2014). *World urbanization prospects, the 2014 revision.* UN Department of Economic and Social Affairs-Population Division.

Ustaoglu, E., & Aydınoglu, A. C. (2020). Suitability evaluation of urban construction land in Pendik district of Istanbul, Turkey. *Land Use Policy*, *99*, 104783. DOI: 10.1016/j.landusepol.2020.104783

Ustaoglu, E., Ortega Lopez, G., & Gutierrez-Alcoba, A. (2023). Building composite indicators for the territorial quality of life assessment in European regions. Combining data reduction and alternative weighting techniques. *Environment, Development and Sustainability*, *27*(3), 6025–6063. Advance online publication. DOI: 10.1007/s10668-023-04116-w

Uy, P. D., & Nakagoshi, N. (2008). Application of land suitability analysis and landscape ecology to urban greenspace planning in Hanoi, Vietnam. *Urban Forestry & Urban Greening*, *7*(1), 25–40. DOI: 10.1016/j.ufug.2007.09.002

Valarmathi, V., Ramkumar, J., Karthikeyan, R., & Doss, S. (2025). Harnessing AI for seismic hazard detection and prediction: Innovations and challenges. In *Emerging AI Applications in Earth Sciences: Challenges, Impact and Analysis* (pp. 97–110). Springer Nature Switzerland. DOI: 10.1007/978-3-031-84583-3_7

Vandermeer, M. C., & Halleux, J. M. (2017). Evaluation of the spatial and economic effectiveness of industrial land policies in northwest Europe. *European Planning Studies, 25*(8), 1454–1475. DOI: 10.1080/09654313.2017.1322042

Varma, K., Srivastava, V., Singhal, A., & Jha, P. K. (2021). Urban and environmental hazards. In Rai, P. K., Singh, P., & Mishra, V. N. (Eds.), *Recent Technologies for Disaster Management and Risk Reduction. Earth and Environmental Sciences Library*. Springer. DOI: 10.1007/978-3-030-76116-5_19

Vera, C. H., Karl, F. D., Richard, F. H., & Daniele, V. (2013). Models and Algorithms for the Integrated Planning of Bin Allocation and Vehicle Routing in Solid Waste Management. *Transportation Science, 48*, 1.

Vicentini, F., Giusti, A., Rovetta, A., Fan, X., He, Q., Zhu, M., & Liu, B. (2009). Sensorized waste collection container for content estimation and collection optimization. *Waste Management (New York, N.Y.), 29*(5), 1467–1472. DOI: 10.1016/j.wasman.2008.10.017 PMID: 19103480

Vondrackova, T., Musilek, J., & Kais, L. (2015). The issue of soft rocks causing problems in foundation engineering. *Proedia Earth and Planetary Science, 15*, 54–59. DOI: 10.1016/j.proeps.2015.08.015

Vrontis, D., Chaudhuri, R., & Chatterjee, S. (2022). Adoption of digital technologies by SMEs for sustainability and value creation: Moderating role of entrepreneurial orientation. *Sustainability (Basel), 14*(13), 1–19. DOI: 10.3390/su14137949

Wang, C. (2025). Multi-Modal Intelligence for Multi-Hazard Response and Urban Resilience (Doctoral dissertation, State University of New York at Stony Brook).

Wang, D. (2023). *Understanding Social Vulnerability to Natural Hazards Through Machine Learning and Dynamic Spatiotemporal Analysis*. Boise State University. DOI: 10.18122/td.2204.boisestate

Wang, D., & Shao, X. (2024). Research of the impact of digital transformation on the production efficiency of manufacturing enterprises: Institution-based analysis of the threshold effect. *International Review of Economics & Finance, 91*(1), 883–897. DOI: 10.1016/j.iref.2024.01.046

Wang, F. (1990). Fuzzy supervised classification of remote sensing maps. *IEEE Transactions on Geoscience and Remote Sensing, 28*(2), 194–201. DOI: 10.1109/36.46698

Wang, H., Shen, Q., & Tang, B.-S. (2015). GIS-based framework for supporting land use planning in urban renewal: Case study in Hong Kong. *Journal of Urban Planning and Development, 141*(3), 05014015. Advance online publication. DOI: 10.1061/(ASCE)UP.1943-5444.0000216

Wang, J., Sun, P., Chen, L., Yang, J., Liu, Z., & Lian, H. (2023). Recent advances of deep learning in geological hazard forecasting. *Computer Modeling in Engineering & Sciences, 137*(2), 1381–1418. DOI: 10.32604/cmes.2023.023693

Wang, T., Kazak, J., Han, Q., & de Vries, B. (2019). A framework for path-dependent industrial land transition analysis using vector data. *European Planning Studies, 27*(7), 1391–1412. DOI: 10.1080/09654313.2019.1588852

Wan, X., Kazmi, S. A. A., & Wong, C. Y. (2022). Manufacturing, exports, and sustainable growth: Evidence from developing countries. *Sustainability (Basel), 14*(3), 1646. DOI: 10.3390/su14031646

Wen, H., Liu, Y., & Liu, Y. (2024). Impact of digitalization on investment and productivity of manufacturing industry: Evidence from China. *SAGE Open*, *14*(3), 21582440241281862. Advance online publication. DOI: 10.1177/21582440241281862

Wen-Tsai, S., Indra, G. T. I., & Sung-Jung, H. (2023). An IoT-Based Aquaculture Monitoring System Using Firebase. *Computers, Materials & Continua*, *76*(2), 2179–2200. DOI: 10.32604/cmc.2023.041022

Wilts, H., Garcia, B. R., Garlito, R. G., Gómez, L. S., & Prieto, E. G. (2021). Artificial intelligence in the sorting of municipal waste as an enabler of the circular economy. *Resources*, *10*(4), 28. DOI: 10.3390/resources10040028

Wimhurst, J. J., Nsude, C. C., & Greene, J. S. (2023). Standardizing the factors used in wind farm site suitability models: A review. *Heliyon*, *9*(5), e15903. DOI: 10.1016/j.heliyon.2023.e15903

World Bank. (2017). Population estimates and projections, https://datacatalog.worldbank.org/dataset/population-estimates-and-projections

World Development Indicators (WDIs), The World Bank (WB). (2026). [Online]. https://databank.worldbank.org/source/world-development-indicators

Worldbank. (2018). What a Waste 2.0: A Global Snapshot of Solid Waste Management to 2050.

Wu, X., & Zou, K. (2025, October). A machine learning-enhanced multi-agent framework for multi-timescale emergency resource allocation in natural disasters. In 2025 2nd International Symposium on AI and Cybersecurity (ISAICS) (pp. 1-6). IEEE. DOI: 10.1109/ISAICS66888.2025.11349790

Wu, M. (2023). *An Intelligent Bibliometric System for Knowledge Association and Hierarchy Discovery*. University of Technology Sydney.

Wu, R., & Cheng, X. (2016). Gender equality in the workplace: The effect of gender equality on productivity growth among the Chilean manufacturers. *Journal of Developing Areas*, *50*(1), 257–274. DOI: 10.1353/jda.2016.0001

Wu, S. J. (2024). Modeling an evaluation framework for adding IoT water-level sensors based on ANN-derived 2D inundation simulations. *Journal of Hydroinformatics*, *26*(9), 2261–2288. DOI: 10.2166/hydro.2024.097

Wu, Y., Li, L., & Zheng, Y. (2024). The impact of digitization in manufacturing on female employment and gender wage gap. *Journal of Asian Economics*, *95*(1), 101821. DOI: 10.1016/j.asieco.2024.101821

Wu, Y., Luo, J., Zhang, X., & Skitmore, M. (2016). Urban growth dilemmas and solutions in China: Looking forward to 2030. *Habitat International*, *56*, 42–51. DOI: 10.1016/j.habitatint.2016.04.004

Xiaoyu, W., Jia, L., Linpeng, Y., & Zhenming, X. (2020). Auto-sorting commonly recovered plastics from waste household appliances and electronics using near-infrared spectroscopy. *Journal of Cleaner Production*, *246*, 118732. DOI: 10.1016/j.jclepro.2019.118732

Xinxing, C., Huaiyang, H., Yuxuan, L., Jiqing, L., & Ming, L. (2022). Robot for automatic waste sorting on construction sites. *Automation in Construction*, *141*, 104387. DOI: 10.1016/j.autcon.2022.104387

Yang, Y., Wang, N., Wang, Z., Ouyang, M., & Wan, C. (2025). From Forecast to Action: A Deep Learning Model for Predicting Power Outages During Tropical Cyclones. arXiv preprint arXiv:2512.06644.

Yang, Q., & Jin, S. (2024). Exploring the impact of digital transformation on manufacturing environment, social responsibility, and corporate governance performance: The moderating role of top management teams. *Sustainability (Basel)*, *16*(11), 4342. DOI: 10.3390/su16114342

Yann, G., Arnab, S., & Paul, C. (2015). A Smart Waste Management with Self-Describing Complex Objects. *International Journal On Advances in Intelligent Systems*, *8*, 1–16.

Yasmin, F., Hasan, M., Sinha, A., & Mondal, S. (2025). AI-driven hazard assessment: a comprehensive study of machine and deep learning approaches. Asian Journal of Civil Engineering, 1-16.

Yelbek, U., Alexandr, N., Timoth, M., Zhanbolat, S., Sungat, A., Alizhan, K., Assel, T.M., Aigul, K.K. (2024). Advancing sanitary surveillance: Innovating a live-feed sewer monitoring framework for effective water level and chamber cover detections. Heliyon, 10, 6, e27395.

Yigitcanlar, T., Sipe, N., Evans, R., & Pitot, M. (2007). A GIS-based land use and public transport accessibility indexing model. *Australian Plants*, *44*(3), 30–37. DOI: 10.1080/07293682.2007.9982586

Yousefi, E., Salehi, E., Zahiri, S. H., & Yavari, A. (2016). Green space suitability analysis using evolutionary algorithm and Weighted Linear Combination (WLC) method. *Space Ontology International Journal*, *5*(4), 51–60.

Zadeh, L. A. (1965). Fuzzy sets. *Information and Control*, *8*(3), 338–353. DOI: 10.1016/S0019-9958(65)90241-X

Zaghba, L., Borni, A., Benbitour, M. K., Fezzani, A., Alwabli, A., Bajaj, M., Dost Mohammadi, S. A., & Ghoneim, S. S. (2024). Enhancing grid-connected photovoltaic system performance with novel hybrid MPPT technique in variable atmospheric conditions. *Scientific Reports*, *14*(1), 8205. DOI: 10.1038/s41598-024-59024-4 PMID: 38589473

Zarin, R., Azmat, M., Naqvi, S. R., Saddique, Q., & Ullah, S. (2021). Landfill site selection by integrating fuzzy logic, AHP and WLC method based on multi-criteria decision analysis. *Environmental Science and Pollution Research International*, *28*(16), 19726–19741. DOI: 10.1007/s11356-020-11975-7

Zeli, W., Heng, L., & Xintao, Y. (2020). Vision-based robotic system for on-site construction and demolition waste sorting and recycling. *Journal of Building Engineering*, *32*, 101769. DOI: 10.1016/j.jobe.2020.101769

Zen, F., & Faizin, M. A. (2021). An overview of digitalization and internationalization MSMEs: A conceptual paper. *Advances in Economics. Business and Management Research*, *193*(1), 195–200.

Zeng, S., Xia, J., Chen, X., Zou, L., Du, H., & She, D. (2020). Integrated land-surface hydrological and biogeochemical processes in simulating water, energy and carbon fluxes over two different ecosystems. *Journal of Hydrology (Amsterdam)*, *582*, 124390. DOI: 10.1016/j.jhydrol.2019.124390

Zhang, F. (2024). Design and Application of Microcontroller-Based Tunnel Construction Environment Monitor. *Tehnicki Vjesnik (Strojarski Fakultet)*, *31*(4), 1060–1068.

Zhang, J., Su, Y., Wu, J., & Liang, H. (2015). GIS based land suitability assessment for tobacco production using AHP and fuzzy set in Shandong province of China. *Computers and Electronics in Agriculture*, *114*, 202–211. DOI: 10.1016/j.compag.2015.04.004

Zhang, J., van Gorp, D., & Kievit, H. (2023). Digital technology and national entrepreneurship: An ecosystem perspective. *The Journal of Technology Transfer*, *48*(1), 1077–1105. DOI: 10.1007/s10961-022-09934-0 PMID: 35602312

Zhang, K., Wang, J., & Wu, Y. (2025). A study of the impact of manufacturing input digitization on firms' organizational resilience: Evidence from China. *Sustainability (Basel)*, *17*(3), 897. DOI: 10.3390/su17030897

Zhang, T., Wang, D., & Lu, Y. (2023). Machine learning-enabled regional multi-hazards risk assessment considering social vulnerability. *Scientific Reports*, *13*(1), 13405. DOI: 10.1038/s41598-023-40159-9 PMID: 37591870

Zhang, X., Fang, C., Wang, Z., & Ma, H. (2013). Urban construction land suitability evaluation based on improved multi-criteria evaluation based on GIS (MCE-GIS): Case of New Hefei City, China. *Chinese Geographical Science*, *23*(6), 740–753. DOI: 10.1007/s11769-013-0609-6

Zhao, Y., Lv, S., & Liu, P. (2024). Advances in earthquake prevention and reduction based on machine learning: A scoping review. *IEEE Access : Practical Innovations, Open Solutions*, *12*, 143908–143929. DOI: 10.1109/ACCESS.2024.3467149

Zheng, Q., Yang, X., Wang, K., Huang, L., Shahtahmassebi, A. R., Gan, M., & Weston, M. V. (2017). Deliminating urban growth boundary through combining land suitability evaluation and cellular automata. *Sustainability (Basel)*, *9*(12), 2213. DOI: 10.3390/su9122213

Zheng, S., Ristovski, K., Farahat, A., & Gupta, C. (2017). Long short-term memory network for remaining useful life estimation. In *Proceedings of the IEEE International Conference on Prognostics and Health Management (ICPHM)* (pp. 88–95). IEEE. DOI: 10.1109/ICPHM.2017.7998311

Zimmerman, H. J. (1991). *Fuzzy set theory and its applications*. Kluwer Academic Publishers. DOI: 10.1007/978-94-015-7949-0

Zolekar, B. R., & Bhagat, V. S. (2015). Multi-criteria land suitability analysis for agriculture in hilly zone: Remote sensing and GIS approach. *Computers and Electronics in Agriculture*, *118*, 300–321. DOI: 10.1016/j.compag.2015.09.016

Zonta, T., da Costa, C. A., da Rosa Righi, R., de Lima, M. J., da Trindade, E. S., & Li, G.-P. (2020). Predictive maintenance in Industry 4.0: A systematic literature review. *Computers & Industrial Engineering*, *150*, 106889. Advance online publication. DOI: 10.1016/j.cie.2020.106889

About the Contributors

Kamalakanta Muduli, is presently working as Associate Professor and Head of the School of Mechanical Engineering, Papua New Guinea University of Technology, Lae, Morobe Province, Papua New Guinea. He is also a Visiting Researcher at University of Derby, UK. He has obtained PhD from School of Mechanical Sciences, IIT Bhubaneswar, Orissa, India. Dr Muduli has over 21 years of academic experience in Universities in India and Papua New Guinea. Dr Muduli is a recipient of ERASMUS+ KA107 award provided by European Union. He has published 106 papers in peer reviewed international journals most of which are indexed in Clarivate analytics, Scopus and listed in ABDC and more than 37 papers in National and International Conferences. He has been also guest edited 10 Special Issues in journals and 21 books published by Springer, Taylor and Francis, MDPI, CRC Press, Wiley Scrivener and Apple Academic Press. Dr Muduli also has supervised four PhD students. His current research interest includes Manufacturing, Sustainable supply chain management, and Industry 4.0 applications in operations and supply chain management. Dr Muduli is a fellow of Institution of Engineers India. He is also a senior member of Indian Institution of Industrial Engineering, member of American Society of Mechanical Engineers(ASME) and Professional member of Industrial Engineering and Operations Management(IEOM) Society.

Sujoy Kumar Jana presently working as Professor and Course Coordinator, MSc in Remote Sensing and GIS at School of Surveying and Land Studies at the Papua New Guinea University of Technology, Papua New Guinea. He having 25 years of extensive work on research, teaching, training and administrative in the university system. He also consults on disaster risk governance and policy, vulnerability and risk assessment, land use land cover mapping, capacity development of decision-makers, river and coastal resilience, critical infrastructure functional risk and urban planning. He has more than 4 books and 50 papers in international and national peer-reviewed journals. His professional activities include roles as editor, editorial board member and reviewer of international journals.

Rashmi Prava Das did her Bachelors in Computer Science and Engineering from North Orissa University and Masters in Computer Science from Birla Institute of Technology, Meshra. She has around 12 years of teaching and research experience in various degree level engineering colleges in Odisha as lecturer, Senior lecturer and Assistant Professor. She did her PhD from CV Raman Global University. She had also supervised few master's thesis. She has published research papers in international journals, conferences and books as chapters.

Indrajit Pal Indrajit Pal is a former member of the Board of Directors for the Global Alliance of Disaster Research Institutes (GADRI), Japan. Dr. Pal is also an Adjunct Professor at Chulalongkorn University Social Research Institute, Thailand. He served as a faculty member at Centre for Disaster Management at Lal Bahadur Shastri National Academy of Administration, Mussoorie, India (A Premier National Institute for training Indian Administrative Services Officers). Dr. Pal holds doctoral and two master's degrees, e.g., Applied Geology and Sociology with urban sociology specialization. Indrajit having more than 16 years of experience on research, teaching, training, advocacy, consultancy primarily focused on Disaster Governance, Risk Management, Incident Command System, Disaster Risk Reduction, Hazard and Risk Assessment, Risk Perceptions, Participatory DRM, Public Health Risk, GIS & Remote Sensing in DRM, Climate Change Adaptation and disaster risk management. Indrajit is reviewer of more than ten international journals and has published number of books and research articles in refereed national and international journals. Indrajit is a member of Board of Directors for the Global Alliance of Disaster Research Institutes (GADRI), Japan

***

Senthil Kumar A.V. is working as Principal, Nehru Institute of Information Technology and Management, Coimbatore, India. He has worked as Professor and Director, PG and Research Department of Computer Applications, Hindusthan College of Arts & Science, Coimbatore, Tamilnadu for 15 years and as Senior Grade Lecturer in CMS College of Science and Commerce for 14 years. To his credit he has industrial experience for five years and teaching experience of 29 years. He has also received his Doctor of Science (D.Sc in Computer Science). He has to his credit 86 Book Chapters, 234 papers in International and National Journals, 85 papers in International Conferences in International and National Conferences, and edited 17 books and 3 Text books. He is as Associate Editor of IEEE Access.

He is an Editor-in-Chief for many journals and Key Member for India, Machine Intelligence Research Lab (MIR Labs). He is an Editorial Board Member and Reviewer for various International Journals. He is also a Committee member for various International Conferences. He has guided 16 Ph.D scholars and guiding 3 Ph. D scholars now.

Arakhita Behera, Assistant Professor in Economics, Department of Social Science and Humanities(DESSH), Regional Institute of Education (NCERT), Ajmer, Rajasthan -305004 (INDIA).

Lotfi Krichen was born in Sfax, TUNISIA. He received the Engineer Diploma, the Doctorate Degree and University Habilitation Diploma (HDR) in Electrical Engineering from the National School of Engineering of Sfax in 1989, 1995 and 2008, respectively. Currently, he is Professor of Electric Machines and Renewable Energy Systems in the Electrical Engineering Department of the National School of Engineering of Sfax. His research interests are Power Electronic Converters, Renewable Energy Systems, Energy Storage Systems, Plug in Electric Vehicle and Smart Grid.

Mouna Rekik was born in Sfax, Tunisia, in 1988. She received her BS, Master's, Ph.D., and HDR degrees in 2011, 2012, 2015, and 2022, respectively, all in electrical engineering from the National Engineering School of Sfax, University of Sfax, Tunisia. She is an Associate Professor in the Department of Electrical Engineering at the National Engineering School of Sfax (ENIS), Tunisia. She is a member of the Electrical Systems and Renewable Energy Laboratory at the University of Sfax and an IEEE member. Her main research interests include power system stability, microgrids, renewable energy systems, plug-in hybrid electric vehicles, and smart grid technologies, artificial intelligence, and circular economy management.

Ajay K. Singh is working as an Associate Professor (Research) in the Department of Humanities and Social Science, Graphic Era (Deemed to be University) Dehradun. He has worked as Assistant Professor (Economics) in the School of Liberal Arts & Management, DIT University Dehradun for 6 years. He did Post-Doctorate Research with EDI of India, Ahmedabad, Gujarat (India). He received MPhil (Economics) from DAVV Indore (India), and PhD (Economics) from IIT Indore (India. He has published several research papers in the diversified area such as climate change, agricultural productivity, assessment of food security; estimation of GFSI, development of environmental sustainability index (ESI) and its association with socio-economic indicators; measurement and determinants of entrepreneurship ecosystem, and dimension of sustainable development and its interlinkages with economic development.

Mariem Smaoui was born in Sfax, Tunisia, in 1987. She received her Engineering degree in 2012 and her Ph.D. degree in 2016, both in Electrical Engineering, from the National Engineering School of Sfax, University of Sfax, Tunisia. She is currently an Assistant Professor in the Department of Electrical Engineering at the Higher Institute of Technological Studies of Sousse, Tunisia. She is a member of the Electrical Systems and Renewable Energy Laboratory at the National Engineering School of Sfax. Her main research interests include the Internet of Things (IoT), renewable energy systems, energy storage systems, microgrids, smart systems, artificial intelligence, and circular economy management.

Mini Srivastava, currently serving as a faculty member in the School of Management and Commerce Studies at Shri Guru Ram Rai University, Dehradun, India. I hold a doctoral degree in Management, and my academic and professional journey has been centered on teaching, research, and academic leadership in the domains of management, innovation, and emerging business practices. My research interests include innovation management, digital transformation, financial technologies (FinTech), sustainable development, and the application of emerging technologies such as artificial intelligence, big data, and blockchain in business processes. I have been actively involved in academic research and have contributed to scholarly publications, conference proceedings, and interdisciplinary research projects focusing on contemporary business challenges and technological advancements. I have participated in and organized national and international conferences, workshops, and faculty development programs, contributing to academic discourse and knowledge dissemination. I am also engaged in curriculum development and academic planning in alignment with NEP 2020, particularly in designing programs for undergraduate and postgraduate management education. My teaching philosophy emphasizes experiential learning, critical thinking, and the integration of theoretical knowledge with practical applications. I am committed to fostering innovation and research-oriented learning among students while contributing to the advancement of academia through continuous research and collaboration.

Gopalakrishnan T is an Assistant Professor in Mechanical Engineering at VISTAS with over 10 years of experience in composite materials, engineering design, and materials engineering. He holds a Ph.D. in Mechanical Engineering (2024) and has authored 34 research papers, multiple book chapters, and patents. An expert in CAD/CAM/FEA, he has received accolades such as the Best Teacher Award (2023) and Faculty Excellence Award (2022). Dr. Gopalakrishnan is passionate about mentoring students and advancing interdisciplinary research in mechanical engineering.

Srinivasan Vasan is an Assistant Professor in the Department of Humanities and Social Sciences at Graphic Era (Deemed to be University), with 14 years of teaching and field expertise. He has served in the Ministry of Health and Family Welfare, as a Guest Faculty, and as a Mobilization Training Specialist with the State Government. Recently, he contributed to a tuberculosis project funded by GFATM and USAID. He has published 60 papers, with 20 more under peer review in UGC-CARE and SCOPUS journals.

Index

A

C

D

E

F

G

H

I

M

P

R

S

T

V

W

www.ingramcontent.com/pod-product-compliance
Lightning Source LLC
LaVergne TN
LVHW081532300926
849161LV00006B/1389
* 9 7 9 8 2 6 0 0 2 3 8 3 9 *